MULTIPERSON DECISION MAKING MODELS USING FUZZY SETS AND POSSIBILITY THEORY

THEORY AND DECISION LIBRARY

General Editors: W. Leinfellner and G. Eberlein

Series A: Philosophy and Methodology of the Social Sciences
Editors: W. Leinfellner (Technical Universtiy of Vienna)
G. Eberlein (Technical University of Munich)

Series B: Mathematical and Statistical Methods
Editor: H. Skala (University of Paderborn)

Series C: Game Theory, Mathematical Programming and Operations Research
Editor: S. H. Tijs (University of Nijmegen)

Series D: System Theory, Knowledge Engineering and Problem Solving
Editor: W. Janko (University of Economics, Vienna)

SERIES B: MATHEMATICAL AND STATISTICAL METHODS

Volume 18

Editor: H. Skala (Paderborn)

Scope

The series focuses on the application of methods and ideas of logic, mathematics and statistics to the social sciences. In particular, formal treatment of social phenomena, the analysis of decision making, information theory and problems of inference will be central themes of this part of the library. Besides theoretical results, empirical investigations and the testing of theoretical models of real world problems will be subjects of interest. In addition to emphasizing interdisciplinary communication, the series will seek to support the rapid dissemination of recent results.

For a list of titles published in this series, see final page.

MULTIPERSON DECISION MAKING MODELS USING FUZZY SETS AND POSSIBILITY THEORY

edited by

JANUSZ KACPRZYK

Systems Research Institute,
Polish Academy of Science, Warsaw, Poland

and

MARIO FEDRIZZI

Institute of Computer Science,
University of Trento, Italy

KLUWER ACADEMIC PUBLISHERS
DORDRECHT / BOSTON / LONDON

Library of Congress Cataloging-in-Publication Data

Multiperson decision making models using fuzzy sets and possibility
theory / edited by Janusz Kacprzyk and Mario Fedrizzi.
p. cm. -- (Theory and decision library. Series B. ; 18)
Includes index.
ISBN 0-7923-0884-0 (alk. paper)
1. Decision-making. 2. Fuzzy sets. I. Kacprzyk, Janusz.
II. Fedrizzi, Mario, 1949- . III. Series: Theory and decision
library. Series B, Mathematical and statistical methods ; v. 18.
T57.95.M85 1990
003'.56--dc20 90-42123

ISBN 0-7923-0884-0

Published by Kluwer Academic Publishers,
P.O. Box 17, 3300 AA Dordrecht, The Netherlands.

Kluwer Academic Publishers incorporates
the publishing programmes of
D. Reidel, Martinus Nijhoff, Dr W. Junk and MTP Press.

Sold and distributed in the U.S.A. and Canada
by Kluwer Academic Publishers,
101 Philip Drive, Norwell, MA 02061, U.S.A.

In all other countries, sold and distributed
by Kluwer Academic Publishers Group,
P.O. Box 322, 3300 AH Dordrecht, The Netherlands.

Printed on acid-free paper

Printed in the Netherlands

TABLE OF CONTENTS

PREFACE

Decision making is certainly a very crucial component of many human activities. It is, therefore, not surprising that models of decisions play a very important role not only in decision theory but also in areas such as Operations Research, Management Science, Social Psychology etc.. The basic model of a decision in classical normative decision theory has very little in common with real decision making: It portrays a decision as a clear-cut act of choice, performed by one individual decision maker and in which states of nature, possible actions, results and preferences are well and crisply defined. The only component in which uncertainty is permitted is the occurence of the different states of nature, for which probabilistic descriptions are allowed. These probabilities are generally assumed to be known numerically, i.e. as single probabilities or as probability distribution functions.

Extensions of this basic model can primarily be conceived in three directions:

1. Rather than a single decision maker there are several decision makers involved. This has lead to the areas of game theory, team theory and group decision theory.

2. The preference or utility function is not single valued but rather vector valued. This extension is considered in multiattribute utility theory and in multicritieria analysis.

3. Finally more than the occurence of the states is uncertain, the uncertainty of the occurence cannot be model-

led by single valued probabilities, the uncertainties are not of a probabilistic nature, or evidences for the assumptions are contradictory or insufficient. For a proper modelling of these "complications" a number of tools or theories have been developed during the last two and a half decades: theories such as evidence theory, possibility theory, fuzzy set theory, areas such as fuzzy logic and approximate reasoning, and terms such as linguistic variables and fuzzy numbers and relations have proven to be very useful for modelling adequately uncertainties of different kinds.

For the modelling of individual decisions in fuzzy environments the article by Bellman and Zadeh on this topic in 1970 in Management Science has been the basis for hundreds of subsequent papers in this area. For multiperson decision making models the origins are more diverse. And this is exactly the area on which this book focusses: the application of fuzzy set theory and possibility theory to model multiperson decisions in which components are uncertain either in a probabilistic or in nonprobabilistic sense.

This book is not written by one person as a monograph or a textbook. It is written by quite a number of scientists who are well-known and prominent in their areas. Sometimes this leads to books which seem to consist of more or less randomly chosen contributions. In this case, however, the editors have to be congratulated on their skill and consequence with which they have selected topics and authors in such a way that a very attractive and well-focussed volume resulted. The five parts of the book lead the reader through the topic area from basic introductions to demanding and sophisticated models. While the first two parts of the book represent attractive introductions or basic theoretical contributions to multiperson decision making and to decision making under fuzziness, each of the last three chapters focusses on one of the types of multiperson decision making which were mentioned above: group decision making, team theory and game theory.

I believe that this book can be used as an excellent introduction to fuzzy multiperson decision making as well as a reference book for this area. May it be of benefit to experts and newcomers to this area as well.

Aachen, February 1990

H.-J. Zimmermann

1

INTRODUCTORY SECTIONS

MULTIPERSON DECISION MAKING: A SELECTIVE REVIEW

Peter C. Fishburn
AT&T Bell Laboratories
Murray Hill, NJ 07974-2070, USA

Abstract: Multiperson decision making is reviewed from the perspective of social choice theory and the theory of elections. The review interweaves abstract theory and practical concerns that deserve consideration in evaluating alternative election methods and in designing good election systems.

Keywords: Social choice, election methods, evaluation.

1. INTRODUCTION

Multiperson decision making, which occurs whenever two or more people actively participate in arriving at a decision, is a ubiquitous human activity. Indeed, even when you or I make decisions on our own that are not multiperson decisions in the strict sense, we almost always take into consideration the situations and values of others, be they family, friends, co-workers, competitors, bosses, neighbors, or people more remote from our daily lives. Many formal approaches to multiperson decision making have been developed over the years, and numerous mathematically oriented methods have been used in their analysis.

My aim in this paper is quite modest in view of the rich variety of approaches and analytical methods of multiperson decision making. It is to review selectively the field of social choice and voting using formulations and techniques that, for the most part, predate fuzzy sets and possibility theory. Examples of fuzzy approaches to social choice that are not reviewed here include Nurmi (1981), Dutta, Panda and Pattanaik (1986) and Dutta (1987). In the paper, set membership follows two-valued logic, and uncertainties characterized numerically conform to standard probability theory.

The paper begins with historical highlights then outlines aspects of election systems that are oriented toward practical concerns. Section 4 provides an overview of Arrowian impossible theorems, and Section 5 reviews election methods and the paradoxes they encounter. Section 6 draws further attention to five factors that are often important in comparing alternative election methods. They are data requirements, computational reasonableness, strategic manipulability, conditions for social choice, and probability of undesirable effects. The final section discusses the role of probability in voting theory.

J. Kacprzyk and M. Fedrizzi (eds.), Multiperson Decision Making Using Fuzzy Sets and Possibility Theory, 3–27.

The emphasis throughout is on the election of one from three or more candidates although other possibilities are not ignored.

Other reviews from the past few decades contain much more information on topics of social choice. They are Riker (1961), Ferejohn and Fiorina (1975), Grofman (1975), Mueller (1976), Niemi and Riker (1976), Plott (1976), Sen (1977), Blair and Pollak (1983) and Fishburn (1984a). Useful books on particular topics will be cited later.

2. HISTORICAL PERSPECTIVE

Three epochs stand out in the development of methods of election and social choice. The first occurred in France in the 1780s, the second in England between 1860 and 1885, and the third largely in the United States between 1950 and 1975. All three presume that each of n voters can rank the m candidates for election from best to worst according to his or her preferences. With voters indexed from 1 to n, we let $x >_i y$ mean that voter i prefers candidate x to y and refer to $(>_1, \ldots, >_n)$ as a *voter preference profile*. We assume here that each $>_i$ is a *linear order* (no ties) on the m candidates, but relax this later.

Suppose the profile $(>_1, \ldots, >_n)$ encodes the voting information considered relevant for an election. Then the problem of social choice is how to aggregate voters' preferences to determine a winning candidate, or several winners if more than one is to be elected, or to determine a consensus ranking of the candidates. When only two candidates compete, there is broad agreement that the winner ought to be determined by simple majority. It is when m exceeds 2 that interesting things happen.

2.1 *The French epoch*

Borda (1781) argued that the *plurality* method, which elects the candidate ranked first by most voters, is flawed because it ignores salient voter preferences when $m \geq 3$. Moreover, it might elect a candidate that would lose to each of the others in pairwise majority comparisons based on the voter preference profile. The method he advocated to correct these defects assigns $m-k$ points to each voter's k^{th}-ranked candidate, $k = 1, \ldots, m$, then determines each candidate's score as the sum over the voters of the points assigned to that candidate. *Borda's method* then chooses the candidate(s) with the greatest score(s).

Borda's method was soon challenged on several counts, including its equal spacing of points assigned to first, second, third, ... ranked candidates, and its failure to elect a majority candidate in some cases. A *majority candidate* is one that would defeat each of the others in pairwise majority comparisons based on $(>_1, \ldots, >_n)$. Condorcet (1785) specified a three-candidate profile with a majority candidate that would lose under *every* Borda-type method that assigns more points to a first choice than a second choice, and more points to a second than a third choice.

Condorcet felt strongly that a majority candidate ought to be elected when one exists, and aggregation methods that satisfy this prescription are referred to today as *Condorcet methods*. He realized that a profile need not have a majority candidate, as when $m = n = 3$ and the profile is $(a >_1 b >_1 c,\ b >_2 c >_2 a,\ c >_3 a >_3 b)$, in which case a has a 2-to-1 majority over b, b has a 2-to-1 majority over c, and c has a 2-to-1 majority over a. In such a case the profile illustrates *Condorcet's effect*, or the *paradox of voting*, or has *cyclical majorities*. Unfortunately, Condorcet's proposal for determining a winner when there is no majority candidate is unclear (Nanson, 1883, p. 229; Black, 1958, pp. 174-177;

Fishburn, 1977; Young, 1988). Part of his proposal was to compare all pairs of candidates and elect the one with the greatest pairwise majority victories, a method rediscovered by Copeland (1951) and sometimes referred to as *Copeland's method.* Problems arise, however, if several candidates have the same maximum number of pairwise majority victories, and it is at this point that Condorcet was obscure.

2.2 *The English epoch*

Other methods for electing one candidate from three or more were forthcoming in the 1800s and included contributions by C. L. Dodgson (a.k.a. Lewis Carroll: see Black, 1958) and Nanson (1883). Nanson's paper is remarkable for its scope and comparative analysis of alternative election methods. He included plurality, the method of having each voter vote for two candidates, Borda's method, plurality runoff procedures and other sequential elimination procedures, and Copeland's method. He highlighted weaknesses of each and proposed his own method designed to overcome them. Given $(>_1, \ldots, >_n)$, *Nanson's method* eliminates all candidates whose Borda scores are less than or equal to the average Borda score. These candidates are then removed from the profile and the process is repeated. The sole candidate not thus eliminated is the winner. If the original profile has a majority candidate, it must win, so Nanson's method is a Condorcet method. If there is no majority candidate and two or more remaining candidates all have the same Borda score, the deadlock can be broken in several ways (Nanson, 1883, p. 233).

The 1800s also saw a flurry of methods for electing several candidates in one election, say to seats in a parliament or on a council. Vote adjustment methods known as quota systems, methods of single transferable vote, and Hare systems (Hare, 1861) were widely discussed around mid-century. One such system proceeds as follows. Given $(>_1, \ldots, >_n)$, suppose k candidates are to be elected, $1 \leq k < m$. Define the *quota* q as the smallest integer such that $q(k+1) > n$. To be elected, a candidate needs at least q first-place votes at some stage of the tally process. The first stage uses the given profile. If one or more candidates each have q or more first-place votes, they are elected; they are then deleted from the profile and the weights of the votes of the voters who had an elected candidate in first place are decreased from 1 to a fraction less than 1 so that the weights of all n voters sum to n, minus q times the number of elected candidates. If no candidate meets or exceeds q by first-place votes at stage 1, the candidate with the fewest first-place votes is eliminated and there is no adjustment in weights of voters. Subsequent stages proceed in a similar way with their reduced profiles and adjusted voter weights until k candidates have been elected. At that point no further elections are possible since the reduced voter weights sum to less than q. The alleged virtues of such a system include equitable treatments of voters and candidates, and effective representation of minorities.

2.3 *The Arrow epoch*

It is widely agreed today that the most important advance in the theory of elections and social choice during the past century was Arrow's discovery (Arrow, 1950, 1963) that a few appealing criteria for social ranking methods are mutually incompatible. Its essential idea is that the problems that arise from Condorcet's paradox of voting, or cyclical majorities, cannot be avoided under any reasonable generalization of majority comparisons.

The crucial technical advance in Arrow's approach that led to his celebrated impossibility theorem was the consideration of a variety of voter preference profiles that might arise in an election among three or more candidates. Earlier work tended to focus on one profile at a time. Arrow, on the other hand, exploited interrelationships among different profiles and showed that if all profiles in a suitably large set for fixed n and $m \geq 3$ satisfy a few basic conditions, then one can deduce a contradiction.

Arrow also observed that if the set of profiles is restricted sufficiently, say to those in which all voters' preferences are single peaked over some fixed linear arrangement of candidates, then his conditions are simultaneously compatible. For single peaked preference profiles, majority comparisons are essentially transitive (Galton, 1907; Black, 1948; Arrow, 1963) and obey Arrow's conditions.

Arrow introduced a new level of mathematics into the subject and profoundly affected later work in social choice theory. Of particular note is the manipulability theorem of Gibbard (1973) and Satterthwaite (1975) that is closely related to Arrow's original theorem. The Gibbard-Satterthwaite theorem says that every method for electing one of $m \geq 3$ candidates that satisfies a few elementary conditions is *manipulable*, i.e., there will be profiles at which some voter by falsifying his true preferences can help to elect a candidate he strictly prefers to the candidate that would have been elected under the 'true' profile. Early writers, including Laplace and Nanson, had commented on the susceptibility of specific methods such as Borda's to so-called strategic voting, but they could scarely have imagined the ubiquity of opportunities for manipulability raised by the theorem of Gibbard and Satterthwaite.

3. ASPECTS OF ELECTION SYSTEMS

Before we return to Arrow's theorem and specific election methods, I outline aspects of election systems to illustrate the breadth of the field and areas of research opportunities. Some aspects that follow are discussed in later sections; a more complete picture of the whole is in Fishburn (1983a).

It is often useful to distinguish between abstract concepts of social choice theory and aspects of election systems one might find in practice. The preceding section's discussion lies more in the former area with its tidy presumption of known voter preference profiles and aggregation procedures – also known as *social choice functions* – that produce a choice, choices, or a social ranking as the case requires.

An actual election system is more appropriately described in somewhat different language. For simplicity, consider an election conducted by a single ballot. We begin with a set of rules, written or unwritten, that guide the entire process, including the nomination procedure, candidate and voter eligibility, the type of ballot to be used, how the ballot must be marked to be counted, and the way the ballots are to be counted or aggregated to determine a winner or collective decision. A moment's reflection shows there is tremendous variety over these aspects in practice. And if one is engaged in the design of practicable election systems, a host of related factors arise, such as: roles of interest groups, effects of polls, voter psychology and ability to understand ballot instructions, candidate strategy, costs of conducting an election, effects on institutional arrangements, and legal provisions for challenges to eligibility, ballots, or outcome.

Following Fishburn (1983a), I list a dozen aspects that include many of the salient factors of election systems:

1. Nomination process or agenda formation
2. Candidate strategy
3. Nature of the electorate
4. Voter psychology
5. Voter strategy
6. Ballots and response procedure
7. Ballot response profile
8. Ballot aggregator
9. Evaluative aspects of aggregation
10. Collective decisions versus social choices
11. Costs and financing
12. Effects on institutions.

A few remarks on each aspect must suffice here to suggest their scope.

The *nomination process or agenda formation* includes rules of candidacy (age, residency, ...; practical feasibility) and often has implications for ballot position, order of voting, and so forth. Nomination rules and other factors can encourage or limit entry (Riker, 1976; Brams and Fishburn, 1983). Depending on voting rules and voter sophistication, a candidate in a sequential-elimination procedure can gain advantage or be at a disadvantage by appearing later in the process (Black, 1958; Farquharson, 1969; Plott and Levine, 1978; Niemi and Rasch, 1987).

The positioning aspect is often a feature of *candidate strategy*, which can also involve stands on policy issues (Riker, 1962; Granberg and Brent, 1980; Downs, 1957; Cox, 1989) and allocation of campaign funds (Brams and Davis, 1974; Aldrich, 1980). Effects of voting systems on the number of candidates that run are discussed by Wright and Riker (1989).

The *nature of the electorate* concerns rules on who can vote and the consequent features of the electorate's demographics, intelligence, and motivation. Many studies have analyzed electorates in detail (Lipset, 1960; Campbell et al., 1960; Key, 1966).

Voter psychology deals with people's likely reactions to different voting systems, the processes by which they form opinions about candidates, and the extent to which their voting behavior can be characterized as rational (Harsanyi, 1969; Shapiro, 1969; Downs, 1957; McKelvey and Ordeshook, 1972). Relatively little seems to be known about voter turnout as a function of election method. A further facet of voter psychology is the extent to which voters form discernable preferences among candidates. Effects of ambiguity of candidates' positions on extents to which voters seek more information are examined experimentally in Collier, Ordeshook and Williams (1989).

Voter strategy involves deliberate decisions by voters to respond in certain ways during the balloting process. Information about other voters' preferences revealed by polls, committee discussions, and so forth, can affect their decisions. Most of the theoretical literature in this area (Farquharson, 1969; Gibbard, 1973; Satterthwaite, 1975; Pattanaik, 1978; Moulin, 1979) presumes a degree of knowledge of other voters' preferences that is unrealistic in most situations. Despite this, a large literature has developed around the theme of motivating voters to vote sincerely (nonstrategically) regardless of what they know about other voters (Groves, 1979; Dasgupta, Hammond and Maskin, 1979; Laffont and Moulin, 1982; McAfee and McMillan, 1988; Reichelstein and Reiter, 1988).

Ballots and response procedure refers to the mechanism by which voters' alleged preferences are revealed or recorded. People may vote differently in secret than in the open: the Australian ballot, which lists candidates and is cast in secret, has been hailed as the major election reform of the nineteenth century. Complicated ballots can discourage voters. Voting machines can simplify matters for some voters and for vote counters while reducing opportunities for fraud. Ballot positions can influence outcomes, and many organizations have designed procedures to minimize position effects.

The *ballot response profile* encodes cast ballots in a manner useful for aggregation. The *ballot aggregator* processes the data in the response profile to arrive at a collective decision in much the same way that a social choice function maps voter preference profiles into social choices within the abstract framework.

Evaluative aspects of aggregation concern properties of the ballot aggregator that support or detract from the aims of the election. Issues of fairness, equity, consistency, and faithfulness to the desires of the electorate provide examples. Does the aggregation process treat candidates equally? Do some voters have more power than others? If you vote for x instead of y, does x's chances of election improve? If a majority candidate exists, what are its chances of being elected?

Evaluative aspects are usually addressed in a *collective decisions versus social choices* mode by bringing conditions or axioms for social choice functions into the picture (Sen, 1970, 1977; Fishburn, 1973; Kelly, 1978; Brams and Fishburn, 1983; Nurmi, 1987; Merrill, 1988). Suppose a social choice function assigns a nonempty subset of candidates, called the *choice set*, to each voter preference profile in its domain. Among the conditions one might desire are

anonymity: the choice set is invariant to permutations of the components of $(>_1, \ldots, >_n)$;

Pareto optimality: if $x >_i y$ for all i, y is not in the choice set;

monotonicity: if x moves up in some $>_i$, all else unchanged, it will remain in the choice set if it was there before the move;

Condorcet's principle: if x is the majority candidate, it is the only candidate in the choice set.

Similar designations might apply to ballot aggregators, or we might ask for the propensity of the aggregator to satisfy such conditions. If we would like to satisfy Condorcet's principle in practice but consider ballots that ask voters to rank all candidates as impractical, what simpler ballot and aggregator will choose the majority candidate in most cases when it exists?

Another concern of collective decisions versus social choices is the design of ballot response mechanisms that will encourage voters to vote sincerely, in a nonmanipulative way.

Costs and financing involves costs of voter registration, costs incurred by candidates and their supporters, and costs borne by organizations or by the public for campaigns, balloting, and ballot counting. In some cases, perceived advantages of multistage runoff elections could be weighed against the costs of multiple ballots.

Effects on institutions refers to political arrangements, interest groups, and party structures that could be affected by the way elections are conducted. Research here has focused on political parties and their changes over time (Downs, 1957; Riker, 1962; Rae, 1967; Butler and Stokes, 1974): see also Wright and Riker (1989). People interested in changing election systems become accustomed to strong resistance from individuals and power groups who believe that changes are inimical to their own interests.

4. IMPOSSIBILITY THEOREMS

We now step back into the social choice mode to illustrate Arrow's impossibility theme through five theorems that range from a result close to his original theorem to the manipulability theorem of Gibbard and Satterthwaite. Many other theorems of a similar nature are recalled in Kelly (1978) and Fishburn (1987). Behind each theorem lies a set of possibility theorems, collections of mutually compatible conditions obtained by weakening one or more of the impossibility theorem's conditions to the point where what remains is consistent.

A few definitions will be helpful. We say that a binary relation $>_0$ on a set T is *acyclic* if it is never true that there are elements $t_1, \ldots, t_m$ in T for which $t_1 >_0 t_2 >_0 \cdots >_0 t_m >_0 t_1$. An acyclic binary relation is a *weak order* if both it and its symmetric complement $\sim_0$, defined by $s \sim_0 t$ if not$(s >_0 t)$ and not$(t >_0 s)$, are transitive. The union of $>_0$ and $\sim_0$ is denoted by $\gtrsim_0$.

We let X denote a *set of candidates* and $N = \{1, \ldots, n\}$ a *set of voters*. Unless stated otherwise, *we assume that each of X and N is a nonempty finite set and* $|X| \geq 3$. In addition, $\mathcal{A}$ denotes a nonempty family of nonempty subsets of X, and F denotes a set of voter preference profiles, i.e., functions $f, g, \ldots$ each of which assigns a binary relation on X to each voter in N. For convenience we let $>_i = f(i)$ and $>_i' = f'(i)$ for $f, f' \in F$.

A *social choice function* with domain $\mathcal{A} \times F$ is a mapping C from $\mathcal{A} \times F$ into the nonempty subsets of X such that, for all $A \in \mathcal{A}$ and all $f \in F$,

$$C(A, f) \subseteq A.$$

Interpretation: When A is the feasible set and voters' preferences are specified by f, $C(A, f)$ is the set of "best" things in A as determined by the choice function C. Given C, we define a binary relation $>_f$ on X for $f \in F$ by

$$x >_f y \text{ if } x \neq y, \ \{x, y\} \in \mathcal{A} \text{ and } C(\{x, y\}, f) = \{x\}.$$

Thus $>_f$ is the binary social preference relation on X induced by C when profile f obtains. The symmetric complement of $>_f$ is denoted by $\sim_f$, and $\gtrsim_f$ is the union of $>_f$ and $\sim_f$. A similar remark applies to $\sim_i$ and $\gtrsim_i$.

Theorem 1 (Arrow and others). *Suppose C is a social choice function on $\mathcal{A} \times F$, $\mathcal{A}$ contains every two-element subset of X, and F is the set of all functions from N into the set of weak orders on X. Then C cannot satisfy all of the following*:

A1 (no dictator). *For every $i \in N$ there exist $\{x, y\} \in \mathcal{A}$ and $f \in F$ such that $x >_i y$ and $y \gtrsim_f x$;*

A2 (social order). *For all $f \in F$, $>_f$ on X is a weak order*;

A3 (Pareto optimality). *For all $x, y \in X$ and all $f \in F$, if $x >_i y$ for all $i \in N$ then $x >_f y$;*

A4 (binary independence). *For all* $x, y \in X$ *and all* $f, f' \in F$, *if* $x >_i y \Leftrightarrow x >_i' y$ *and* $y >_i x \Leftrightarrow y >_i' x$ *for all* $i \in N$, *then* $x >_f y \Leftrightarrow x >_{f'} y$.

Thus, if the choice function is always required to order the candidates (A2), obey Pareto optimality (A3), and base the social comparison between x and y only on voters' preferences between those two (A4), then some voter is a "dictator": there is an i such that society prefers x to y whenever i prefers x to y. The standard proof of Theorem 1 is combinatorial. Its first step shows that, when A2-A4 hold, some voter dictates social preference between some two alternatives; the second step shows that this voter's power for that pair spreads over all pairs of candidates, thus violating A1.

The social ordering axiom A2 goes well beyond the prevention of social preference cycles. When it is weakened to acyclicity, other conditions must be strengthened to restore impossibility.

Theorem 2 (Mas-Colell and Sonnenschein, 1972). *Suppose the hypotheses of Theorem* 1 *hold and* $|N| \geq 4$. *Then C cannot satisfy all of the following*: A3; A4;

A1* (no vetoer). *For every* $i \in N$ *there exist* $\{x, y\} \in \mathcal{A}$ *and* $f \in F$ *such that* $x >_i y$ *and* $y >_f x$;

A2* (acyclicity). *For all* $f \in F$, $>_f$ *on X is acyclic*;

A5 (strong monotonicity). *For all* $i \in N$, *all* $x, y \in X$, *and all* $f, f' \in F$, *if* $>_j = >_j'$ *for all* $j \neq i$, *either* $(y >_i x,\ x \sim_i' y)$ *or* $(x \sim_i y,\ x >_i' y)$, *and* $x \gtrsim_f y$, *then* $x >_{f'} y$.

Theorems 1 and 2 rely on interconnections between voter preference profiles to demonstrate incompatibility. It is possible, however, to obtain enough structure within a single profile to achieve a similar conclusion.

Theorem 3 (Parks, 1976; Pollak, 1979). *Suppose the hypotheses of Theorem* 1 *hold except that F consists of a single profile f*, $F = \{f\}$, *such that for every map* f' *from N into the set of weak orders on a three-element set* $\{t_1, t_2, t_3\}$ *there exist* $x_1, x_2, x_3 \in X$ *such that, for all* $i \in N$ *and all* $j, k \in \{1, 2, 3\}$, $x_j >_i x_k \Leftrightarrow t_j >_i' t_k$. *Then C on* $\mathcal{A} \times \{f\}$ *cannot satisfy all of the following*: A1; A2; A3;

A6 (intraprofile consistency). *For all* $x, y, a, b \in X$, *if* $x >_i y \Leftrightarrow a >_i b$ *and* $y >_i x \Leftrightarrow b >_i a$ *for all* $i \in N$, *then* $x >_f y \Leftrightarrow a >_f b$.

Since the interprofile independence condition A4 of Theorem 1 is irrelevant under the hypotheses of Theorem 3, we replace it by the intraprofile axiom A6 that requires similar social preferences for similar voter preference patterns on pairs of candidates.

Theorems 1-3 focus on social choices from two-candidate feasible sets. Our other two theorems deal with choices from larger sets.

Theorem 4 (Grether and Plott, 1982). *Suppose C is a social choice function on* $\mathcal{A} \times F$, $|N| \geq 2$, $|X| \geq 4$, $\mathcal{A}$ *contains every subset of X with m or more elements, and F is the set of all functions from N into the set of weak orders on X, with m fixed*, $3 \leq m < |X|$. *Then C cannot satisfy all of the following*:

B1 (no dictator). *For every* $i \in N$ *there exist* $A \in \mathcal{A}$ *and* $f \in F$ *such that* $C(A, f) \not\subseteq \{x \in A : x \gtrsim_i y \text{ for all } y \in A\}$;

B2 (weak axiom of revealed preference). *For all* $f \in F$ *and all* $A, B \in \mathcal{A}$, *if* $A \subset B$ *and* $A \cap C(B, f) \neq \emptyset$, *then* $C(A, f) = A \cap C(B, f)$;

B3 (Pareto optimality). *For all* $x \in A \in \mathcal{A}$ *and all* $f \in F$, *if* $x >_i y$ *for all* $i \in N$, *then* $y \notin C(A, f)$;

B4 (independence). *For all* $A \in \mathcal{A}$ *and all* $f, f' \in F$, *if* $x >_i y \Leftrightarrow x >'_i y$ *for all* $x, y \in A$ *and all* $i \in N$, *then* $C(A, f) = C(A, f')$.

The conditions are numbered here to correspond to the same-numbered conditions in Theorem 1. Given profile f, B2 says that the choice set for A consists of its elements in the choice set of its superset B, provided that $A \cap C(B, f)$ is not empty. The others are straightforward extensions of their Theorem 1 counterparts.

Theorem 3 is based on a single *profile*. Our final theorem is based on a single *feasible set*, namely X. Note also that it requires every choice set to be a singleton.

Theorem 5 (Gibbard, 1973; Satterthwaite, 1975). *Suppose* C *on* $\{X\} \times F$ *is a social choice function,* $|N| \geq 2$, F *is the set of all functions from* N *into the set of weak orders on* X, $|C(X, f)| = 1$ *for every* $f \in F$, *and* $|\cup_F C(X, f)| \geq 3$. *Then* C *cannot satisfy both of the following:*

B1* (no dictator). *For every* $i \in N$ *there exist* $f' \in F$ *and* $x, y \in \cup_F C(X, f)$ *such that* $x >'_i y$ *and* $C(X, f') = \{y\}$;

B5 (nonmanipulability). *For all* $f, f' \in F$ *and all* $i \in N$, *if* $>_j = >'_j$ *for all* $j \neq i$, *then* $C(X, f) \gtrsim_i C(X, f')$.

The final condition presumes that all voters but i have the same preferences in f' as in f. Its conclusion, where $\gtrsim_i$ refers to f, says that voter i cannot get a preferred outcome by falsifying his preferences. In other words, C is not unilaterally manipulable by any voter at any profile.

5. ELECTION METHODS AND PARADOXES

A large number of election methods have been propounded in the past two centuries. While early writers recognized problems with others' methods, it remained for Arrow and others to show that problems are inevitable when three or more candidates are involved. In a word, there is no completely satisfactory election method.

The design of good election methods therefore involves tradeoffs among factors. This section and the next explore this theme further. I shall view election methods from the social choice perspective and pay attention to their presumed data from voters and how they aggregate data to arrive at social choices.

This section outlines four categories of election methods, describes their properties, gives a few axiomatic characterizations, and identifies some of their foibles. The latter are called *paradoxes* when they reveal unanticipated possibilities or problems. The next section says more about data, computation, and axiomatic conditions. For simplicity I focus on the election of one candidate from $m \geq 3$ without worrying about how ties are broken. Various two-candidate election methods, including simple majority and hierarchical representative systems, are axiomatized in May (1952), Murakami (1968) and Fishburn (1973, 1983b). Many of the paradoxical possibilities for $m \geq 3$ are described in Fishburn (1974a), Niemi and Riker (1976), Fishburn and Brams (1983), and Saari (1984, 1987). Gehrlein (1983) gives a thorough review of Condorcet's paradox of voting.

5.1 *Scoring methods*

Scoring methods operate on ranked voter data by assigning different points to different positions in the rankings. Points for each candidate are summed over the voters and the winner (or social ranking) is determined from point totals. Voter rankings might be linear orders, weak orders, or perhaps truncated orders, as when each voter is asked to rank his or her top three choices.

Borda's method, the oldest and still most popular scoring method, has been characterized axiomatically by Young (1974), Hansson and Sahlquist (1976) and Nitzan and Rubinstein (1981). More general scoring methods are axiomatized by Smith (1973) and Young (1975a). A feature of their axioms is to allow n to vary over all positive integers while m is fixed ($m \geq 2$). Young's formulation considers the set Π of all functions π that map the $m!$ linear orders on X into the nonnegative integers with $\pi > 0$ for some order. The *summary profile* π tells how many voters have each of the linear orders on X as their preference ranking. With $C(\pi)$ the choice set for π, two of Young's (1975a) conditions are

consistency: $C(\pi) \cap C(\pi') \neq \varnothing \Rightarrow C(\pi+\pi') = C(\pi) + C(\pi')$;

neutrality: if σ is a permutation on X that maps π into π_σ in the natural way, then $C(\pi_\sigma) = \sigma(C(\pi))$.

The latter condition is a fairness axiom for candidates. Consistency says that if disjoint voting groups with n and n' members have a common candidate in their choice sets, then the choice set of the combined group of $n+n'$ voters consists of the common choices of the original groups.

Consistency and neutrality imply a lexicographic scoring method. When they are joined by

continuity: $[C(\pi) = \{x\},\ \pi' \in \Pi] \Rightarrow C(K\pi+\pi') = \{x\}$ for all sufficiently large integers K,

C is determined by a *simple score vector* $w = (w_1, \ldots, w_m)$ of real numbers: for each π, $C(\pi)$ contains the candidates with the maximum point total when w_k points are assigned to the kth-ranked candidate in each voter's order. Monotonicity implies $w_1 \geq \cdots \geq w_m$, nonconstancy or Pareto optimality implies $w_1 > w_m$, and Pareto optimality or a stronger monotonicity condition gives $w_1 > w_2 > \cdots > w_m$.

Two Borda paradoxes are noted in Fishburn (1974a, 1974b), given $m \geq 3$. First, there are profiles π with the following property: Borda's method linearly orders the candidates; if the first (or last) candidate is deleted and we recompute the Borda ordering for the other $m-1$ *after* this deletion, the new ordering is the reverse of the original ordering of those $m-1$. Second, there are profiles that have Borda winner x, but x is a Borda loser (recomputed after deletions) for *every* proper subset of X that contains x and at least one other candidate, *except* for one subset with two members. The first of these paradoxes is extensively generalized by Saari (1982).

Further anomalies arise from different score vectors applied to a profile without deletions. Let w and w' be strictly decreasing score vectors for m candidates, normalized by $w_1 = w_1' = 1$ and $w_m = w_m' = 0$. When $w \neq w'$, Fishburn (1981) shows that there is a profile for which the decreasing order of w point totals is the reverse of the decreasing order of w' point totals. Saari (1984) generalizes this by dropping monotonicity for w and w' and allowing any linear orders on X for the point-total orders obtained with w and w'.

It has also been observed that Borda's method is superior in several ways to all other scoring methods based on monotonic simple score vectors. Suppose linear orders are assigned randomly and independently to voters with equal probability for each of the $m!$ orders. Given $m=3$, Gehrlein and Fishburn (1978a) compute the probability that the winner for score vector $w=(1,\lambda,0)$ coincides with the majority candidate, given its existence. They show that in the limit as $n\to\infty$ the coincidence probability is maximized by Borda's method ($\lambda=1/2$), where it equals approximately 0.90119. Related results appear in Gehrlein and Fishburn (1978b, 1981). Under the same random assignment assumption and with nonincreasing nonconstant score vectors $w=(w_1,\ldots,w_m)$ and $v=(v_1,\ldots,v_{m-1})$, let $P_m(w,v)$ be the limit in n of the probability that the w winner from X equals the v winner after a loser, chosen at random, is deleted. Then $P_m(w,v)$ for each $m\geq 3$ is maximized when w and v are Borda vectors (Gehrlein et al., 1982). Hence Borda's method is comparatively robust against changing the winner when a loser is deleted.

5.2 *Condorcet methods*

The Copeland and Nanson methods are but two of many proposals that elect the majority candidate when it exists. Henriet (1985) gives an axiomatization of Copeland's method. Nanson (1883, p. 217) discusses a modification of his method (called "Nanson's function" in Fishburn, 1977) that deletes only the lowest Borda score candidate(s) at each stage. *Black's method* (Black, 1958) elects the majority candidate when it exists and otherwise chooses the Borda winner.

To define other Condorcet methods, let F be the set of n-tuples $f=(>_1,\ldots,>_n)$ of *linear* orders on X. As before, $C(A,f)$ is the choice set from $A\subseteq X$ when f obtains. Some methods, like Copeland's, depend only on the simple majority relation M_f defined on X for each $f=(>_1,\ldots,>_n)$ by

$$xM_fy \quad \text{if} \quad |\{i\in N: x>_iy\}| > |\{i\in N: y>_ix\}|\,.$$

To define three of these, let $U_f(x,A)=\{y\in A: yM_fx\}$, x's *up set* in A; $D_f(x,A)=\{y\in A: xM_fy\}$, x's *down set* in A; and let $D_f^+(x,A)$ be all y in A such that there is an M_f path from x to y in A, i.e., x's *extended down set* in A.

Schwartz's method (Schwartz, 1972): $C(A,f)=\{x\in A: x\in D_f^+(y,A)$ & $y\notin D_f^+(x,A)$ for no $y\in A\}$.

Fishburn's method (Fishburn, 1977): $C(A,f)=\{x\in A: U_f(y,A)\subset U_f(x,A)$ for no $y\in A\}$.

Miller's method (Miller, 1980, 1983): $C(A,f)=\{x\in A: D_f(x,A)\subset D_f(y,A)$ for no $y\in A\}$.

In crude terms, x is in Schwartz's choice set if it is in A's "top cycle", in Fishburn's choice set if nothing else in A has a "smaller" up set, and in Miller's choice set if nothing else in A has a "larger" down set. Miller's choice set, often referred to as A's *uncovered set*, has been examined extensively by others (Shepsle and Weingast, 1984; Banks, 1985; McKelvey, 1986; Feld et al., 1987; Hartley and Kilgour, 1987). A recent proposal in this area is *Dutta's method* (Dutta, 1988) which takes $C(A,f)$ as the necessarily unique minimal $B \subseteq A$ such that B equals B's uncovered set and, for all $x \in A \setminus B$, x is not in the uncovered set of $B \cup \{x\}$. Dutta characterizes his method axiomatically and notes that it is always a subset of Miller's choice set.

Other methods that depend only on M_f are *pairwise majority sequential elimination methods* often used in practice. When $A = \{a_1, \ldots, a_k\}$ and the *voting order* is $a_1, a_2 \cdots a_k$, we compare a_1 to a_2, keep the majority winner, compare it to a_3, keep the majority winner, ..., through $k-1$ binary comparisons. Assuming that ties might arise and that they are not broken in a prespecified manner, $C(A,f)$ would be the set of candidates in A that survive to the end under some way of breaking majority ties during the process. The union of such $C(A,f)$ over *all* voting orders contains $a \in A$ if and only if the other $k-1$ members of A can be ordered in some way, say $x_2 \cdots x_k$, so that $a(M_f \cup T_f)x_2\ (M_f \cup T_f)x_3 \cdots (M_f \cup T_f)x_k$, where T_f denotes majority tie.

Another class of Condorcet methods requires preference counts for binary comparisons. Let

$$N_f(x, y) = |\{i \in N : x >_i y\}| .$$

Since Borda scores can be computed from these counts, the Nanson and Black methods are in this class. Let $M(x, A, f) = \min\{N_f(x, y) : y \in A \setminus \{x\}\}$, the smallest count for x against a member of A. The best of the worst wins under

Condorcet's maximin method: $C(A,f) = \{x \in A : M(x, A, f) \geq M(y, A, f)$ for all $y \in A\}$.

Given (A, f) let $\mathcal{L}$ be the set of linear orders L on A that maximize the sum of all $N_f(x, y)$ for which xLy and $x, y \in A$.

Kemeny's method (Kemeny, 1959; Slater, 1961): $C(A,f) = \{x \in A :$ for some $L \in \mathcal{L}$, xLy for all $y \in A \setminus \{x\}\}$.

Kemeny's method chooses candidates at the tops of the orders on A that maximize "agreement" with f on A. Suppose that, in the variable electorate formulation of the preceding subsection with X as the feasible set, we define a choice function $C(X, \pi) = C(\pi)$ by the top candidates in a set $\mathcal{L}(\pi)$ of linear orders on X, as done for Kemeny's method. Young and Levenglick (1978) prove that Kemeny's method is the unique choice function of this type whose underlying $\mathcal{L}$ function satisfies

neutrality: if σ is a permutation on X that maps π into π_σ, then $\mathcal{L}(\pi_\sigma) = \sigma(\mathcal{L}(\pi))$;

consistency: $\mathcal{L}(\pi) \cap \mathcal{L}(\pi') \neq \varnothing \Rightarrow \mathcal{L}(\pi + \pi') = \mathcal{L}(\pi) \cap \mathcal{L}(\pi')$;

Condorcet: $N_\pi(x, y) > N_\pi(y, x) \Rightarrow$ if $L = \cdots yx \cdots$, then $L \notin \mathcal{L}(\pi)$; $N_\pi(x, y) = N_\pi(y, x) \Rightarrow [\cdots xy \cdots \in \mathcal{L}(\pi) \Leftrightarrow \cdots yx \cdots \in \mathcal{L}(\pi)]$.

The third and final class of Condorcet methods requires information about f that goes beyond the binary preference counts. We mention two of these. Call $x \in A$ a *Condorcet alternative for* A (with respect to f) if yM_fx for no $y \in A$. Let $\alpha(x, A, f)$ be the minimum number of inversions in the linear orders of $f = (>_1, \ldots, >_n)$ that make x a Condorcet alternative for A, and let $\beta(x, A, f)$ be the number of components (possibly 0) in the largest subsequence g of f for which x is a Condorcet alternative for A (with respect to g). To smooth out integer effects let

$$\alpha^*(x, A, f) = \lim_{K \to \infty} \alpha(x, A, Kf) / K,$$

$$\beta^*(x, A, f) = \lim_{K \to \infty} \beta(x, A, Kf) / K,$$

where Kf is f replicated K times.

Dodgson's method (Dodgson, 1876; Black 1958): $C(A, f) = \{x \in A : \alpha^*(x, A, f) \leq \alpha^*(y, A, f) \text{ for all } y \in A\}$.

Young's method (Young, 1975b): $C(A, f) = \{x \in A : \beta^*(x, A, f) \geq \beta^*(y, A, f) \text{ for all } y \in A\}$.

I use Dodgson's name for the first of these since he may have originated inversion counts as a route to a social choice in the absence of a majority candidate. The use of limits allows both methods to satisfy

homogeneity: $C(A, Kf) = C(A, f)$ for all $K \in \{1, 2, \ldots\}$.

Paradoxes arise from some Condorcet methods. When $|A| \geq 4$ and C adheres to Schwartz's method or a sequential elimination method, we can have $y \in C(A, f)$ when there is an $x \in A$ that *every* voter prefers to y, in violation of Pareto optimality (Fishburn, 1974a, 1977). We say that *monotonicity is violated* if $x \in C(A, f)$ and, after x moves up in some $>_i$ to give f', all else unchanged, $x \notin C(A, f')$. Dodgson's method violates monotonicity for $|A| \geq 4$ but not $|A| = 3$ (Fishburn, 1977, 1982). Smith (1973) shows that Nanson's method (modified to delete the lowest Borda score) violates monotonicity for $|A| \geq 3$. He shows that this is true also for a range of scoring methods that proceed by sequential elimination. Fishburn (1982) shows that pairwise majority sequential elimination methods violate monotonicity when $|A| \geq 4$.

Another paradox, dubbed the *no show paradox* by Fishburn and Brams (1983), is endemic to Condorcet methods. Using the variable electorate formulation, Moulin (1988) assumes that choice sets are singletons. His version of the no show paradox occurs whenever the addition of a new voter changes the winner to a candidate the new voter prefers less than the candidate who would have won had he not voted. Moulin proves that this cannot happen for a Condorcet method when $|A| \leq 3$ but must be possible for every Condorcet method when $|A| \geq 4$ and $n \geq 25$. Holzman (1989) extends the analysis. Moulin also discusses a related *twin paradox* whereby the addition of a voter with preferences identical to another voter causes the election of a candidate less preferred by both. This can occur for every Condorcet method when $|A| = m \geq 4$ and $n \geq 25 + \binom{m}{2}$.

Tideman (1987) shows how many Condorcet methods and others can be perversely affected by *clones*, i.e., candidates who are similar to other candidates. A Condorcet method based on counts N_f that is "independent of clones" is described in Tideman (1987) and Zavist and Tideman (1989).

5.3 *NonCondorcet sequential elimination methods*

We have seen that some sequential elimination methods are Condorcet methods. Other are not. The most widely used are plurality runoff schemes and single transferable vote methods (quota systems) described in Section 2. Many can be recast as sequential scoring methods.

The *plurality runoff method* asks each voter to vote for one candidate on the first ballot then conducts a simple majority election between the two with the most first ballot votes. Fishburn and Brams (1983) illustrate paradoxes for this method for $m=3$, including violations of monotonicity, the *multiple-districts paradox* (Young, 1974) in which the same candidate wins in each district but loses the combined-districts election, and the no show paradox. Their version of the latter says that the addition of identical ranked ballots with candidate x in *last* place (by the potential no-shows) can change the winner from another candidate to x. The monotonicity paradox was described earlier by Smith (1973), Doron and Kronick (1977) and Doron (1979), and further analysis of no show appears in Ray (1986).

The description of no show just given assumes that the two stages of plurality runoff are based on a single ranked ballot. A related paradox, recognized as early as 1910 (Report, 1910; Meredith, 1913), occurs for quota systems with ranked ballots when two or more of $m \geq 4$ candidates are to be elected. It happens when one of the elected candidates would have been a loser if more people who ranked him first had voted. This and a related paradox involving truncated voter rankings are also illustrated in Fishburn and Brams (1983). For more on truncation effects see Fishburn and Brams (1984).

5.4 *Unranked methods*

We conclude this section with comments on election methods that only ask voters to check some number of names on the ballot. For tally purposes, each name checked is awarded a full vote. *Single-stage methods* use one ballot and elect the candidate with the most votes. *Two-stage methods* use a similar first ballot to decide which $k \geq 2$ candidates are to go onto a second ballot from which the winner is determined.

Each single-stage method is described by a nonempty subset S of $\{1, 2, \ldots, m\}$ and the rule that a ballot will be counted if and only if s names are checked for some $s \in S$. The plurality method has $S=\{1\}$, Nanson (1883) mentions $S=\{2\}$, and "vote for no more than 3" has $S=\{1, 2, 3\}$. Two-stage methods can be described by a pair (S_1, S_2) of sets of positive integers, where S_j applies to ballot j, and the number $k \geq 2$ of candidates that go onto the second ballot. The most common is the plurality runoff method described earlier.

Although unranked methods are not Condorcet methods, extensive computer simulations and theoretical studies have assessed their propensities for electing the majority candidate when it exists. The most salient results of simulation studies conducted by W. V. Gehrlein and the present author during the 1970s are summarized in Fishburn and Gehrlein (1982), which provides references to the earlier work. Also see Gehrlein (1983). Among practical systems, plurality is rather bad. For example, when $m=5$, it elects the majority candidate, given existence, about 60% of the time. By comparison, approximate percentages are 70 for $S=\{2\}$ and $S=\{1, 2, 3\}$, and 86 for Borda's method. For $m=5$ and two-stage methods that have $k=2$, the comparable figures are 82, 90 and 99 (for a runoff between the highest two Borda scores).

Theoretical assessments of the plurality method are included in Gehrlein and Fishburn (1978b). With $m=3$ and random assignments of preference orders to voters, they show that the limit ($n \to \infty$) probability that the plurality winner is the same as the Borda winner is

$$123/144 + [3/(4\pi^2)]\{\sin^{-1}(\sqrt{13}/4)\}^2,$$

or approximately 0.758338. A more complex expression is obtained for the limit probability that the plurality winner is inferior to each of the others on the basis of simple majorities. Its numerical value is about 0.033843.

Approval voting, a term coined by Robert Weber, is the single-stage method with $S=\{1,2,\ldots,m\}$. It arose from several sources in the 1970s (Ottewell, 1977; Kellett and Mott, 1977; Weber, 1977; Brams and Fishburn, 1978; Morin, 1980), was axiomatized by Fishburn (1978a, 1978b) and Sertel (1988a) [see also Sertel (1988b)], field tested by The Institute of Management Sciences in 1985 (Fishburn and Little, 1988), and has been adopted for the elections of five professional societies. Extensive information and analyses appear in Brams and Fishburn (1983) and Merrill (1988). Our studies show that, among other things, it does fairly well at electing majority candidates, is immune to strategic manipulation for $m=3$ and is very difficult to manipulate by insincere voting for $m \geq 4$, and tends to elect candidates who are well supported both by voters who vote for only one or two candidates and by those who vote for many (Brams and Fishburn, 1988; Brams, 1988).

Although a voter can vote for any number of candidates in an approval voting election, several people (Weber, 1977; Merrill, 1979; Brams and Fishburn, 1983, Section 5.5) have shown that an optimal voting strategy is to vote for all candidates that lie above the "midpoint in preference" between one's most preferred and least preferred candidates. Data from the TIMS field test suggest that many people do this.

6. COMPARING ELECTION METHODS

Since the late 1700s many books and an untold number of reports and articles have been devoted to comparing election methods and arguing the virtues of new methods designed to correct perceived defects in older methods. What have we learned from this, and where do we go from here?

As noted earlier, tradeoffs among factors must be considered in comparing and designing election systems. Of the many aspects spoken of earlier, I focus on five that strike me as some of the most relevant to comparative analyses:

1. Data requirements
2. Computational reasonableness
3. Strategic manipulability
4. Conditions for social choice
5. Probability of undesirable effects.

The following remarks reinforce and supplement things said in previous sections. They intend to raise questions as much as to provide definitive answers.

6.1 *Data requirements*

It is important to understand the electorate and its motivation and ability to respond to ballot instructions in a meaningful way. What is fine for a university group may be unsuitable for a large diverse electorate. While this seems obvious, I mention it because many proposed methods have been devised by technically sophisticated people who may be more mathematically than practically oriented.

I have deliberately omitted methods that elicit data on preferences or utilities that go beyond rankings. Even rankings may presume far more than is reasonable. What sense would it make in an election among ten candidates to ask voters for complete rankings when they know nothing about most of the ten? Voter indecisiveness in polls suggests that many people have no clear first choice, not to mention second, third, ... choices.

When it is reasonable to elicit voter rankings, many aggregation methods are possible, and other factors must be considered in deciding among the alternatives. However, even when rankings are feasible, there may be reasons for looking at unranked methods. In such cases, methods such as plurality, approval voting, and plurality with a runoff come into play.

6.2 *Computational reasonableness*

This factor involves both the ease of summarizing ballot data and the computation needed to determine the choice set or the winning candidate.

Ballot summarization is easiest for plurality and not much harder for approval voting. This partly explains why the plurality method is so widely used and why approval voting has attracted considerable attention. Borda and other simple scoring methods ask for tabulation of the number of first choices, second choices, and so forth for each candidate. However, Borda can also make do with the same majority comparison counts required by Condorcet maximin and Kemeny. The simplest Condorcet methods (Copeland, Schwartz, Fishburn, Miller) need only pairwise majority comparisons without the counts, and these might be elicited by direct pairwise votes if m is small. The most complex Condorcet methods (Dodgson, Young) can require the entire summary profile π as defined earlier. If m is large and n is very large, it could be onerous to calculate and record π.

Given the summarized ballot data, the computation of the choice set is trivial for plurality and approval voting, straightforward for scoring methods and the simpler Condorcet methods, and may be prohibitively difficult for the most complex Condorcet methods. The possibility of ties, or nonsingleton choice sets, can complicate matters. Nanson (1883) argued against the modification of his method that deletes only the lowest Borda score candidate on each round because of the calculations required by the latter. He also saw Condorcet's maximin method as computationally impractical in comparison with Copeland's method.

The advent of computers obviously changes the picture. It has also given rise to theories of computational complexity (Garey and Johnson, 1979) that have been applied to social choice problems. In the variable electorate mode, Bartholdi, Tovey and Trick (1989a) prove that it is NP-hard to compute Dodgson and Kemeny winners. Also see Kelly (1988a, 1988b) on the same theme. A somewhat different complexity theme is pursued by Lewis (1985) and Gottinger (1987).

One other aspect of computability that could be important is whether a voter or candidate with some facility at arithmetic could easily compute the winner, given the summarized ballot data.

6.3 *Strategic manipulability*

Most voting systems intend voters to report their preferences truthfully. Some, however, encourage insincere voting or strategic misrepresentation of preferences, and this is usually viewed as a defect of those systems. Plurality voting is a good example. If a voter's first choice has no chance, he or she will often vote for another candidate to avoid a wasted vote. Brams and Fishburn (1983, Chapter 9) estimate that about 75 percent of the voters in the 1980 United States presidential election who preferred John Anderson voted instead for one of the two front runners, Ronald Reagan and Jimmy Carter. Anderson won about 7 percent of the popular vote, but would have gotten more than 20 percent if all voters had voted sincerely.

The Borda method also encourages misrepresentation. If a voter has a clear favorite, ranked first, he may rank the strongest competitor last to decrease its chances even though he does not prefer it least. Under plurality with a runoff, some voters may vote insincerely on the first ballot to promote a candidate who could be defeated by their favorite in a runoff.

Other procedures, including many Condorcet methods, are complex enough to discourage misrepresentation even when voters know a lot about other voters' preferences. However, Gärdenfors' theorem (Gärdenfors, 1976) indicates manipulation possibilities for all Condorcet methods. The computational complexity analysis of Bartholdi, Tovey and Trick (1989b) shows that a tie-breaking variant of Copeland's method is very difficult to manipulate by misrepresentation.

While approval voting seems fairly robust against strategic misrepresentation for voter gain, it does encourage voters to think strategically. Whether a voter votes for his second choice, second and third choices, ..., in addition to his first choice, may depend on his perceptions of their chances as well as his preferences.

6.4 *Conditions for social choice*

A conservative estimate of the number of conditions proposed as desirable properties for election methods or social choices functions since the time of Borda and Condorcet is about 100. Their sheer number and variety are indicative of the challenge one meets when comparing existing methods or designing new methods.

Several writers, including Fishburn (1977), Richelson (1979, 1981), Nurmi (1987), Tideman (1987) and Banks and Bordes (1988) provide cross classifications of election methods and desirable conditions that tell which methods satisfy which conditions. Their tables organize and give coherence to the comparison process.

Most people want elections that are fair to voters and candidates and that elect candidates who are best for the purposes for which they are to be elected. But fairness and related concepts are multifaceted, and different people view them differently. For example, some people regard the election of a majority candidate when one exists as the preeminent feature of a good election method while other do not. If you prefer x to y but the presence of your sincere vote could change the winner from x to y, you might wonder about the fairness of the method. Nevertheless, every Condorcet method exhibits this anomaly in certain situations.

My point is not that Condorcet methods are generally unfair or inimical to the desires of the electorate. It is rather that every election method, no matter how reasonable and fair it seems, can encounter paradoxes and problems. The task for people who evaluate alternative methods is partly to determine which methods are least susceptible to the more noxious paradoxes and problems.

6.5 *Probability of undesirable effects*

Of all paradoxical concerns, those associated with Condorcet's paradox of voting have received far more attention than the others taken together. Since most of this is reviewed in Gehrlein (1983), I shall make only a few remarks.

Several studies (Garman and Kamien, 1968; Niemi and Weisberg, 1968; Gehrlein and Fishburn, 1976, 1979) assess the probability of the existence of a majority candidate based on randomly generated preference profiles. Beginning with Riker (1958) and Niemi (1970), others have sought cases in practice in which Condorcet's paradox might have arisen. Recent studies on this theme, which also address the question of whether the majority candidate would have been elected by various methods, include Coombs, Cohen and Chamberlin (1984), Fishburn (1986) and Fishburn and Little (1988). Theoretical assessments of the likelihoods that various election methods will elect the majority candidate are summarized in Fishburn and Gehrlein (1982) and Gehrlein (1983).

The latter studies are based on computer simulation. The same method was used in Fishburn (1974a) to look for instances in which a Pareto dominated candidate is elected under pairwise majority sequential elimination and to assess the propensities of paradoxes associated with Borda's method. Generally speaking, the probabilities of the paradoxes examined were quite small. For example, of 12,000 instances with $m \in \{4, 5, 6, 7\}$ and $n \in \{4, 11, 21\}$, one instance of a Pareto dominated winner was observed.

On a more theoretical line Ray (1986) shows that in a three-candidate contest the probability of the no show paradox is high when the winner under single transferable vote (plurality with a runoff) differs from the plurality winner. He says that this considerably weakens the attractiveness of plurality with runoff as a alternative to the basic plurality method.

The recent book by Merrill (1988) is highly recommended for further study on the comparative issue. It includes a number of results based on simulation that address effects of strategic voting as well as nonstrategic behavior.

7. PROBABILITY IN VOTING THEORY

Probability theory has played a role in the theory of elections well beyond its use in addressing likelihoods of paradoxes and outcomes. Historical examples are found in Condorcet (1785) and Laplace (1812). I conclude with recent uses of probability in voting theory that were not discussed earlier.

Several authors, including Kreweras (1965), Zeckhauser (1969), Barbera and Sonnenschein (1978) and Fishburn (1973, Chapter 18; 1978c; 1987, Chapter 8) consider probability distributions or lotteries on candidates for making social choices. Some of this work focuses on impossibility theorems (Barbera and Sonnenschein, 1978; Fishburn, 1987) and some on properties satisfied or violated by social choice lottery methods.

We illustrate the latter with a proposal of Kreweras (1965) that is generalized in Fishburn (1984b). The voting data are, for each pair of candidates $\{x, y\}$, the number of voters who express a preference for x over y minus the number who express the opposite preference. No assumption is made about whether voters have transitive preferences. The choice set is a nonempty set of lotteries over candidates, no one of which is inferior to another on the basis of a simple bilinear form based on the data. In many cases the choice set assigns probability 1 to a particular candidate. This happens whenever there is a majority candidate. More generally, if the set of candidates can be partitioned into A and B such that more voters prefer a to b than b to a whenever $a \in A$ and $b \in B$, then every lottery in the choice set assigns probability 0 to B. In addition, if there is a way to assign weak orders to voters that explain the data by sincere expressions of preference, and if x Pareto dominates y by this assignment, then y has zero probability of becoming the winner.

Probabilities have also been used at the level of voters' choices to replace traditional assumptions of voter preference rankings and unambiguous preferences by individual choice probabilities. Examples are Intriligator (1973) and Fishburn and Gehrlein (1977). The first of these takes $m \geq 2$ and bases social lotteries on individuals' choice probabilities. The second focuses on $m = 2$ and uses ordinary votes to construct social choice lotteries on the two candidates.

Electoral uncertainties both by candidates about voters' desires and voters about candidates' positions on issues have been considered in spatial models of elections. A recent example is Enelow and Hinich (1989). Their paper analyzes situations at the level of the electorate as a whole but gives references to earlier work on individual uncertainty, and nicely summarizes the main differences between results obtained from deterministic and probabilistic spatial theories.

REFERENCES

Aldrich J. H. (1980) Before the Convention: Strategies and Choices in Presidential Nominating Campaigns. University of Chicago Press, Chicago.

Arrow K. J. (1950) A difficulty in the concept of social welfare. J. Polit. Econ. 58, 328-346.

Arrow K. J. (1963) Social Choice and Individual Values, 2nd edition. Wiley, New York.

Banks J. S. (1985) Sophisticated voting outcomes and agenda control. Social Choice and Welfare 1, 295-306.

Banks J. S., and G. A. Bordes (1988) Voting games, indifference, and consistent sequential choice rules. Social Choice and Welfare 5, 31-44.

Barbera S., and H. Sonnenschein (1978) Preference aggregation with randomized social orderings. J. Econ. Theory 18, 244-254.

Bartholdi J., C. A. Tovey and M. A. Trick (1989a) Voting schemes for which it can be difficult to tell who won the election. Social Choice and Welfare 6, 157-165.

Bartholdi J., C. A. Tovey and M. A. Trick (1989b) The computational difficulty of manipulating an election. Social Choice and Welfare (in press).

Black D. (1948) On the rationale of group decision-making. J. Polit. Econ. 56, 23-34.

Black D. (1958) The Theory of Committees and Elections. Cambridge University Press, Cambridge.

Blair D. H., and R. A. Pollak (1983) Rational collective choice. Sci. Amer. 249, 88-95.
Borda, Jean-Charles de (1781) Mémoire sur les élections au scrutin. Hist. Acad. Royale Scis.
Brams S. J. (1988) MAA elections produce decisive winners. Focus 8, 1-2.
Brams S. J., and M. D. Davis (1974) The 3/2's rule in presidential campaigning. Amer. Polit. Sci. Rev. 68, 113-134.
Brams S. J., and P. C. Fishburn (1978) Approval voting. Amer. Polit. Sci. Rev. 72, 831-847.
Brams S. J., and P. C. Fishburn (1983) Approval Voting. Birkhauser, Boston.
Brams S. J., and P. C. Fishburn (1988) Does approval voting elect the lowest common denominator? PS: Polit. Sci. and Politics 21, 277-284.
Butler D., and D. Stokes (1974) Political Change in Britian, 2nd edition. Macmillan, London.
Campbell A., P. E. Converse, W. E. Miller and D. E. Stokes (1960) The American Voter. Wiley, New York.
Collier K., P. C. Ordeshook and K. Williams (1989) The rationally uninformed electorate: some experimental evidence. Public Choice 60, 3-29.
Condorcet, Marquis de (1785) Essai sur l'application de l'analyse à la probabilité des décisions rendues à la pluralité des voix. Paris.
Coombs C. H., J. L. Cohen and J. Chamberlin (1984) An empirical study of some election systems. Amer. Psychologist 39, 140-157.
Copeland A. H. (1951) A 'reasonable' social welfare function. Mimeographed, University of Michigan Seminar on Applications of Mathematics to the Social Sciences.
Cox G. W. (1989) Undominated candidate strategies under alternative voting rules. Mathematical and Computer Modelling 12, 451-459.
Dasgupta P., P. Hammond and E. Maskin (1979) The implementation of social choice rules: some general results on incentive compatibility. Rev. Econ. Studies 46, 185-216.
Dodgson C. L. (1876) A method of taking votes on more than two issues. Clarendon Press, Oxford. See Black (1958).
Doron G. (1979) The Hare voting system is inconsistent. Polit. Studies 27, 283-286.
Doron G., and R. Kronick (1977) Single transferable vote: an example of a perverse social choice function. Amer. J. Polit. Sci. 21, 303-311.
Downs A. (1957) An Economic Theory of Democracy. Harper and Row, New York.
Dutta B. (1987) Fuzzy preferences and social choice. Math. Soc. Sciences 13, 215-229.
Dutta B. (1988) Covering sets and a new Condorcet choice correspondence. J. Econ. Theory 44, 63-80.
Dutta B., S. C. Panda and P. K. Pattanaik (1986) Exact choice and fuzzy preferences. Math. Soc. Sciences 11, 53-68.
Enelow J. M., and M. J. Hinich (1989) A general probabilistic spatial theory of elections. Public Choice 61, 101-113.
Farquharson R. (1969) Theory of Voting. Yale University Press, New Haven.
Feld S. L., B. Grofman, R. Hartley, M. Kilgour and N. Miller (1987) The uncovered set in spatial voting games. Theory and Decision 23, 129-155.
Ferejohn J. A., and M. P. Fiorina (1975) Purposive models of legislative behavior. Amer. Econ. Rev. 65, 407-414.
Fishburn P. C. (1973) The Theory of Social Choice. Princeton University Press, Princeton.

Fishburn P. C. (1974a) Paradoxes of voting. Amer. Polit. Sci. Rev. 68, 537-546.

Fishburn P. C. (1974b) On the sum-of-ranks winner when losers are removed. Discrete Math. 8, 25-30.

Fishburn P. C. (1977) Condorcet social choice functions. SIAM J. Appl. Math. 33, 469-489.

Fishburn P. C. (1978a) Symmetric and consistent aggregation with dichotomous voting, in J.-J. Laffont (Ed.): Aggregation and Revelation of Preferences. North-Holland, Amsterdam, pp. 201-218.

Fishburn P. C. (1978b) Axioms for approval voting: direct proof. J. Econ. Theory 19, 180-185. Corrigendum 45 (1988), p. 212.

Fishburn P. C. (1978c) Acceptable social choice lotteries, in H. W. Gottinger and W. Leinfellner (Eds.): Decision Theory and Social Ethics, Issues in Social Choice. Reidel, Dordrecht, pp. 133-152.

Fishburn P. C. (1981) Inverted orders for monotone scoring rules. Discrete Appl. Math. 3, 27-36.

Fishburn P. C. (1982) Monotonicity paradoxes in the theory of elections. Discrete Appl. Math. 4, 119-134.

Fishburn P. C. (1983a) Dimensions of election procedures: analyses and comparisons. Theory and Decision 15, 371-397.

Fishburn P. C. (1983b) A new characterization of simple majority. Econ. Letters 13, 31-35.

Fishburn P. C. (1984a) Discrete mathematics in voting and group choice. SIAM J. Alge. Discrete Methods 5, 263-275.

Fishburn P. C. (1984b) Probabilistic social choice based on simple voting comparisons. Rev. Econ. Studies 51, 683-692.

Fishburn P. C. (1986) Empirical comparisons of voting procedures. Behav. Sci. 31, 82-88.

Fishburn P. C. (1987) Interprofile Conditions and Impossibility. Harwood Academic, Chur.

Fishburn P. C., and S. J. Brams (1983) Paradoxes of preferential voting. Math. Mag. 56, 207-214.

Fishburn P. C., and S. J. Brams (1984) Manipulation of voting by sincere truncation of preferences. Public Choice 44, 397-410.

Fishburn P. C., and W. V. Gehrlein (1977) Towards a theory of elections with probabilistic preferences. Econometrica 45, 1907-1924.

Fishburn P. C., and W. V. Gehrlein (1982) Majority efficiencies for simple voting procedures: summary and interpretation. Theory and Decision 14, 141-153.

Fishburn P. C., and J. D. C. Little (1988) An experiment in approval voting. Management Sci. 34, 555-568.

Galton F. (1907) One vote, one value. Nature 75, 414.

Gärdenfors P. (1976) Manipulation of social choice functions. J. Econ. Theory 13, 217-228.

Garey M., and D. Johnson (1979) Computers and Intractability: A Guide to the Theory of NP-Completeness. Freeman, San Francisco.

Garman M., and M. Kamien (1968) The paradox of voting: probability calculations. Behav. Sci. 13, 306-316.

Gehrlein W. V. (1983) Condorcet's paradox. Theory and Decision 15, 161-197.

Gehrlein W. V., and P. C. Fishburn (1976) The probability of the paradox of voting: a computable solution. J. Econ. Theory 13, 14-25.

Gehrlein W. V., and P. C. Fishburn (1978a) Coincidence probabilities for simple majority and positional voting rules. Social Sci. Research 7, 272-283.

Gehrlein W. V., and P. C. Fishburn (1978b) Probabilities of election outcomes for large electorates. J. Econ. Theory 19, 38-49.

Gehrlein W. V., and P. C. Fishburn (1979) Proportions of profiles with a majority candidate. Computers and Mathematics with Applications 5, 117-124.

Gehrlein W. V., and P. C. Fishburn (1981) Scoring rule and majority agreements for large electorates with arbitrary preferences. Math. Soc. Sciences 2, 23-33.

Gehrlein W. V., B. Gopinath, J. C. Lagarias and P. C. Fishburn (1982) Optimal pairs of score vectors for positional scoring rules. Appl. Math. and Optimization 8, 309-324.

Gibbard A. (1973) Manipulation of voting schemes: a general result. Econometrica 41, 587-601.

Gottinger H. W. (1987) Choice and complexity. Math. Soc. Sciences 14, 1-17.

Granberg D., and E. Brent (1980) Perceptions of issue positions of presidential candidates. Amer. Scientist 68, 617-625.

Grether D. M., and C. R. Plott (1982) Nonbinary social choice: an impossibility theorem. Rev. Econ. Studies 49, 143-149.

Grofman B. (1975) A review of macro election systems, in R. Wildenmann (Ed.): Sozialwissenschaftliches Jahrbuch für Politik 4. Günter Olzog, Munich, pp. 303-352.

Groves T. (1979) Efficient collective choice when allocation is possible. Rev. Econ. Studies 46, 227-241.

Hansson B., and H. Sahlquist (1976) A proof technique for social choice with variable electorate. J. Econ. Theory 13, 193-200.

Hare T. (1861) The Election of Representatives, Parliamentary and Municipal: A Treatise. Longman, Green, London.

Hansanyi J. C. (1969) Rational-choice models of political behavior vs. functionalist and conformist theories. World Politics 21, 513-538.

Hartley R., and D. M. Kilgour (1987) The geometry of the uncovered set in the three-voter spatial model. Math. Soc. Sciences 14, 175-183.

Henriet D. (1985) The Copeland choice function. An axiomatic characterization. Social Choice and Welfare 2, 49-63.

Holzman R. (1989) To vote or not to vote: what is the quota? Discrete Appl. Math. 22, 133-141.

Intriligator M. (1973) A probabilistic model of social choice. Rev. Econ. Studies 40, 553-560.

Kellett J., and K. Mott (1977) Presidential primaries: measuring popular choice. Polity 11, 528-537.

Kelly J. S. (1978) Arrow Impossibility Theorems. Academic Press, New York.

Kelly J. S. (1988a) Social choice and computational complexity. J. Math. Econ. 17, 1-8.

Kelly J. S. (1988b) 5. Computational complexity. Social Choice and Welfare 5, 313-317.

Kemeny J. (1959) Mathematics without numbers. Daedalus 88, 577-591.

Key V. O. (1966) The Responsible Electorate. Harvard University Press, Cambridge, MA.

Kreweras G. (1965) Aggregation of preference orderings, in S. Sternberg et al. (Eds.): Mathematics and Social Sciences 1. Mouton, Paris.

Laffont, J.-J., and H. Moulin, Eds. (1982) Special issue on implementation. J. Math. Econ. 10, 1-145.

Laplace, Marquis de (1812) Théorie Analitique des Probabilités. Paris.
Lewis A. A. (1985) On effectively computable realizations of choice functions. Math. Soc. Sciences 10, 43-80.
Lipset S. M. (1960) Political Man. Doubleday, New York.
Mas-Colell A., and H. Sonnenschein (1972) General possibility theorems for group decisions. Rev. Econ. Studies 39, 185-192.
May K. O. (1952) A set of independent necessary and sufficient conditions for simple majority decisions. Econometrica 20, 680-684.
McAfee R. P., and J. McMillan (1988) Multidimensional incentive compatibility and mechanism design. J. Econ. Theory 46, 335-354.
McKelvey R. D. (1986) Covering, dominance and institution free properties of social choice. Amer. J. Polit. Sci. 30, 283-315.
McKelvey R. D., and P. Ordeshook (1972) A general theory of the calculus of voting, in J. Herndon (Ed.): Mathematical Applications in Political Science, VI. University Press of Virginia, Charlottesville, pp. 32-78.
Meredith J. C. (1913) Proportional Representation in Ireland. Dublin.
Merrill S. (1979) Approval voting: a 'best buy' method for multicandidate elections? Math. Mag. 52, 98-102.
Merrill S. (1988) Making Multicandidate Elections more Democratic. Princeton University Press, Princeton.
Miller N. R. (1980) A new 'solution set' for tournaments and majority voting. Amer. J. Polit. Sci. 24, 68-96.
Miller N. R. (1983) The covering relation in tournaments: two corrections. Amer. J. Polit. Sci. 27, 382-385.
Morin R. A. (1980) Structural Reform: Ballots. Vantage Press, New York.
Moulin H. (1979) Dominance solvable voting schemes. Econometrica 47, 1337-1351.
Moulin H. (1988) Condorcet's principle implies the no show paradox. J. Econ. Theory 45, 53-64.
Mueller D. C. (1976) Public choice: a survey. J. Econ. Literature 14, 395-433.
Murakami Y. (1968) Logic and Social Choice. Routledge and Kegan Paul, London.
Nanson E. J. (1883) Methods of elections. Transactions and Proceedings Royal Society Victoria 19, 197-240.
Niemi R. G. (1970) The occurrence of the paradox of elections in university elections. Public Choice 8, 91-100.
Niemi R. G., and B. E. Rasch (1987) An extension of Black's theorem on voting orders to the successive procedure. Public Choice 54, 187-190.
Niemi R. G., and W. H. Riker (1976) The choice of voting systems. Sci. Amer. 234, 21-27.
Niemi R. G., and H. F. Weisberg (1968) A mathematical solution for the probability of the paradox of voting. Behav. Sci. 13, 317-323.
Nitzan S., and A. Rubinstein (1981) A further characterization of Borda ranking method. Public Choice 36, 153-158.
Nurmi H. (1981) Approaches to collective decision making with fuzzy preference relations. Fuzzy Sets and Systems 6, 249-259.
Nurmi H. (1987) Comparing Voting Systems. Reidel, Dordrecht.
Ottewell G. (1977) The arithmetic of voting. In Defense of Variety 4, 42-44.
Parks R. P. (1976) An impossibility theorem for fixed preferences: a dictatorial Bergson-Samuelson welfare function. Rev. Econ. Studies 43, 447-450.

Pattanaik P. K. (1978) Strategy and Group Choice. North-Holland, Amsterdam.
Plott C. R. (1976) Axiomatic social choice theory: an overview and interpretation. Amer. J. Polit. Sci. 20, 511-596.
Plott C. R., and M. E. Levine (1978) A model of agenda influence on committee decisions. Amer. Econ. Rev. 68, 146-160.
Pollak R. A. (1979) Bergson-Samuelson social welfare functions and the theory of social choice. Quart. J. Econ. 93, 73-90.
Rae D. W. (1967) The Political Consequences of Electoral Laws. Yale University Press, New Haven.
Ray D. (1986) On the practical possibility of a 'no show paradox' under the single transferable vote. Math. Soc. Sciences 11, 183-189.
Reichelstein S., and S. Reiter (1988) Game forms with minimal message spaces. Econometrica 56, 661-692.
Report of the Royal Commission Appointed to Enquire into Electoral Systems (1910). HMSO, London.
Richelson J. T. (1979) A comparative analysis of social choice functions, I, II, III: a summary. Behav. Sci. 24, 355.
Richelson J. T. (1981) A comparative analysis of social choice functions, IV. Behav. Sci. 26, 346-353.
Riker W. H. (1958) The paradox of voting and Congressional rules for voting on amendments. Amer. Polit. Sci. Rev. 52, 349-366.
Riker W. H. (1961) Voting and the summation of preferences: an interpretive bibliographical review of selected developments during the last decade. Amer. Polit. Sci. Rev. 55, 900-911.
Riker W. H. (1962) The Theory of Political Coalitions. Yale University Press, New Haven.
Riker W. H. (1976) The number of political parties: a reexamination of Duverger's law. Comparative Politics 9, 93-106.
Saari D. G. (1982) Inconsistencies of weighted summation voting systems. Math. Oper. Research 7, 479-490.
Saari D. G. (1984) The ultimate in chaos in weighted voting systems. Advances Appl. Math. 5, 286-308.
Saari D. G. (1987) The source of some paradoxes from social choice and probability. J. Econ. Theory 41, 1-22.
Satterthwaite M. A. (1975) Strategy-proofness and Arrow's conditions: existence and correspondence theorems for voting procedures and social welfare functions. J. Econ. Theory 10, 187-217.
Schwartz T. (1972) Rationality and the myth of the maximum. Noûs 6, 97-117.
Sen A. K. (1970) Collective Choice and Social Welfare. Holden-Day, San Francisco.
Sen A. K. (1977) Social choice theory: a re-examination. Econometrica 45, 53-89.
Sertel M. R. (1988a) Characterizing approval voting. J. Econ. Theory 45, 207-211.
Sertel M. R. (1988b) A non-dictatorial compromise. Social Choice and Welfare 5, 19-29.
Shapiro M. J. (1969) Rational political man: a synthesis of economic and socio-psychological perspectives. Amer. Polit. Sci. Rev. 63, 1106-1119.
Shepsle K. A., and B. R. Weingast (1984) Uncovered sets and sophisticated voting outcomes with implications for agenda control. Amer. J. Polit. Sci. 28, 49-74.
Slater P. (1961) Inconsistencies in a schedule of paired comparisons. Biometrica 48, 303-312.

Smith J. H. (1973) Aggregation of preferences with variable electorate. Econometrica 41, 1027-1041.
Tideman T. N. (1987) Independence of clones as a criterion for voting rules. Social Choice and Welfare 4, 185-206.
Weber R. J. (1977) Comparison of voting systems. Mimeographed.
Wright S. G., and W. H. Riker (1989) Plurality and runoff systems and numbers of candidates. Public Choice 60, 155-175.
Young H. P. (1974) An axiomatization of Borda's rule. J. Econ. Theory 9, 43-52.
Young H. P. (1975a) Social choice scoring functions. SIAM J. Appl. Math. 28, 824-838.
Young H. P. (1975b) Extending Condorcet's rule. Mimeographed, The City University of New York.
Young H. P. (1988) Condorcet's theory of voting. Amer. Polit. Sci. Rev. 82, 1231-1244.
Young H. P., and A. Levenglick (1978) A consistent extension of Condorcet's election principle. SIAM J. Appl. Math. 35, 285-300.
Zavist T. M., and T. N. Tideman (1989) Complete independence of clones in the ranked pairs rule. Social Choice and Welfare 6, 167-173.
Zeckhauser R. (1969) Majority rule with lotteries on alternatives. Quart. J. Econ. 83, 696-703.

FUZZY SET THEORY AS A THEORY OF VAGUENESS

Vilém Novák

Czechoslovak Academy of Sciences, Mining Institute,

A. Římana 1768, 708 00 Ostrava - Poruba, Czechoslovakia

Abstract: The paper is a survey of the main notions and results obtained in fuzzy set theory. The presentation is based on the results in fuzzy logic. The connection of fuzzy set theory with it is demonstrated, formal definition of a structure of membership degrees is presented and few elements of fuzzy logic are succintly mentioned. Then the definitions of basic operations and notions of fuzzy set theory including their properties are presented. Finally, the relation of fuzzy set theory to the notion of possibility is mentioned.

Keywords: Fuzzy sets, Fuzzy logic, Truth value, Possibility theory.

1. INTRODUCTION

A quarter of century had passed since Lotfi A. Zadeh published his seminal paper (Zadeh, 1965) on a special kind of functions which he began to call *fuzzy sets*. In that time, nobody dared to guess that a new mathematical discipline and even a new way ot thinking is science has been founded. Fuzzy sets caused turning of attention of scientists and philosophers to the phenomenon of vagueness which is inherent in human regarding of world. This phenomenon was recognised not to be a consequence of imperfection of measurements but, on the contrary, to be something which makes human being possible to record complex and relevant phenomena and to disregard unimportant, marginal ones. It is by no means accidental that vagueness is one of the nost outstanding features of the semantics of natural language. The reason consists in the fact that the expressions of natural language

J. Kacprzyk and M. Fedrizzi (eds.), Multiperson Decision Making Using Fuzzy Sets and Possibility Theory, 28–42.

reflect the way of human recording of phenomena in the world. *Vagueness* is one of the forms of *indefiniteness*. The second form of indefiniteness is *uncertainty*. This is present when an event is to set in, while vagueness appears when we want to record a given phenomenon independently of its setting in. Therefore, probability theory, which can be considered to be a mathematical theory of uncertainty, cannot be confused with fuzzy set theory (which is sometimes the case).

Fuzzy set theory is one of the theories enabling us to grasp the phenomenon of vagueness in a formal way. There are many successful applications of its in various fields and even in industry. Analogously as classical set theory, fuzzy set theory stems from the assumption that groupings of objects can also be viewed as objects. These new objects, however, can be determined unsharply, i.e. they can be vague. The question of unsharp determination reduces to the question whether a given object, say x, belongs to the given grouping, say A. In symbols, we ask whether

$$x \in A \tag{1}$$

(i.e. x is an element of A) is true. If the grouping A is vague then we are not able in general to answer "yes" or "no". Thus, we may seek some kind of a scale on which we could express a degree of truth of (1). We naturally come to many-valued logic which is usually called *fuzzy logic*. In the next section, we succintly present the demands made on the scale of truth values and few important concepts of fuzzy logic.

2. FUZZY LOGIC

2.2 The structure of truth values

The scale of truth values plays a crucial role in fuzzy set theory. There are many studies devoted to the intuitive and formal requirements on its structure (see e.g. Bellman and Giertz, 1973, Negoita and Ralescu, 1975, Termini et al., 1984, Yager, 1982, and others). In general, it must be a complete lattice. There are good reasons to assume it is a *residuated lattice*, i.e. a complete, infinitely distributive lattice

$$\mathcal{L} = \langle L, \vee, \wedge, \otimes, \rightarrow, \mathbf{1}, \mathbf{0} \rangle \tag{2}$$

where $\mathbf{0}$, $\mathbf{1}$ are the smallest and the greatest elements, respectively, and $\otimes$, $\rightarrow$ are the binary operations of (bold) multiplication and residuation respectively with the following properties:

(a) $\langle L, \otimes, \mathbf{1} \rangle$ is a commutative monoid.
(b) The operation $\otimes$ is isotone in both variables and $\rightarrow$

is antitone in the first variable and isotone in the second one.

(c) The adjunction property

$$a \otimes b \leq c \quad \text{iff} \quad a \leq b \longrightarrow c$$

holds for every a, b, $c \in L$.

The reasons for accepting of such a structure are analysed in Pavelka (1979), Novák and Nekola (1983), Novák and Pedrycz (1988), and Novák (1989).

As a special case, most often used in applications, the degrees of truth are assumed to be based on the interval of real numbers L = ⟨0, 1⟩, where we put $\vee$:= max, $\wedge$:= min,

$$a \otimes b = 0 \vee (a + b - 1), \tag{3}$$

and

$$a \longrightarrow b = 1 \wedge (1 - a + b) \tag{4}$$

for every a, $b \in \langle 0, 1\rangle$. The use of this interval is very natural but it is sometimes the cause of confusion of truth values with probabilities. Let us stress that this is only a formal similarity caused by using the same sets of measures. However, as we have mentioned in Section 1, the probability reflects the indefiniteness connected with setting in of some event. The elements of the above defined structure represent measures of the indefinitenes connected with recording of a phenomenon, no matter whether it does or does not set in.

In the sequel, we also use the operations of *biresiduation*

$$a \longleftrightarrow b := (a \longrightarrow b) \wedge (b \longrightarrow a) \tag{5}$$

and *negation*

$$\neg a := a \longrightarrow \mathbf{0}. \tag{6}$$

Using (3) and (4), we obtain

$$a \longleftrightarrow b = 1 - |a - b|$$

and

$$\neg a = 1 - a$$

for every a, $b \in \langle 0, 1\rangle$.

2.2 Some fundamental concepts of fuzzy logic

Fuzzy logic, similarly as any kind of logic, has its syntax and semantics. The syntax of fuzzy logic consists of rules for the formulae formation, of a list of axioms, i.e. a list of selected formulae a priori assumed to be true (in some degree), and of three rules of inference using which new formulae are derived. The formulae are formed in a usual way from atomic formulae which are connected by the connectives $\vee$ (disjunction), $\wedge$ (conjunction), $\Rightarrow$ (implication), & (bold conjunction), $\{o_j;\ j \in J\}$ (a set of additional *n*-ary connectives), and quantified using $\exists$ (existential quantifier) and $\forall$ (general quantifier). The possibility to use the additional connectives is a specific feature of fuzzy logic.

Among the rules of inference, the rule of *modus ponens* is most important:

$$r_{MP} : \frac{A,\ A \Rightarrow B}{B} \quad \left(\frac{a,\ c}{a \otimes c} \right) \tag{7}$$

In (7), A, $A \Rightarrow B$ are formulae and a, c are the respective truth values. In words, knowing the formulae A and $A \Rightarrow B$ with the respective truth values a and c, we derive the formula B with the truth value $a \otimes c$. The expression in the brackets is a semantic part and the other expression in (7) is a syntactic part of the rule r_{MP}. Note that in the case of $L = \langle 0,\ 1 \rangle$, the operation $\otimes$ cannot be replaced by $\wedge$ (cf. Pavelka, 1979 or Novák and Pedrycz, 1988).

The notion of a *proof* can also be defined. In fuzzy logic, every proof has a value which is derived from the values of the axioms using the semantic part of the rules of inference.

The *semantics* of fuzzy logic deals with the notion of a *model* $\mathcal{D}$. Any formula A has a truth value $\mathcal{D}(A) \in L$ in a given model $\mathcal{D}$. The model is a mathematical abstraction representing our idea what a given formula in fact means. The evaluation of formulae in the model is realised using the operations $\vee$, $\wedge$, $\longrightarrow$, $\otimes$. For example,

$$\mathcal{D}(A \Rightarrow B) = \mathcal{D}(A) \longrightarrow \mathcal{D}(B).$$

Let us remark, that fuzzy logic presented here has the important property of completeness, i.e. the global syntactic and semantic evaluation of formulae coincide. For details and theorems concerning fuzzy logic see Pavelka (1979), Novák (1989) and Novák (1990).

3. FUZZY SET THEORY

3.1 Fuzzy approach to the phenomenon of vagueness

The phenomenon of vagueness is given to us when regarding the world. For many years, science attempted to cover this phenomenon striving for still greater and greater exactness in all respects. However, vagueness is present in any human view of the world and thus, an entirely exact view does not exist. It became indispensable to find a way how to grasp vagueness using some formal tool. Though there are several attempts to develop such a tool, two mathematical theories became elaborated in the best way and get still more popularity, namely the alternative set theory and fuzzy set theory. The former is not dealt with in this paper and the reader is referred to Vopěnka (1979), or Termini et al. (1984).

In general, vagueness is encountered when trying to separate a grouping of objects. However, this may not be possible to be done exactly. Assume we are given a basic grouping of objects in which we want to separate a new one. The former grouping can be considered to be given sharply, i.e. it may be a set. The newly separated grouping can be unsharp. A classical example is the problem how to separate a grouping of numbers which fit the vague phenomenon of "being tall" (we consider people). Then the basic set is, e.g. the interval of real numbers $\langle 40, 300\rangle$ (in cm).

How can an unsharp grouping be characterised? It is natural for the human mind to use a scale which brings us to what can be called *fuzzy approach*: any element of the above mentioned basic set is assigned an element from the scale which measures, how we understand the element to be a member of the unsharp grouping. With respect to the previous sections, we may formalize fuzzy approach as follows: Let U be a basic set of elements and L a scale. Let A be an unsharp grouping of objects taken from U. Then each element of A is characterised by an element of L, i.e. we write

$$A = \{\langle x, c\rangle, x \in U, c \in L\}. \tag{8}$$

The element $c \in L$ can be considered to be a degree of truth of the proposition "x is an element of A". Hence, L is the support of the lattice $\mathcal{L}$ of truth values (2). Since the assignement of the elements $c \in L$ in (8) is unique, A is a function

$$A: U \longrightarrow L. \tag{9}$$

This function is called a *fuzzy set*. The element $A(x) = c \in L$ is usually written as

$$Ax \in L$$

and it is called the *degree of membership* of x in A. If $Ax = \mathbf{1}$ then x belongs to A. If $Ax = \mathbf{0}$ then x does not belong to A.

Many authors (also in this volume) use the symbol $\mu_{\mathbf{A}}$ (sometimes $f_{\mathbf{A}}$, $\chi_{\mathbf{A}}$, etc.) for the function (9) and call it the *membership function* of the fuzzy set A. The symbol A is then used only as a symbol for the unsharp grouping taken as an object. In our notation, the fuzzy set is identified with the membership function which seems to be more transparent and typographically simpler.

3.2 Fundamental concepts of fuzzy set theory

In this section we present some selected concepts of fuzzy set theory. In detail they can be found in the extensive literature, e.g. Negoita and Ralescu (1975), Dubois and Prade (1980), Novák (1989) and on many other places.

The fuzzy set is often written in the form of (8) or in the succint form as

$$A = \bigcup_{x \in U} Ax/x$$

where $Ax \in L$ is a degree of membership of the element $x \in U$. If A is a fuzzy set in U then we write $A \lessapprox U$. The set of all the fuzzy sets in U is denoted by $\mathcal{F}(U)$.

The fuzzy set $\{Ax/x\}$ is called a *fuzzy singleton*. A *support* of a fuzzy set A is a classical set

$$\mathrm{Supp}(A) = \{x;\ Ax \neq \mathbf{0}\}.$$

Similarly, a *kernel* of a fuzzy set A is a classical set

$$\mathrm{Ker}(A) = \{x;\ Ax = \mathbf{1}\}.$$

A fuzzy set A is *normal* if $\mathrm{Ker}(A) \neq \emptyset$.

The a-cut of A is a classical set

$$A_a = \{x;\ Ax \geq a\}.$$

Let $A, B \lessapprox U$. The basic operations with fuzzy sets are:

union

$$C = A \cup B \quad \text{iff } Cx = Ax \vee Bx$$

intersection

$$C = A \cap B \quad \text{iff } Cx = Ax \wedge Bx$$

bold intersection

$$C = A \otimes B \quad \text{iff } Cx = Ax \otimes Bx$$

residuation

$$C = A \ominus B \quad \text{iff } Cx = Ax \longrightarrow Bx$$

complement

$$\bar{A} = A \oplus \emptyset.$$

In the case of $L = \langle 0, 1 \rangle$, we obtain

$$(A \otimes B)x = 0 \vee (Ax + Bx - 1)$$

$$\bar{A}x = \neg Ax = 1 - Ax$$

for every $x \in U$.

A fuzzy set $A \subsetapprox U$ is a *fuzzy subset* of $B \subsetapprox U$, $A \subseteq B$, if

$$Ax \leq Bx$$

holds for every $x \in U$.

A specific feature of fuzzy sets if the possibility to define additional operations on the basis of the operations on L. Given an n-ary operation o: $L^n \longrightarrow L$, it must fulfil the following *fitting condition* (Pavelka, 1979; Novák and Pedrycz, 1988): There are $k_1, \ldots, k_n > 0$ such that

$$(a_1 \leftrightarrow b_1)^{k_1} \otimes \ldots (a_n \leftrightarrow b_n)^{k_n} \leq$$

$$\leq o(a_1, \ldots, a_n) \leftrightarrow o(a_1, \ldots, a_n) \qquad (10)$$

holds for every a_i, $b_i \in L$, i = 1, ... n, where the power is taken with respect to the operation $\otimes$.

Using o, we may define an n-ary operation $O{:}\mathcal{F}(U)^n \longrightarrow \mathcal{F}(U)$ assigning a fuzzy set $C \subsetapprox U$ to every n-tuple of fuzzy set A_1, ..., $A_n \subsetapprox U$ by putting

$$C = O(A_1, \ldots, A_n) \quad \text{iff} \quad Cx = o(A_1x, \ldots, A_nx) \qquad (11)$$

for every $x \in U$. Then it is easy to prove:

Theorem. Let $\{o_d;\ d \in I\}$ be a set of fitting operations enriching the lattice $\mathcal{L}$ (2) which are the bases of the corresponding operations $\{O_d;\ d \in I\}$ due to (11). Then

$$\langle \mathcal{F}(U), \cup, \cap, \otimes, \oplus, \emptyset, U, \{O_d;\ d \in I\}\rangle$$

is a residuated lattice enriched by the operations $\{O_d;$ $d \in I\}$.

This theorem gives rise to the definition of new operations on fuzzy sets and demonstrates their properties.

Some examples of the additional operations are ($L = \langle 0, 1\rangle$)

algebraic product

$$C = A \,.\, B \text{ iff } Cx = Ax \,.\, Bx$$

algebraic sum

$$C = A \oplus B \text{ iff } Cx = Ax + Bx - Ax \,.\, Bx$$

bounded difference

$$C = A \ominus B \text{ iff } Cx = 0 \vee (Ax - Bx),$$

and many others.

Let U be a linearly ordered set. A fuzzy set $A \subsetneqq U$ is *convex* if

$$Ax_2 \geq Ax_1 \wedge Ax_3$$

holds for every $x_1, x_2, x_3 \in U$, $x_1 \leq x_2 \leq x_3$.

Let $U = U_1 \times \ldots \times U_n$ and $A_i \subsetneqq U_i$, $i = 1, \ldots n$. A *Cartesian product* of fuzzy sets $A_1, \ldots, A_n$ is a fuzzy set

$$(A_1, \ldots, A_n)\langle x_1, \ldots, x_n\rangle = A_1x_1 \wedge \ldots \wedge A_nx_n.$$

A fuzzy set

$$R \subsetneqq U_1 \times \ldots \times U_n$$

is called a *fuzzy relation*. Let $R \subsetneqq U \times V$, $S \subsetneqq V \times W$ be fuzzy relations. Then the *composition* of fuzzy relations is a fuzzy relation

$$T = R \circ S \subsetneqq U \times W$$

defined by

$$(R \circ S)\langle x, z\rangle = \bigvee_{y \in V} (R\langle x, y\rangle \wedge S\langle y, z\rangle). \qquad (12)$$

The operation $\wedge$ in (12) can be replaced by $\otimes$. Then we obtain a *weak composition* of fuzzy relations and denote it by $R \underset{x}{\circ} S$.

It is possible to define naturally the inverse, domain and range of fuzzy relations. Important is also the problem of solution of fuzzy relational equations: given fuzzy relations R and T, or S and T, the task is to find the relation S so that $R \circ S = T$ ($R \underset{x}{\circ} S = T$) held true. This problem was first solved by Sanchez (1976).

The following typology of fuzzy sets is often used in applications: Let $\leq$ be a linear ordering in the universe U

and $A \subsetneq U$ be a fuzzy set. Then A is an S^+-fuzzy set if

$$x \preceq y \text{ implies } Ax \leq Ay,$$

and it is an S^--fuzzy set if

$$x \preceq y \text{ implies } Ax \geq Ay$$

for every $x, y \in U$. If there is a point $y_o \in U$ such that

$$Ax \leq Ay_o \text{ and } Az \leq Ay_o$$

holds for every $x \preceq y_o$ and $y_o \preceq z$ then A is a Π-fuzzy set.

A powerful principle of fuzzy set theory is the *extension principle* which makes possible to extend any operation on U to operation with fuzzy sets.

Let $*: U \times U \longrightarrow U$ be a binary operation on U and $A, B \subsetneq U$ be fuzzy sets. Then the extension of "$*$" on the fuzzy sets A, B is the fuzzy set $C = A * B$ with the membership function

$$Cz = \begin{cases} \bigvee \{Ax \wedge Bx;\ x, y \in U \text{ and } z = x * y\} \\ 0 \quad \text{otherwise} \end{cases}$$

for every $z \in U$.

A convex normal fuzzy set $A \subsetneq \mathbb{R}$ is called a *fuzzy number*. Using the extension principle, all the common operations with fuzzy numbers are defined. For details see Dubois and Prade (1978, 1980, 1988), Ramík and Římánek (1985), Novák (1989) and others.

3.3 The use of fuzzy sets in approximate reasoning

Approximate reasoning is a way of reasoning which uses vague concepts. The use of fuzzy sets in it was first proposed by Zadeh (1972, 1975).

3.3.1 A model of natural language semantics

The applications of fuzzy sets are often connected with the use of natural language. Fuzzy sets seem to be a suitable base for the construction of the model of the semantics of certain syntagms (couples of words connected according to the grammatical rules). The most often form of syntagms is the form

adjective - noun,

for example, *a tall man*, *a big house*, etc. Let $\mathcal{A}$ be a syntagm and U a universe. Then the meaning of $\mathcal{A}$ is a fuzzy set $A \subsetneqq U$,

$$M(\mathcal{A}) = A.$$

The membership degree $Ax \in L$ expresses the degree to which an element fits the vague phenomenon being denoted by the syntagm $\mathcal{A}$ and formalised by the fuzzy set A.

The syntagms can be connected by conjunctions *and*, *or*. The meaning of such expressions is then defined using the operations on fuzzy sets. However, the situation is rather complicated from the linguistic point of view and, therefore, we cannot simply assign a unique operation to a conjunction.

Very interesting and quite successful is the model of the meaning of the so called *linguistic modifiers*, i.e. special kind of adverbs. In fuzzy set theory, the following modifiers are quite well elaborated: *very*, *highly*, *more or less*, *roughly* and *rather*.

In general, the meaning of a linguistic modifier $\mathbf{m}$ is a couple of functions

$$M(\mathbf{m}) = \langle \zeta_{\mathrm{m}}, \nu_{\mathrm{m}} \rangle$$

where $\zeta_{\mathrm{m}}: U \longrightarrow U$ is a shifting function and $\nu_{\mathrm{m}}: L \longrightarrow L$ acts on the membership degrees. The meaning of the linguistic expression $\mathbf{m}\mathcal{A}$ where $\mathcal{A}$ is a syntagm is

$$M(\mathbf{m}\mathcal{A}) = \nu_{\mathrm{m}} \circ A \circ \zeta_{\mathrm{m}} ,$$

i.e. the degree of membership of $x \in U$ is

$$M(\mathbf{m}\mathcal{A})x = \nu_{\mathrm{m}}(A(\zeta_{\mathrm{m}}(x))).$$

This makes us possible to construct the meaning of expressions such as *a very young boy*, *more or less big value* etc.

The function ν_{m} may have the following, experimentally derived, forms:

$$\nu_{\mathbf{very}}(a) = a^2$$

$$\nu_{\mathbf{more\ or\ less}}(a) = 2a - a^2$$

$$\nu_{\mathbf{highly}}(a) = a^3$$

$$\nu_{\mathbf{roughly}}(a) = -a^4 + 4a^3 - 6a^2 + 4a$$

$$\nu_{\text{rather}}(a) = \begin{cases} 2a^4 & \text{for } 0 \le a \le 1/\sqrt{2} \\ -2a^4 + 4a^2 - 1 & \text{for } 1/\sqrt{2} < a \le 1 \end{cases}$$

for any $a \in \langle 0, 1 \rangle$.

The function ζ_m is put equal to *Id* (identical function) by some authors. However, this is linguistically incorrect. For example, ζ_{very} was exprerimentally estimated as follows: Let $U = \mathbb{R}$,

$$\|Ker(A)\| = |\sup(Ker(A)) - \inf(Ker(A))|$$

and

$$s_A = \begin{cases} \inf(Ker(A)) & \text{for } A \text{ being an } S^+\text{-fuzzy set} \\ \sup(Ker(A)) & \text{for } A \text{ being an } S^-\text{-fuzzy set} \\ \text{approximate center of } Ker(A) & \text{for } A \text{ being a } \Pi\text{-fuzzy set} \end{cases}$$

Then

$$\zeta_{\text{very}}(x) = x + (-1)^k.d.\|Ker(A)\|$$

where $k = 1$ if A is an S^+-fuzzy set or A is a Π-fuzzy set and $x \le s_A$, otherwise $k = 2$. The constant d was estimated to fall into the interval $\langle 0.25, 0.40 \rangle$. More details can be found in Novák (1989).

3.3.2 The compositional rule of inference

The compositional rule of inference is a tool having the most interesting applications. It is based on the rule of modus ponens mentioned in the preceding section.

Let $\mathcal{A}$ be a linguistic expression. Then the meaning of the proposition

$$p = \text{"}X \text{ is } \mathcal{A}\text{"} \tag{13}$$

is

$$M(p) = M(\mathcal{A}) = A, \tag{14}$$

$A \subsetneq U$. Let z be a proposition

$$z = \text{"IF } X \text{ is } \mathcal{A} \text{ THEN } Y \text{ is } \mathcal{B}\text{"} \tag{15}$$

where $M(\mathcal{B}) = B \subsetneqq U$. Then the meaning of z is

$$M(z) = R = A^{*} \circledast B^{*} \qquad (16)$$

where A^{*}, $B^{*} \subsetneqq U \times V$ are fuzzy relations defined by

$$A^{*}\langle x, y\rangle = Ax$$

$$B^{*}\langle x, y\rangle = By$$

for every $x \in U$ and $y \in V$ (A^{*}, B^{*} are so called cylindrical extensions of A, B, respectively).

The compositional rule of inference is a formula

$$B = A' \blacksquare R$$

where A' is the meaning of the proposition "X is $\mathcal{A}'$" and the symbol $\blacksquare$ denotes a composition of fuzzy relations. The expression $\mathcal{A}'$ can be a modification of $\mathcal{A}$ e.g. by a linguistic modifier. The composition $\blacksquare$ depends on the model of the meaning (15). If (16) is used then $\blacksquare := \underset{\times}{\circ}$ (cf. Section 3.2). This is a direct generalization of the rule of modus ponens (7). If (15) is a linguistic expression for the existence of a causal relation between A and Y then we can put

$$R = A \times B$$

and then use $\blacksquare := \circ$ (12). This case is simpler for computations and it is often used in applications.

4. POSSIBILITY THEORY

Zadeh (1978a) discussed first the concept of possibility, and founded *possibility theory* based on fuzzy set theory.

The notion of possibility is connected with the question of logical contradiction. If, during the deliberation about some event we come to the contradiction with the reason, then such an event is impossible. But, any event which already set in, cannot lead to a contradiction, i.e. it is trivially possible.

We may seek for a measure of realisability which will be called a possiblity measure. We assume the range of its values to be also the interval $\langle 0, 1\rangle$.

There is a relation between probability and possibility which can be formally expressed by the following relation:

$$P(A) \le \Pi(A) \qquad (17)$$

where A is an event, $P(A)$ is its probability and $\Pi(A)$ is the measure of its possibility. The relation (17) should be interpreted as follows: If an event A may set in with the probability $P(A)$ then it is possible with the measure at least as big as $P(A)$. However, if an event is possible with a measure $\Pi(A) > 0$ then it need not set in, i.e. the probability $P(A)$ can be even equal to 0.

As we have stressed before, human regarding of the world always contains some vagueness. Hence, the measure of possibility in general should have more than two values since inherently vague events are not either possible or impossible. This leads to fuzzy sets.

Let U be a set and X a variable which attains values in U. The range of the values of X is a certain subset $A \subseteq U$. However, since the vagueness is inherent, A should rather be a fuzzy set, i.e.

$$A \lessapprox U.$$

The variable X induces a possibility distribution $\Pi_{\mathbf{x}}$ which is *postulated* to be equal to A, i.e.

$$\Pi_{\mathbf{x}} = A. \qquad (18)$$

This relation is called the *fundamental postulate* of the possibility theory. It can also be written in the following form: Let $\mathcal{A}$ be a linguistic expression with the meaning $M(\mathcal{A}) = A \lessapprox U$. Then the proposition

"X is $\mathcal{A}$"

induces the possibility distribution (18). The *possibility measure* of the value $X = x$, $x \in U$, is

$$\pi_{\mathbf{x}}(x) = Ax, \qquad (19)$$

i.e. it is a measure of the possibility of the event "x is $\mathcal{A}$".

For example, if X is a variable whose values are numbers then we may ask about the possibility of the proposition "x is a small number" where $x \in \mathbb{R}$ is a value of the variable X. By the fundamental postulate, we obtain from (19) that the measure of this possibility is equal to the membership degree of x in the fuzzy set representing the meaning of the syntagm *small number*.

Let $B \lessapprox U$ be a fuzzy set. Then the possibility measure of the event "the value of X belongs to B" is

$$\pi(B) = \bigvee_{x \in U} (Bx \wedge \Pi_{\mathbf{x}}(x)).$$

Let A, $B \subsetneq U$. From the above definition we obtain

$$\pi(A \cup B) = \pi(A) \vee \pi(B)$$

$$\pi(A \cap B) = \pi(A) \wedge \pi(B).$$

It is also possible to introduce n-ary possibility distributions considering the universe $U = U_1 \times \ldots \times U_n$.

Since $\mathcal{A}$ is an expression of natural language, the rules for the computation of the meaning of $\mathcal{A}$ outlined in the preceding section can be naturally extended to the possibility distributions. Hence, we can compute the measure of the possibility of the events such as "x is a very small number", "x is small and y is rather big" etc. The reader is referred, e.g. to Zadeh (1978a, b), Dubois and Prade (1988), Yager (1982) and others.

5. CONCLUSIONS

In this paper, we briefly outlined the main concepts of fuzzy set theory and possibility theory and their philosophical background. Because of the lack of space, our presentation could not be very exhaustive. It should rather be taken as several introducing comments on the basis of which the reader can get acquainted with the cited literature. Of course, the literature on fuzzy sets is much more extensive that that contained in the references. Some additional references can be found in the other papers of this book.

REFERENCES

Bellman R., and Giertz M. (1973) On the analytic formalism of the theory of fuzzy sets. Inf. Sci. 5, 149-156.

Dubois D., and Prade H. (1978) Operations on fuzzy numbers. Int. J. Systems Sci. 9, 613-626.

Dubois D., and Prade H. (1980) Fuzzy Sets and Systems. Theory and Applications. Academic Press, New York.

Dubois D., and Prade H. (1988) Possibility Theory - An Approach to Computerized Processing of Uncertainty. Plenum Press, New York.

Novák V. (1989) Fuzzy Sets and Their Applications. Adam Hilger, Bristol.

Novák V. (1990) On the syntactico-semantical completeness of first-order fuzzy logic I, II. Kybernetika (to appear).

Novák V., and Nekola J. (1983) Basic operations with fuzzy sets from the point of fuzzy logic., in E. Sanchez and Gupta (Eds.): Fuzzy Information, Knowledge Representation and Decision Analysis, Pergamon Press, Oxford.

Novák V., and Pedrycz W. (1988) Fuzzy sets and t-norms in the light of fuzzy logic. Int. J. Man-Machine Stud. 29, 113-127.

Negoita C. V., and Ralescu D. A. (1975) Application of Fuzzy Sets to System Analysis. Birkhäuser, Stuttgart.

Pavelka J. (1979), On fuzzy logic I, II, III. Zeit. Math. Logik Gurndl. Math. 25, 45-52, 119-134, 447-464.

Ramík J., and Římánek J. (1985) Inequality relation between fuzzy numbers and its use in fuzzy optimisation. Fuzzy Sets and Systems 16, 123-138.

Sanchez E. (1976) Resolution of composite fuzzy relation equations. Inf. Control 30, 38-48.

Skala H. J., Termini S., and Trillas E. , Eds. (1984) Aspects of Vaguenes. Theory and Decision vol 39. D. Reidel, Dordecht.

Vopěnka P. (1979) Mathematics In the Alternative Set Theory. Teubner, Leipzig.

Yager R. R., Ed. (1982) Fuzzy Set and Possibility Theory. Recent Developments. Pergamon Press, New York.

Zadeh L. A. (1965) Fuzzy sets. Inf. Control 8, 338-353.

Zadeh L. A. (1975) The concept of a linguistic variable and its application to approximate reasoning I, II, III. Inf. Sci. 8, 199-257, 301-357; 9, 43-80.

Zadeh L. A. (1978a) Fuzzy sets as a basis for a theory of possibility. Fuzzy Sets and Systems 1, 3-28.

Zadeh L. A. (1978b) PRUF - a meaning representation language for natural languages. Int. J. Man-Machine Stud. 10, 395-460.

VAGUE NOTIONS IN THE THEORY OF VOTING

Hannu Nurmi*, Mario Fedrizzi and Janusz Kacprzyk*****

* *Department of Political Science, University of Turku, SF - 20500 Turku, Finland*

** *Institute of Informatics, University of Trento, Via Rosmini 42, 38100 Trento, Italy*

*** *Systems Research Institute, Polish Academy of Sciences, ul. Newelska 6, 01-447 Warsaw, Poland*

Abstract. The social choice theory is well-known for its dramatic incompatibility results. We argue that these results are not very robust under plausible «perturbations» of their underlying assumptions. In particular, we give an overview of some results on probabilistic decision schemes and the aggregation of fuzzy preference relations. It turns out that with somewhat more vague basic concepts the theory of social choice can be of assistance in the design of good procedures for group decision making.

Keywords: decision scheme, fuzzy tournament, social choice function, axiomatic social choice theory.

1. INTRODUCTION

From the very beginning the study of *social choice* has been characterized by negative results. In what is nowadays considered to be the first specifically social-choice-focused treatise, published in 1781, de Borda attacked the plurality voting method for the intuitively counter-intuitive outcomes that it might lead to in special circumstances. Borda's contemporary Marquis de Condorcet also pointed to particular difficulties in interpreting voting outcomes in the presence of majority voting cycles (see, e.g., De Grazia, 1953; Sommerlad & McLean, 1989). Similarly, some hundred years latter Dodgson found many puzzling features in the voting systems used in his immediate surroundings, the Oxford colleges (see, e.g., Black, 1958; Riker, 1961). Also in the late 1800's Nanson (1882)

J. Kacprzyk and M. Fedrizzi (eds.), Multiperson Decision Making Using Fuzzy Sets and Possibility Theory, 43–52.

discussed problems related to many methods that were used in various contexts. All these authors suggested remedies to the problems they had encountered, but new puzzles kept popping up.

What distinguishes the modern *social choice theory* from what has been found by its precursors are results indicating that not only do we encounter problems when using common voting procedures but we are necessarily going to encounter them no matter which systems we choose. The modern social choice theory is notorious for a particular type of negative results often called *impossibility* or perhaps more aptly *incompatibility* results that suggest that looking for a system that has a set of desirable choice-theoretic properties necessarily excludes systems that satisfy another set of often equally plausible properties. Classic results in this field is of course Arrow's (1963) theorem (see also Kelly, 1978). Let us have a brief look at the point of departure adopted in the modern social choice theory.

Consider a set N of n individuals and a set A of alternatives. N is typically assumed to be finite, while A may well be infinite. Indeed, in the spatial theory of voting it is common to assume that A is a convex subset of $\mathbb{E}^m$, the m-dimensional Euclidean space (see, e.g., Enelow & Hinich, 1984). Let 2^A be the set of all subsets of A. Each individual i in N is assumed to have a binary *preference relation* R_i over A that is both connected and transitive. Denote individual i's all possible binary relations over A by $\mathbb{R}_i$. A social *social choice function* C can now be defined as follows

$$C: A \times \mathbb{R}_1 \times \ldots \times \mathbb{R}_n \rightarrow 2^A.$$

Thus it assigns to each n-tuple, or profile, of individual preference relations a subset of alternatives, viz. those considered «*socially best*».

In Arrow's classic work the function studied was not exactly of the above type, but the so-called *social welfare function.* The domain of the latter consists of the individual preference relations, while its range consists of all possible binary relations, that is

$$W: \mathbb{R}_1 \times \ldots \times \mathbb{R}_n \rightarrow \mathbb{R}$$

where W is the social welfare function and $\mathbb{R}$ denotes all possible (collective) preference relations.

In some results – notably in the Gibbard-Satterthwaite theorem – a special case of the social welfare function is focused upon, viz. a singleton-valued social choice function or *social decision function* (Gibbard, 1973; Satterthwaite, 1975).

2. TWO RESULTS

Apart from Arrow's theorem probably the best-known result in the social choice theory is the results achieved independently by Gibbard and Satterthwaite. It states that all universal and non-trivial social decision functions defined for at least three alternatives are either manipulable or dictatorial. *Universality* means here that the function is defined for all preference profiles and alternative sets. *Non-triviality*, in turn, requires that for each alternative there must be a preference profile such that this particular alternative is chosen.

The results thus states that given these two intuitively obvious requirements one cannot have *non-manipulable* decisions functions unless one accepts *dictatorial* functions. Gärdenfors (1976) has generalized this result to cover, in addition to social decision functions, also the *social choice functions*. Obviously as far as the purpose of voting is to uncover individual preferences, the manipulability of choice or decision function seems counter-productive. In the next section we shall see that somewhat more vague basic concepts can be used to avoid the incompatibilities involved in the Gibbard-Satterthwaite theorem.

Another result that has been discussed in the theory of voting is McKelvey's (1976, 1979) theorem which is related to the presence and coverage of the majority cycles in higher dimensional policy spaces. McKelvey focuses on the m-dimensional Euclidean space $\mathbb{E}^m$. Each dimension of the space can be interpreted as a property as, e.g., price, durability or length. Each alternative considered by the voting body can be viewed as a point in the space. Each voter is assumed to have an optimal point or outcome in the space. Moreover, each voter has a utility function U_i defined on $\mathbb{E}^m$ such that

$$U_i(x) = f_i \,||\, x - x'_i \,||, \qquad \text{for all } x \text{ in } \mathbb{E}^m.$$

Here x'_i denotes individual i's optimum point, $||\,.\,||$ is the Euclidean norm and f_i is monotonically decreasing.

Under these assumption McKelvey proves that in the case where the *core* is empty – i.e. for all each alternative x there is at least one other alternative y such that y is preferred by majority to x – one can reach any point in $\mathbb{E}^m$ from any other point via a majority trajectory. That is, an *agenda* or a set of outcomes can be built from any starting point to any other point such that any outcome between these two has the property that it defeats the preceding one with a majority of votes assuming that all voters vote according to their preferences.

Prima facie this is a startling results as it shows that the majority rule per se does not guarantee any «responsiveness» of the voting outcomes to the preferences of voters under the stated assumptions. Indeed, the outcomes do not necessarily stay within the limits of Pareto optimality. On closer scrutiny one may question the plausibility of some of McKelvey's assumption as, e.g., a myopic voting behaviour or a rather specific assumption concerning the voters' utility functions. In his 1979 article McKelvey, however, generalizes considerably the scope of his theorem by replacing the Euclidean norm assumption with one that only requires that the preferences of voters have a continuous utility representation in $\mathbb{E}^m$.

3. PROBABILISTIC SOCIAL CHOICE

One way of reacting to the above results and other of similar negative nature is to investigative their underlying assumptions. We shall do that later. But one could also ask whether the mechanism one tries to construct – be it a social welfare function, social choice function or social decision function – is the right one. More specifically it can be argued that if good deterministic preference aggregation devices are impossible to work out, one should look for good indeterministic ones. The traditional way of dealing with indeterminism in a precise way is to resort to probability theory.

Suppose that instead of the above mentioned aggregation functions we would like to have an intuitively plausible *probabilistic choice mechanism*, viz. a mechanism that given individual preference relations would result in a choice probability vector indicating for each alternative its probability of being chosen. The choice itself would be determined by a random device having the corresponding probability distribution over its possible outcomes (see, e.g., Intriligator, 1973). Obviously social decision functions are special cases of probabilistic ones in that they result in a special types of probability distributions, viz. those assigning to one alternative the choice probability equal to 1.

More precisely, a *decision scheme* is a function d from the set of preference profiles (i.e. n-tuples of preference relations) over A to the set of measures over A with a fixed weight equal to 1 (Barbera, 1979). A measure over A, in turn, is a function assigning a non-negative number m(a) to each a in A. The weight of the measure, finally, is the sum of the numbers assigned to the elements of A.

Quite interesting results about the properties of the decision schemes have been achieved by Barbera (1979), Dutta (1980) and Intriligator (1973, 1982). In particular, their strategic properties seem entirely different from those of social decision functions. Following Barbera we shall distinguish three different decision schemes: (i) point voting schemes, (ii) supporting size schemes, and (iii) simple schemes.

Point voting schemes are characterized by positional vectors of the type $C = \langle c_1, \ldots, c_k \rangle$ such that $c_1 \geq c_2 \geq \ldots \geq c_k$,
where K is the cardinality of A. When a point voting scheme is used, the probability of alternative a_j being chosen, P_j, is determined as follows

$$P = \sum_{i \in N} c_{r(i,j)}$$

where r(i, j) is the rank given to a_j by voter i. Thus, probability P_j is the sum of scores given to a_j by all the voters. An example is here a modification of the classic Borda count where instead of getting k-1, k-2, etc. points for each first, second, etc. rank, an alternative gets (k-1)/M, (k-2)/M, etc. points for each first, second etc. rank. Here $M = [(k-1) + (k-2) + \ldots + (1)]\, n$.

In the *supporting size decision schemes*, on the other hand, the starting point is the matrix of pairwise comparison of alternatives. The entry (i, j) indicates the number of voters preferring a_i to a_j. If all preferences are strict, then obviously the entry (j, i) is obtained by subtracting the entry (i, j) from n. In these schemes the probability of choosing alternative a_i is determined by the entries of the i-th row of the pairwise comparison matrix. More formally,

$$P_i = \sum_{\substack{i \varepsilon N \\ i \neq j}} s_{ij}$$

where s_{ij} is a suitable function of the number of voters preferring a_i to a_j; s_{ij} must be chosen so that the sum of P_i's equals unity. Thus, e.g.,

$$s_{ij} = n_{ij}/nk$$

where n_{ij} is the number of voters preferring a_i to a_j.

The simple decision scheme, finally, are schemes that can be viewed as both point voting and supporting size decision schemes. More formally, a simple voting scheme determines the choice probability of a_j as follows.

$$P_j = \sum_{l \varepsilon A} \sum_{i \varepsilon N} t\ v^i_{jl} + k\ .$$

Here t and k are constants and

v^i_{jl} = 1 if individual i prefers a_j to a_l
= 0, otherwise.

An example of a simple voting scheme is a modified Borda count. When k = 3 and n = 5, it can be represented as both a supporting size decision scheme with the weight 5/15 assigned to a unanimous support for an alternative against the other ones, weight 4/15 assigned to 4 voters' support, weight 3/15 to 3 voters' support, 2/15 to 2 voters' support, 1/15 to 1 voters' support and 0 to no support. As a point voting scheme it can be represented with the weights 2/15, 1/15 and 0 assigned to the first, second and third rank, respectively. Thus, this version of the Borda count is a simple voting scheme.

While the Gibbard-Satterthwaite theorem tells us that all plausible social decision functions are either manipulable or dictatorial, Barbera's (1979) theorems show that this is not the case if one settles for somewhat more vague constructs, viz. the decision schemes which are probabilistic generalization of the social decision function. Before going into the theorems a few words about the terminology would be helpful.

Firstly, a social decision function is *manipulable* iff there is a situation (i.e. a set of alternatives and a preference profile) in which it is beneficial for a voter to act as if his preference order were R_i' while it really is R_i ($R_i' \neq R_i$, of course). Secondly, a decision function or scheme is *strategy-proof* if is never (i.e. in no situation) manipulable by any individual voter. In other words, when the decision function or scheme is strategy-proof, the outcome resulting from all voters revealing their true preferences is a *Nash equilibrium*, i.e. no voter can benefit from a unilateral deviation from revealing his true preference once the others do so.

Barbera (1979) shows that the point voting, supporting size and simple decision schemes are all strategy-proof. He shows this by proving a set of characterization theorems for the above three types of schemes. For each scheme there is a set of necessary and sufficient conditions that contains strategy-proofness. In fact, the characterization theorems only differ with respect to one condition, a probabilistic independence condition. For our purposes it suffices to observe that strategy-proofness in the sense of non-manipulability by individuals is not uncommon among probabilistic voting procedures.

Although Barbera's theorems show that the manipulability of social choice procedures is not so pervasive as the Gibbard-Satterthwaite theorem has led many authors to believe, it should be interpreted correctly. It deals with the strategy-proofness in the sense of non-manipulability by individuals, not by groups.

Indeed, Dutta (1980) shows that if coalitions are allowed, every point voting or supporting size scheme is manipulable. Nonetheless, Barbera's results show that it is worthwhile to study the robustness of results obtained in deterministic settings.

Moreover, the fact that many of the negative results of the social choice theory are crucially dependent on the deterministic setting they deal with is shown by the results of Intriligator (1973). He gives an axiomatic characterization of a relatively straightforward decision scheme, viz. the *average rule*. Instead of individual preference relations he starts from *individual probability vectors*

$$P_i = \langle P_{i1}, \ldots, P_{ik} \rangle$$

where P_{i1} indicates the probability that voter i would choose alternative 1 if he had to make a choice from among the alternatives in A. It is assumed that the components of the vector satisfy the usual requirements of probabilities, viz. that $P_{i1} \geq 0$, for all 1 in A, and $\sum_i P_{i1} = 1$.

The average rule defines the probability P_j that alternative j is chosen by the group as

$$P_j = (1/n) \sum_{i \in N} P_{i,j}, \text{ for all } j \text{ in } A \quad .$$

In order words, the choice probabilities are simply averages of the individual choice probabilities. This rule has a number of plausible properties, *inter alia*, the citizens' sovereignty or non-imposition, Pareto optimality and independence. The *citizen's sovereignty* is a requirement that for all choice probability vectors there is a n-tuple of individual probability vectors that yields the former vector. The *Pareto optimality* requires that if all individuals assign to alternative 1 at least as large choice probability as to alternative j, and at least one voter gives the former a strictly larger choice probability, then the group as a whole should assign to alternative 1 a larger choice probability than to alternative j. The *independence condition* is the requirement that the choice probability of 1 depends on no other considerations than the individual choice probabilities of 1. This condition is much more straightforward than its analog in Arrow's theorem. As the average rule obviously also satisfies the unrestricted domain condition, we observe that we have a rule that satisfies analogs of all conditions that Arrow imposed on social welfare functions in his theorem. Of course, we do not end up with a social welfare function, not even a social choice function, but a decision scheme.

We have thus seen that with nondeterministic notions one can avoid some negative results of the social choice theory. Of course, in practical decision making it may be difficult to justify the use of probabilistic schemes. Especially if the decision setting is unique, i.e. non-repetitive, and one alternative has to be chosen, the probabilistic properties of schemes may lose much of their intuitive appeal. Let us now turn to other ways of employing vague concepts in finding plausible solution concepts for voting games.

4. SOLUTIONS TO VOTING GAMES

McKelveys's results discussed in Section 2 suggest that the *amendment procedure* – i.e. pairwise comparison of alternatives with a majority rule – does not necessarily converge to any plausible subset of alternatives. Indeed, the end results may not even be in the set of Pareto-undominated alternatives. What, however, if instead of the majority rule, some other "qualified" majority would be required in the pairwise comparisons? Kramer (1977) shows that this may be a way out of the arbitrariness of the majority rule. Specifically, he demonstrates that if one adopts an *α-majority rule* with α determined by the so-called min-max number, then the social preference relation P_α determined as x P_α y iff x is preferred to y by at least αn voters contains no cycles and has a maximal element. The required majority, αn, has to be determined anew for each configuration of the voters' optima. Of course, one could choose a very large α to cover all possible cases, but this would lead to very indecisive choice rules since, obviously, if $\alpha = 1$, we end up with the Pareto set. Thus it would be useful to have as small a value of α as possible without conflicting with the acyclicity requirement. The min-max number is defined as follows. First, for each x in R_m let v(x) be the maximum number of votes against x, i.e. $v(x) = \max_y n(y, x)$. Then the min-max set is $M = \{z \varepsilon R^m \mid v(z) = \min_x v(x)\}$. Finally, the min-max number is $m^* = v(x)$, for any z in M.

By Kramer's theorem the trajectory of α-majority outcomes will enter the min-max set and remain there once the set is reached. Unfortunately, the min-max set may also be very large and hence the α-majority rule be rather indecisive. Moreover, the value of α has to determined for each new decision situation.

The α-majority rule can be viewed as a generalization or "fuzzification" of the simple majority rule. Similar analogs of "crisp" solution concepts have been suggested in the literature (see, e.g., Nurmi, 1981; Kacprzyk & Nurmi, 1989; Nurmi & Kacprzyk, 1989; also Kacprzyk & Fedrizzi, 1988).

These concepts are based on two different approaches: (i) some start form the notion of individual preference relations and work out a collective fuzzy preference relation, and (ii) some start from the assumption that the individual preference relations are of fuzzy nature and these approaches are in fact aggregations of those relations. The latter approaches represent similar departures from the underlying "given" preassumptions of the traditional social choice theory as Intriligator's assumption concerning the individual probabilistic choice functions. To see how the approaches of type (ii) can be utilized in finding plausible solution concepts in voting games, let us consider the concept of a *fuzzy tournament* (see Nurmi & Kacprzyk, 1989, for further discussion).

Suppose that each individual i in the group N has a *fuzzy preference relation* R_i over A, the set of alternatives. If the cardinality of A is small enough, this fuzzy relation can be represented by a matrix $R_i = [r^i_{jk}] = [\mu_{R(i)}(a_j, a_k)]$; $i \varepsilon N$; $a_j, a_k \varepsilon A$. Here μ_{R_i} : $A \times A \to [0,1]$ is the membership function of the fuzzy relation R_i. We assume that R_i is reciprocal, i.e. $r^i_{jk} + r^i_{kj} = 1$, and $r^i_{kk} = 0$.

The values in R can be interpreted so that 1 (0, respectively) indicates a definite preference of the element represented by the row (column) over the element represented by the column (row), while the value 0.5 indicates indifference; the situations in-between are represented by numbers between 0 and 1.

The *fuzzy individual preference relations* can be utilized in defining plausible subsets of the Pareto set as solutions to voting games. Let us define the α– *dominance relation* on A as follows.

$$a_j D a_k <=> r^i_{jk} \geq a, \quad \text{for all } i \varepsilon N, \text{ and } j, k \varepsilon A.$$

In other words, an alternative α– dominates another one iff the degree to which all voters prefer the fomer to the latter is at least α . If α > 0.5, the set of α– undominated alternatives coincides with the Pareto set. Obviously, by decreasing the value of α we get smaller, and thus more decisive, solution sets because $\alpha 1 < \alpha 2 < 0.5 => UD(\alpha 1) \leq UD(\alpha 2) \leq UD(0.5)$, where UD (α) denotes the (sub) set of α– undominated alternatives.

Obviously if one looks for as decisive a solution concept as possible, then one ought to choose the value of α which gives the smallest set UD (α). This set is, however on *unanimity* consideration, nor on majoritarian principles. More *majoritarian* in spirit is the concept of an *α- uncovered set,* which is a fuzzification of Miller's (1977, 1980) uncovered set. We start from the *α– covering relation*, over A, MC_α defined as follows: x MC_α y iff x 0.5 majority dominates y and, moreover, whenever y α– majority dominates some alternative z, so does x. The relation of α– majority dominance, in turn, is defined as that x α– majority dominates y iff at least a majority of voters has a preference for x over y to degree of at least α .

The *α– uncovered set*, UMC_α , consists of those alternatives that are not α– covered by any other alternative. We observe that if α1 > α 2, then x $MC_{\alpha 1}$ y = > x $MC_{\alpha 2}$ y and, hence, $UMC_{\alpha 1} \leq UMC_{\alpha 2}$ for all x, y, ε A. An obvious solution concept in the majoritarian spirit would be a nonempty UMC_α with the minimal value of α.

5. CONCLUDING REMARKS

The assumption underlying the results of the social choice theory are at least prima facie plausible. After all, it is reasonable to assume that the voters who are seriously interested in the act of voting should be able to form a well-structured opinion about the alternatives at hand. It is, however, not entirely obvious that the traditional assumption of complete and transitive preference relations is the most plausible way of representing individual opinions. Intriligator's choice probability assumption is certainly worth considering as an alternative assumption concerning voter opinions. Similarly, the fuzzy preference relations could be used as a starting point in the design and evaluation of group decision procedures.

Also the constructs investigated in the social choice theory can be criticized as being too «deterministic». If instead of social choice or social decision functions we would be interested in decision schemes that result in choice probability distributions over the alternatives, then many incompatibility results could be avoided. Of course, the problem of how to end up with one alternative once the choice probabilities are found, still remains. An obvious way would be to let a random device with appropriate outcome distribution to determine which alternative is to be chosen. In settings where long-run frequencies do not make sense, the utilization of random devices is however difficult to justify. The probabilistic decision schemes should be viewed as perturbations of their deterministic counterparts in a similar way as fuzzy systems theory concepts purport to be generalizations of their nonfuzzy counterparts. This way of looking at the results achieved in nondeterministic social choice theory would seem to suggest that the classic incompatibility results are not necessarily robust social laws, but perhaps rather artifacts implied by rather special assumptions and goals of the theorists.

REFERENCES

Arrow, K.J. (1963) *Social Choice and Individual Values*, 2nd ed. Wiley, New York.

Barbera, S. (1979) Majority and Positional Voting in a Probabilistic Framework, *The Review of Economic Studies XLVI*, 379-389.

Black, D. (1958) *The Theory of Committees and Elections*. Cambridge University Press, Cambridge.

DeGrazia, A. (1953) Mathematical Derivation of an Election System, *Isis 44*, 42-51.

Dutta, B. (1980) Strategic Voting in a Probabilistic Framework, *Econometrica 48*, 448-456.

Enelow, J.M. and M.J. Hinich (1984) *The Spatial Theory of Voting: An Introduction*. Cambridge: Cambridge University Press.

Gärdenfors, P. (1976) Manipulation of Social Choice Functions, *Journal of Economic Theory 13*, 217-228.

Gibbard, A. (1973) Manipulation of Voting Schemes: A General Result, *Econometrica 41*, 587-601.

Intriligator, M.D. (1973) A Probabilistic Model of Social Choice, *Review of Economic*

Studies 40, 553-560.

Intriligator, M.D. (1982) Probabilistic Models of Choice, *Mathematical Social Sciences 2*, 157-166.

Kacprzyk, J. and M.Fedrizzi (1988) A «soft» measure of consensus in the setting of partial (fuzzy) preferences. European Journal of Operational Research 34, 316-325.

Kacprzyk, J. and H. Nurmi (1989) Linguistic Quantifiers and Fuzzy Majorities for More Realistic and Human-Consistent Group Decision Making, in G. Evans, W. Karwowski and M. Wilhelm (eds.), *Fuzzy Methodologies for Industrial and Systems Engineering*, Amsterdam: Elsevier, pp. 267-281.

Kelly, J.S. (1978) *Arrow Impossibility Theorems*. New York: Academic Press.

Kramer, G.H.(1977) A Dynamical Model of Political Equilibrium, *Journal of Economic Theory* 12, 310-334.

McKelvey, R.D. (1976) Intransitivities in Multidimensional Voting Models and Some Implications for Agenda Control, *Journal of Economic Theory 12*, 472-482.

McKelvey, R.D. (1979) General Conditions for Global Intransitivities in Formal Voting Models, *Econometrica 47,* 1085-1112.

Miller, N.R. (1977) Graph-Theoretical Approaches to the Theory of Voting, *American Journal of Political Science 21*, 769-803.

Miller, N.R. (1980) A New Solution Set for Tournaments and Majority Voting, *American Journal of Political Science 24*, 68-96.

Nanson, E.. (1882) Methods of Elections, *Transactions and Proceedings of the Royal Society of Victoria*, Art. XIX, 197-240.

Nurmi, H. (1981) Approaches to Collective Decision Making with Fuzzy Preference Relations, *Fuzzy Sets and System 6*, 249-259.

Nurmi, H. and J. Kacprzyk (1989) On Fuzzy Tournaments and Their Solution Concepts in Group Decision Making, European Journal of Operational Research forthcoming.

Riker, W.H. (1961) Voting and the Summation of Preferences, *American Political Science Review 60*, 900-911.

Satterthwaite, M. (1975) Strategy-Proofness and Arrow's Conditions, *Journal of Economic Theory 10,* 187-217.

Sommerlad, F. and I. McLean, Transl. (1989) The Political Theory of Condorcet, *University of Oxford, Social Studies Faculty Centre Working Paper 1/89.*

2

General Issues Related to Decision Making Under Fuzziness

AGGREGATION OF POSSIBILITY MEASURES

Didier DUBOIS Henri PRADE
Institut de Recherche en Informatique de Toulouse
Laboratoire Langages et Systèmes Informatiques
Université Paul Sabatier, 118 route de Narbonne
31062 TOULOUSE Cedex (FRANCE)

Abstract : Possibility measures are not closed under convex combinations, since the nestedness property of focal sets is not preserved under this aggregation mode. This paper intends to investigate the eventwise aggregations of possibility measures that preserve the consonance property of these set-functions. It is proved that these operations are defined as the union (via the maximum) of some monotonic pointwise transformations of the fuzzy sets that characterize the possibility measures ; especially the previously introduced weighted maximum operations are among the natural solutions. This work also suggests to view a fuzzy set as a disjunctive aggregation of possibly dissonant imprecise, individual opinions, the weights reflecting the degrees of membership of individuals in the group. The extension of these results to decomposable set-functions is considered.

Keywords : Possibility theory ; consensus ; probability ; weighted combination.

1. INTRODUCTION

The unique operation that can combine probability measures in an *eventwise* manner, and provide a probability measure as a result is generally a linear convex combination (McConway, 1981 ; Lehrer and Wagner, 1981). "Eventwise" means that the aggregated measure of an event depends on the probability values of this event only. This noticeable result is basically due to the additivity of probability measures. So, convex combination of probabilities, also called convex mixtures play a crucial role in multiperson decision models, where each person expresses his opinion as a probability distribution on a set of alternatives ; see French (1985) for a survey on this topic. Convex mixtures are also important in utility theory (Herstein and Milnor, 1953). Moreover most families of upper and lower probabilities can be defined as convex sets of probability measures (e.g. Dubois and Prade, 1988b).

Instead of representing a person's opinion about a set of alternatives in terms of his probability of selecting one or another, it looks as natural to model the possibility of his choosing each alternative. In other words, each person indicates the subset of alternatives he would *not* reject, where the propensity to reject is a matter of degree. This view is in accordance with Shackle's view of decision processes being guided by the possibility of options rather by their (often unattainable) probabilities (Shackle, 1961).

J. Kacprzyk and M. Fedrizzi (eds.), Multiperson Decision Making Using Fuzzy Sets and Possibility Theory, 55–63.

With this idea in mind, the problem is to construct one possibility measure as the consensual aggregation of several individual possibility measures. The aggregation problem is then carried over from the probabilistic setting to the one of possibility theory. Wagner (1988) recently investigated the same problem in the setting of belief functions (Shafer, 1976). He proved that the unique eventwise aggregation function is once again a convex mixture. In spite of the fact that possibility measures are consonant plausibility measures, Wagner's result does not apply to possibility measures if we wish to keep the closure property requesting that the aggregation of possibility measures is again a possibility measure. Indeed, the family of possibility measures over a set is not convex (e.g. Dubois and Prade, 1988b).

This paper solves the problem of aggregating possibility measures in an eventwise fashion while preserving the consonance property. It is proved that these operations are characterized by taking the union of some monotonic pointwise transformations of the fuzzy sets underlying the possibility measures. After stating the main result, some comments are made in order to interpret the type of combination process at work in this aggregation. Then it is indicated how to extend the result to decomposable measures (Dubois and Prade, 1982 ; Weber, 1984), thus putting together our results and those of the probabilistic setting into a common framework.

2. CONSENSUS OF POSSIBILITY MEASURES

Let Ω be a finite set, and Π a set-function from 2^{Ω} in [0,1]. Π is called a possibility measure if and only if the following properties are satisfied (Zadeh, 1978 ; Dubois and Prade, 1988a)

$$\Pi(\emptyset) = 0 \;\; ; \;\; \Pi(\Omega) = 1 \tag{1}$$

$$\Pi(A \cup B) = \max(\Pi(A), \Pi(B)) \tag{2}$$

In this paper Ω is a set of potential decisions. Let π be the possibility distribution underlying Π, i.e. $\pi(\omega) = \Pi(\{\omega\})$, $\forall\, \omega \in \Omega$. $\pi(\omega)$ evaluates the extent to which decision ω looks feasible for some individual, i.e. to what extent he is not liable of rejecting it. $\pi(\omega) = 1$ when nothing prevents ω from being chosen, and $\pi(\omega) = 0$ if ω is completely ruled out. $\Pi(A)$ evaluates to what extent A contains at least one potentially acceptable decision. This interpretation is in accordance with (2) since $\Pi(A) = \max\{\pi(\omega) | \omega \in A\}$ (or $\sup\{\pi(\omega) | \omega \in A\}$ if Ω is not finite provided that (2) is extended to non-countable unions).

Let I be a finite set of n individuals, and $\{\Pi_i \mid i \in I\}$ be a collection of possibility measures reflecting the opinions of the individuals in I. A consensus function is one that maps $\Pi_1, \Pi_2, \ldots, \Pi_n$ to another possibility measure Π that summarizes the opinion of the group. The following conditions can be requested in order to characterize possibility consensus functions

<u>A1</u> $\forall\, A \subseteq \Omega, \forall\, \Pi_1, \ldots, \Pi_n$, there exists a function $C_A : [0,1]^n \to [0,1]$ such that

$$\Pi(A) = C_A(\Pi_1(A), \Pi_2(A), \ldots, \Pi_n(A))$$

<u>A2</u>(c) $\forall\, A \subseteq \Omega$, if $\Pi_1(A) = \Pi_2(A) = \ldots = \Pi_n(A) = c$ then $\Pi(A) = c$.

Axiom A1 is in the spirit of the axiom of irrelevance of alternatives (i.e. $\Pi(A)$ does not depend on events other than A) ; axiom A2(c) is a unanimity rule at level c. If A2(c) is true $\forall\, c \in [0,1]$, then the consensus function is idempotent. A1 and A2(c) are exact counterparts of well-accepted axioms in probabilistic theories of consensus (McConway, 1981 ; Lehrer and Wagner, 1981).

We are now in a position to prove the main results of the paper.

Proposition 1 : If the consensus function is defined according to axiom A1, and if A2(0) holds, then $\forall$ A, B $\subseteq \Omega$, $C_A = C_B$, provided that $|\Omega| \geq 3$.

Proof : For any $(\alpha_1, \alpha_2, \ldots, \alpha_n) \in [0,1]^n$, and any proper non-empty subsets A, B of Ω, such that A is a proper subset of B (this requires $|\Omega| \geq 3$), choose $\Pi_1, \Pi_2, \ldots, \Pi_n$ such that $\Pi_i(A) = \alpha_i$, $\forall i$, $\Pi_i(B - A) = 0$, $\forall i$. Because $A \subset B \subset \Omega$, we can have $\alpha_i < 1$ since, then, $\Pi_i(\bar{B}) = 1$ is feasible, where $\bar{B}$ is the complement of B. Clearly this implies $\Pi_i(B) = \Pi_i(A) = \alpha_i$. Then since Π is a possibility measure $\Pi(B) \geq \Pi(A)$, which writes $C_B(\alpha_1,\ldots, \alpha_n) \geq C_A(\alpha_1,\ldots, \alpha_n)$. Moreover $\Pi(B) = \Pi(A \cup B) \leq \Pi(A) + \Pi(B - A)$ (i.e. Π is subadditive). Now from A2(0), $\Pi(B - A) = C_{B-A}(0, \ldots, 0) = 0$. Hence $C_B = C_A$. Besides if $A \cap B \neq \emptyset$ then $C_{A\cap B} = C_A = C_B$. Hence $\forall \omega \in \Omega$, $\forall A \subset \Omega$, $\omega \in A$ implies $C_{\{\omega\}} = C_A$. Particularly if $B = \{\omega,\omega'\}$, $C_{\{\omega\}} = C_B = C_{\{\omega'\}}$. Q.E.D.

As a consequence the consensus function can be pointwisely defined on $[0,1]^n$, and is denoted C.

Corollary : The consensus function satisfies A2(1).

Proof : If $\Pi_1(A) = \Pi_2(A)\ldots \Pi_n(A) = 0$, then from A2(0), $C(\Pi_1(A), \Pi_2(A), \ldots, \Pi_n(A)) = 0$ i.e. $\Pi(A) = 0$. As a consequence $\Pi(\bar{A}) = \Pi_1(\bar{A}) =\ldots = \Pi_n(\bar{A}) = 1$ i.e. $C(1, 1, \ldots, 1) = 1$. Q.E.D.

Note that on 2-element sets, proposition 1 is not valid since if $\Omega = \{\omega_1,\omega_2\}$, $C_{\{\omega_1\}}(\alpha_1, \ldots, \alpha_n) = 1 - \Pi_{i=1,n} (1 - \alpha_i)$, $C_{\{\omega_2\}}(\alpha_1, \ldots, \alpha_n) = \max(\alpha_1, \ldots, \alpha_n)$ define a consensus function that obeys A1 and A2(0). Moreover $C_\Omega(1, 1, \ldots, 1) = 1$.

Proposition 2 : Any possibilistic consensus function is monotonically non-decreasing.

Proof : This is obvious because possibility measures are monotonically non-decreasing under inclusion, and if $A \supseteq B$, $\Pi(B) \geq \Pi(A)$ translates into $C(\beta_1, \ldots, \beta_n) \geq C(\alpha_1, \ldots, \alpha_n)$ letting for all i, $\beta_i = \Pi_i(B)$, $\alpha_i = \Pi_i(A)$ for some possibility measure Π_i. Q.E.D.

Proposition 3 : Possibilistic consensus functions are all of the form $C(\alpha_1, \ldots, \alpha_n) = \max(f_1(\alpha_1), f_2(\alpha_2), \ldots, f_n(\alpha_n))$ where $f_1, \ldots, f_n$ are functions from [0,1] to [0,1] such that $f_i(0) = 0$, $\forall i$, and $f_i(1) = 1$ for some i.

Proof : Let $A_1, \ldots, A_n$ be n (possibly non-distinct) subsets of Ω, where $n = |I|$. By definition

$$\Pi(\textstyle\bigcup_{j=1,n} A_j) = C(\Pi_1(\bigcup_{j=1,n} A_j), \Pi_2(\bigcup_{j=1,n} A_j), \ldots, \Pi_n(\bigcup_{j=1,n} A_j))$$

which develops into

$$\max_{j=1,n} C(\alpha_{1j}, \alpha_{2j}, \ldots, \alpha_{nj}) = C(\max_{j=1,n} \alpha_{1j}, \ldots, \max_{j=1,n} \alpha_{nj}) \quad (3)$$

letting $\alpha_{ij} = \Pi_i(A_j)$. This is a functional equation that characterizes admissible consensus functions. Let $\beta_i = \max_{j=1,n} \alpha_{ij}$. Obviously, the right-hand side of (3) only depends on $(\beta_1, \beta_2, \ldots, \beta_n)$. If each value α_{ij} is modified into

$$\alpha'_{ij} = 0 \text{ if } i \neq j$$
$$= \beta_j \text{ if } i = j$$

then the left-hand side of (3) remains invariant, i.e. for any $(\beta_1, \beta_2, \ldots, \beta_n) \in [0,1]^n$

$$C(\beta_1, \beta_2, \ldots, \beta_n) = \max_{j=1,n} C(0, \ldots, \beta_j, \ldots, 0) \quad (4)$$

where β_j is put at the jth place of C. Define $f_j(x)$ as the image by C of the n-tuple with zeros only, except x in the jth position. (4) leads to

$$C(\beta_1, \beta_2, \ldots, \beta_n) = \max_{j=1,n} f_j(\beta_j) \quad (5)$$

Now because of axiom A2(0), $C(0, \ldots, 0) = 0$ leads to $\forall j, f_j(0) = 0$ and because of A2(1), $C(1, \ldots, 1)$ leads to $f_j(1) = 1$ for at least one j. Q.E.D.

This result contrasts with the one obtained by McConway, Lehrer and Wagner for probability measures, by which the only admissible probability consensus function is

$$Q(\beta_1, \beta_2, \ldots, \beta_n) = \Sigma_{j=1,n} \lambda_j\beta_j \text{ with } \Sigma\lambda_j = 1, \lambda_j \geq 0.$$

However the similarity is striking : the sum is turned into a maximum, and the weight of importance is turned into a modification function. In fact, the weight λ_j is similar to $f_j(1)$. Note also that the functions f_j can be chosen monotonically non-decreasing, due to the non-decreasingness of C. Strictly speaking, however, only the useful part of the f_j's in (5) must be non-decreasing. Moreover, the above results also hold if Ω is not finite.

The function f_i modifies the shape of the fuzzy set underlying Π_i, i.e. $\forall \omega$, $\pi_i(\omega)$ becomes $f_i(\pi_i(\omega))$. Indeed $\max_{i=1,n} f_i(\Pi_i(A)) = \max_{i=1,n} f_i[\sup_{\omega\in A} \pi_i(\omega)] = \max_{i=1,n} \sup_{\omega\in A} f_i(\pi_i(\omega))$ provided that f_i is non-decreasing. This modification occurs on the support of π_i only since $f_i(0) = 0$. However $f_i \circ \pi$ can be subnormalized, i.e. $\sup f_i \circ \pi = f_i(1) < 1$ (for some i, but not for all).

An example of admissible possibility consensus function is the weighted maximum (Dubois and Prade, 1986) i.e.

$$C(\beta_1, \beta_2, \ldots, \beta_n) = \max_{j=1,n} \min(\lambda_j, \beta_j) \quad (6)$$

where $\max_j \lambda_j = 1$. This is obtained for $f_j(x) = \min(\lambda_j, x)$. More generally one might consider modifier functions of the form $\lambda_j * f$ where $*$ is a triangular norm (Schweizer and Sklar, 1983), and f is an increasing function such that $f(0) = 0$ and $f(1) = 1$. This form enables the subnormalization effect to be separated from the shape modification effect.

These results point out that if Π_1 and Π_2 are possibility measures then $\max(\Pi_1,\Pi_2)$ is still a possibility measure, but $\min(\Pi_1,\Pi_2)$ is not.

Note that we have stated axiom A1 in terms of possibility measures. It is possible to state it in terms of necessity measures, i.e. set-functions N defined by $\forall A, N(A) = 1 - \Pi(\bar{A})$. But it is easy to show that if C is a possibility consensus function, then $K(x_1, x_2, ..., x_n) = 1 - C(1 - x_1, 1 - x_2, ..., 1 - x_n)$ is a necessity consensus function. For n = 2, we have

$$\Pi(A) = C(\Pi_1(A), \Pi_2(A)) \text{ and}$$

$$\begin{aligned} \forall A,\ N(A) = K(N_1(A), N_2(A)) &= K(1 - \Pi_1(\bar{A}), 1 - \Pi_2(\bar{A})) \\ &= 1 - C(\Pi_1(\bar{A}), \Pi_2(\bar{A})) \\ &= 1 - \Pi(\bar{A}) \end{aligned}$$

For instance, if $C(\Pi_1,\Pi_2) = \max(\Pi_1,\Pi_2)$, then $K(N_1,N_2) = \min(N_1,N_2)$. Moreover if $C(\Pi_1,\Pi_2) = \max(\min(\lambda_1,\Pi_1), \min(\lambda_2,\Pi_2))$, then

$$K(N_1,N_2) = \min(\max(1 - \lambda_1, N_1),\ \max(1 - \lambda_2,N_2)),$$

i.e. a weighted minimum (Dubois and Prade, 1986) is performed on the necessity measures, corresponding to a weighted maximum on the possibility distributions ; indeed $\min_i(\max(1 - \lambda_i, N_i(A)) = \min_i \max(1 - \lambda_i, \inf_{\omega\notin A}[1 - \pi_i(\omega)]) = \min_i \inf_{\omega\notin A}[1 - \min(\lambda_i,\pi_i(\omega))] = \inf_{\omega\notin A} [1 - \max_i \min(\lambda_i, \pi_i(\omega)]$.

3. SOME INTERPRETIVE REMARKS

The fact that the convex mixture is the only admissible consensus operation for belief functions suggests that any belief function can be interpreted as the result of an opinion aggregation procedure. Namely if $(\mathcal{F},m)$ is a body of evidence on Ω, i.e. m is a basic probability assigment on 2^Ω such that $\sum_{A\neq\emptyset} m(A) = 1$, and $\mathcal{F} = \{A \mid m(A) > 0\}$ is a set of focal elements, m(A) can be interpreted as the weight of relative importance of an individual selecting A as the set of admissible options for him, provided that we allow non-distinct focal elements. Indeed it is possible that several individuals $i_1, ..., i_k$ choose the same A as the set of admissible options according to them.

Similarly, the result pertaining to possibility measures suggests that any fuzzy set F on Ω, with membership function μ_F is the result of a possibilistic consensus procedure on the α-cuts of F, say $F_{\alpha_1}, F_{\alpha_2}, ..., F_{\alpha_m}$ with $\alpha_1 = 1 \geq \alpha_2 ... \geq \alpha_m$, where α_i stands for $f_i(1)$, $\forall i$. Indeed the well-known representation formula (Zadeh, 1971) :

$$\forall\omega,\ \mu_F(\omega) = \max_{i=1,n} \min(\mu_{F_{\alpha_i}}(\omega), \alpha_i)$$

extends to $\forall A$, $\Pi(A) = \max_{i=1,n} \min(\Pi_i(A), \alpha_i)$ where Π_i is the (crisp) possibility measure based on F_{α_i} (i.e. $\pi_i = \mu_{F_{\alpha_i}}$). This is clearly a special instance of possibilistic consensus function. Here α_i evaluates to what extent the individual selecting F_{α_i} is a possible decider, i.e. to what extent his proposal is to be part of the consensual set of options. In that particular case, the higher the weight of importance of the decider, the smaller the selected subset. Note that this consensus rule is very respectful of individuals due to its disjunctive character. Clearly, the weighted union of a collection of classical subsets $F_1, ..., F_m$ of a universe Ω gives birth to a normalized fuzzy set (i.e. which contains at least on element ω with a degree of membership 1) in the general case where the F_i's are not nested. However, given the fuzzy set defined by $\mu_F(\omega) = \max_i \min(\lambda_i, \mu_{F_i}(\omega))$ where λ_i assesses the importance of the individual choosing the classical subset F_i, it is not possible in general, from μ_F and the λ_i's to find out the original F_i's. Indeed the α-cuts of F are generally larger than the F_i's.

In (Arrazola et al., 1989) a good example of aggregation of possibility distributions by a weighted union is given. Each possibility distribution represents a fuzzy estimate of the value of a parameter (e.g. the price of a second hand car) obtained by analogical reasoning (e.g. from the known price of a similar car), while the weight reflects the quality of the analogy. Any conclusion obtained by analogy is only plausible and does not exclude another distinct conclusion. Thus a consensus aggregation is natural in this case (in the example, the car the price of which we want to estimate may be more or less similar to cars having different prices, then the weighted union of the different prices offers a good summarization of the possible values of the price of the car we are interested in). This contrasts with the conclusions obtained from general laws applying to the same item, which are naturally combined in a conjunctive manner. For instance if we know that the car c_0 both belongs to categories $\mathfrak{C}_1$ and $\mathfrak{C}_2$ and that the price of any car in category $\mathfrak{C}_i$ is in the subset A_i, then the price of c_0 should be in $A_1 \cap A_2$ (where nothing prevents A_i to be a fuzzy set).

The disjunctive modes of aggregation considered in this paper contrast with Dempster rule of combination (Shafer, 1976) and fuzzy conjunction connectives, which correspond to generalized set intersections (e.g. Dubois and Prade, 1988b). Their range of applicability is certainly not consensus formation since these rules badly deal with conflicting information. They seem to be more adapted to what French (1985) calls the "expert problem", by which several experts are asked to estimate the *real* value of some parameter. This is not a matter of voting (i.e. picturing the set of options proposed by the group) but rather a problem of finding where the truth lies. In that case, when experts diverge, the problem is then to find who is wrong rather than building consensus. Moreover agreement of independent experts must lead to confirmation effect, which is in accordance with Dempster rule especially, but not with A2(c).

4. EXTENSION TO DECOMPOSABLE MEASURES

A decomposable measure on a finite set Ω is a set-function g such that

$$\forall\ A, B, A \cap B = \emptyset \Rightarrow g(A \cup B) = g(A) * g(B) \tag{7}$$

for some operation $*$ on [0,1] (Dubois and Prade, 1982 ; Weber, 1984). In fact operation $*$ is a triangular co-norm in the sense of Schweizer and Sklar (1983)), i.e. a monotonically non-decreasing semi- group operation with identity 0 and such that $a * 1 = 1, \forall\ a \in [0,1]$. The problem of consensus functions can be stated for decomposable measures. It is more general than both the probabilistic and the possibilistic settings since probability measures are special cases of decomposable measures ($a * b = \min(1, a + b)$), and so are possibility measures ($a * b = \max(a,b)$). Here we do not intend to solve the consensus problem completely, but just to give preliminary results.

Let $g_1, g_2, \ldots, g_n$ be n $*$-decomposable measures. Let g be the result of a consensus formation, supposed to be also a $*$-decomposable measure. We reconsider axioms A1 and A2(0) where Π is changed into g. It is easy to prove that, again the consensus function can be defined by means of a mapping C, $[0,1]^n \to [0,1]$ that does not depend on the considered event (Proposition 1). The proof is similar, but instead of using the subaddivitity property, we use the $*$-additivity, i.e. $g(A \cup B) = g(A) * g(B - A)$ where, due to $g_i(B - A) = 0, \forall i$ by construction, and due to A1(0), $g(A \cup B) = g(A) * 0 = g(A)$. Of course C satisfies A2(1), and must be monotonically non-decreasing.

The problem of characterizing admissible consensus functions is open. It comes down to solving the functional equation (3) where max is changed into a triangular co-norm. For n = 2 it reads : find the functions C such that $\forall\ x, y, z, t \in [0,1]$

$$C(x,y) * C(z,t) = C(x * z, y * t) \quad (8)$$

When $* = C$ this equation is called the bissymetry property, and (8) is a particular case of the generalized bissymetry equation (Aczel, 1966, p. 330). Here we only give examples of solutions to (8). The following proposition deals with a class of decomposable measures different from possibility and probability.

Proposition 4 : When $*$ is a strict co-norm with additive generator φ,

i) $C(x,y) = \varphi^{-1}(a\varphi(x) + b\varphi(y))$ is a solution of (8), and is a consensus function for $*$-decomposable measures, that satisfies A1, and A2(c) for $c \in \{0,1\}$, where $a \geq 0$, $b \geq 0$, $\max(a,b) > 0$.

ii) C moreover satisfies the unanimity rule (A2(c), $\forall c \in [0,1]$) whenever $a + b = 1$.

Proof : A strict co-norm is of the form $x * y = \varphi^{-1}(\varphi(x) + \varphi(y))$ where φ is an increasing bijection from $[0,1]$ to $[0,+\infty)$. Let us check that C defined as in i) is a solution of (8)

$$\begin{aligned}\varphi(C(x,y) * C(z,t)) &= \varphi(C(x,y)) + \varphi(C(z,t)) \\ &= a\varphi(x) + b\varphi(y) + a\varphi(z) + b\varphi(t) \\ &= a(\varphi(x) + \varphi(z)) + b(\varphi(y) + \varphi(t)) \\ &= a\varphi(x * z) + b\varphi(y * t) = \varphi(C(x * z, y * t))\end{aligned}$$

hence $C(x,y)$ is a solution of (8). If g is a $*$-decomposable measure, then let $A \cap B = \emptyset$ and $g(A \cup B) \triangleq C(g_1(A \cup B), g_2(A \cup B)) = C(g_1(A), g_2(A)) * C(g_1(B), g_2(B))$ hence $C(0,0)=0$ is obvious letting $A = B = \emptyset$; $C(1,1) = 1$ is true once $\max(a,b) > 0$. Now if $a + b = 1$, C is a generalized arithmetic mean (Aczel, 1966) and is idempotent. Q.E.D.

An example of $*$-decomposable measure that corresponds to Proposition 4 is defined by the axiom $A \cap B = \emptyset \Rightarrow g(A \cup B) = g(A) + g(B) - g(A) \cdot g(B)$, and the family of consensus rules laid bare in Proposition 4 is defined by

$$C(x_1, x_2, \ldots, x_n) = 1 - \prod_{i=1,n} (1 - x_i)^{a_i} \text{ for } a_i > 0 \quad (9)$$

whose generator is $\varphi(x) = -\mathrm{Ln}(1 - x)$ and where $\prod$ stands for the product. The unicity of this rule can be conjectured. Technical details on this point are left for further research. Note that the case of probability measures correspond to nilpotent co-norms, i.e. of the form $x * y = \varphi^{-1}(\min(1, \varphi(x) + \varphi(y)))$ where φ is an increasing bijection from $[0,1]$ to $[0,1]$ where the normalization condition ($g(\Omega) = 1$) is of the form

$$\Sigma_{\omega\in\Omega}\, \varphi(g(\{\omega\})) \geq 1 \quad (10)$$

When (10) holds with equality, $\varphi \circ g$ is a probability measure. For these set-functions, Proposition 4 also holds, and $C(x,y) = \varphi^{-1}(\min(1, a\varphi(x) + b\varphi(y))$ as a family of admissible consensus functions as well. However when (10) holds with equality, the condition $a + b = 1$ is forced. Indeed

$$\Sigma_{\omega\in\Omega}\, \varphi(g(\{\omega\}) = a\Sigma_{\omega\in\Omega}\, \varphi(g_1(\{\omega\})) + b\Sigma_{\omega\in\Omega}\, \varphi(g_2(\{\omega\})) = a + b = 1$$

which explains results obtained with consensus of probability measures.

Remember that ∗-decomposable measures based on $x * y = \min(1, x + y + \lambda xy)$ are belief or plausibility functions for $\lambda \geq -1$ (see for instance Dubois and Prade, 1982). The above results indicate that for each value of λ, we get a family of belief functions closed under a consensus operation of the form defined in Proposition 4, for instance (9) when $\lambda = -1$. This consensus operation is different from the convex mixture found by Wagner (1988) for belief and plausibility functions. The same remarks apply to the subclass of possibility measures, with their own consensus functions. Moreover consensus functions adapted to subclasses of belief functions cannot be applied to general belief functions due to the unicity of convex mixtures as consensus functions for general belief functions. Conversely convex mixtures do not preserve the properties of subclasses of belief functions (except probability measures). These remarks point out that the unicity result of Wagner (1988) should not be used as an argument against other consensus functions, but they are supporting the idea that instead of choosing a mathematical framework first, and then be stuck on a consensus operation, one may as well choose a type of consensus operation adapted to a situation, and then select the mathematical model that can accept it.

5. CONCLUSION

Results on consensus rules for probability measures have been extended to possibility measures and decomposable set-functions, thus providing a richer and more flexible framework for modeling multiperson decision-making. Note that the admissible consensus rules are more restricted in the case of probability measures than with other set-functions, but the added degrees of freedom for ∗-decomposable measures pertain to the modeling of individual importance in the group. Moreover the possibilistic consensus rule is significantly different from the other ones. While the probabilistic rules express averaging, the possibilistic rules are disjunctive, hence even more respectful of individual opinions (but less decisive, as a consequence). The fact that possibilistic consensus functions are defined up to a monotonic transformation of possibility distributions involved in the aggregation recalls that possibility distributions express ordinal rather than cardinal information, as also found when constructing a possibility distribution from a subjective ordering between events modeling "at least as possible as" (Dubois, 1986).

Lastly the basic axiom A1 can be challenged because it is very strong. Accepting it only for singleton sets leads to consider consensus problems in the framework of fuzzy set theory, i.e. pointwise (instead of eventwise) operations for fuzzy set aggregation viewed as social choice functions ; see Fung and Fu (1975), Montero de Juan (1987), Choleva (1985) and Dubois and Koning (1989) on this approach.

References

Aczel J. (1966) Lectures on Functional Equations and Applications. Academic Press, New York.

Arrazola I., Plainfossé A., Prade H., and Testemale C. (1989) Extrapolation of fuzzy values from incomplete data bases. Information Systems 14(6), to appear.

Choleva W. (1985) Aggregation of fuzzy opinions : an axiomatic approach. Fuzzy Sets and Systems 17, 249-258.

Dubois D. (1986) Belief structures, possibility theory and decomposable confidence measures on finite sets. Computers and Artificial Intelligence (Bratislava) 5, 403-416.

Dubois D., and Koning J.L. (1989) Social choice axioms for fuzzy set aggregation. Fuzzy Sets and Systems, Special Issue on 'Aggregation and Best Choice with Imprecise Opinions', to appear.

Dubois D., and Prade H. (1982) A class of fuzzy measures based on triangular norms - A general framework for the combination of uncertain information. Int. J. of General Systems 8(1), 43-61.

Dubois D., and Prade H. (1986) Weighted minimum and maximum in fuzzy set theory. Information Sciences 39, 205-210.

Dubois D., and Prade H. (1988a) Possibility Theory : an Approach to Computerized Processing of Uncertainty. Plenum Press, New York.

Dubois D., and Prade H. (1988b) Modeling uncertainty and inductive inference. Acta Psychologica 68, 53-78.

French S. (1985) Group consensus probability distributions : a critical survey, in J.M. Bernardo, M.H. DeGroot, D.V. Lindley and A.F.M. Smith (Eds.) : Bayesian Statistics 2, North- Holland, Amsterdam, pp. 183-202.

Fung L.W., and Fu K.S. (1975) An axiomatic approach to rational decision making in a fuzzy environment, in L.A. Zadeh, K.S. Fu, K. Tanaka and M. Shimura (Eds.) : Fuzzy Sets and their Applications to Cognitive and Decision Processes, Academic Press, New York, pp. 227-256.

Herstein I.N., and Milnor J.W. (1953) An axiomatic approach to measurable utility. Econometrica 21, 291-297.

Lehrer K., and Wagner C.G. (1981) Rational Consensus in Science and Society. D. Reidel Publ. Co., Boston.

McConway K. (1981) Marginalization and linear opinion pools. J. Amer. Statistical Assoc. 76, 410-414.

Montero de Juan F.J. (1987) Aggregation of fuzzy opinions in a non-homogeneous group. Fuzzy Sets and Systems 25, 15-20.

Schweizer B., and Sklar A. (1983) Probabilistic Metric Spaces. North-Holland, Amsterdam.

Shackle G.L.S. (1961) Decision, Order and Time in Human Affairs. Cambridge University Press, Cambridge, U.K..

Shafer G. (1976) A Mathematical Theory of Evidence. Princeton University Press, Princeton.

Wagner C.G. (1988) Consensus for belief functions and related uncertainty measures. Report ORNL/TM-10748, Oak Ridge National Laboratory, Oak Ridge, Tenn..

Weber S. (1984) $\perp$ – decomposable measures and integrals for archimedean t-conorms $\perp$. J. Math. Anal. Appl. 101, 114-138.

Zadeh L.A. (1971) Similarity relations and fuzzy orderings. Information Sciences 3, 177-200.

Zadeh L.A. (1978) Fuzzy sets as a basis for a theory of possibility. Fuzzy Sets and Systems 1, 3-28.

MODELLING VALUED PREFERENCE RELATIONS

Sergei Ovchinnikov
San Francisco State University
San Francisco, CA 94132, U.S.A.

ABSTRACT

A theory of a valued weak preference relation is described. The transitivity property of a strict preference relation and indifference relation associated with a weak preference relation is established.

Keywords: Preference Relations, Transitivity Property

1. INTRODUCTION

Preference relations play an important role in decision-making, especially, in the context of measurement theory (see, for example, [4]). A weak preference relation R on the set of alternatives A is defined as a binary relation satisfying properties:

1) $xRy \vee yRx$,

2) $(xRy \wedge yRz) \rightarrow xRz$,

for all x, y, and z in A. (We use standard notations for logical connectives "and", "or", "implies", and negation "not".) Therefore, R is 1) complete and 2) transitive binary relation on A. Note, that completeness implies reflexivity, i.e.,

3) $(\forall x)\ xRx$

for all $x \in A$.

There are two binary relations associated with a given weak preference relation. An indifference relation I is defined by

$$xIy \quad \text{iff} \quad xRy \wedge yRx,$$

for all x and y in A, and a strict preference relation P is defined by

$$xPy \quad \text{iff} \quad xRy \wedge \sim yRx,$$

for all x and y in A.

In classical theories binary relations P, I, and R are considered as models for intuitive concepts of strict preference, indifference and preference-or-indiference.

J. Kacprzyk and M. Fedrizzi (eds.), Multiperson Decision Making Using Fuzzy Sets and Possibility Theory, 64–70.

The following theorem is a classical result in the theory of preference relations.

Theorem. Suppose R is a weak preference relation on A. Then
i) I is a symmetric and transitive binary relation on A, i.e.,

$$xIy \rightarrow yIx,$$

and

$$(xIy \wedge yIz) \rightarrow xIz,$$

for all x, y, z in A.
ii) P is an antisymmetric and transitive binary relation on A, i.e.,

$$xPy \rightarrow \sim yPx,$$

and

$$(xPy \wedge yPz) \rightarrow xPz,$$

for all x, y, z in A.

It turns out that, in some instances, classical binary relations are not adequate models for intuitive concepts of preference and indifference. For example, an individual could prefer x to y to a certain 'degree' or a probability of preference is introduced. To resolve these diffculties a notion of a valued binary relation is introduced. A valued binary relation R on A is a function $R: A \times A \rightarrow [0,1]$; in other words, R is a function of two variables on A with codomain [0, 1]. Fuzzy binary relations [1] and probabilistic relations [2] are examples of valued preference relations.

In this paper, we present an approach to the theory of valued preference relations based on multiple-valued logic. The paper is organized as follows:

In section 2 triangular norms (t−norms), conorms, and negation functions are introduced. These functions are used as models for logical connectives in section 3. In the same section 3 we introduce a theory for preference relations and establish some properties of these relations. The classical results about preference and indifference relations are generalized in section 4.

2. T-NORMS AND NEGATION FUNCTIONS

This section contains a brief overview of t−norms, t−conorms, and negation functions. Proofs and references are found in [7] and [5].

A *t−norm* T is defined as a function T: $[0,1]\times[0,1] \rightarrow [0,1]$ satisfying the following properties:
(i) $T(x,1) = x$,
(ii) $T(x,y) \leq T(z,u)$, if $x \leq z$ and $y \leq u$,
(iii) $T(x,y) = T(y,x)$, and
(iv) $T(x,T(y,z)) = T(T(x,y),z)$,
for all x,y,z and u in [0,1].

An *Archimedian* t−norm is a t−norm satisfying
(v) $T(x,x) < x$, for all x in (0,1).

Let $\mathbb{R}^+ = [0,+\infty]$. A t−norm is continuous and Archimedian if and only if

$$T(x,y) = f(g(x) + g(y))$$

where

a) g is a continuous and strictly decreasing function from [0,1] to $\mathbb{R}^+$, such that $g(1) = 0$,

b) f is a continuous from $\mathbb{R}^+$ onto [0,1], such that $f(x) = g^{-1}(x)$ on $[0, g(0)]$, and

$f(x) = 0$ for all $x \geq g(0)$, or, equivalently,

$$T(x,y) = g^{-1}(\min(g(x) + g(y)), g(0))$$

We say that a t−norm T has *zero divisors* if it satisfies

(vi) $T(x,y) = 0$ for some positive x and y.

If T is a continuous Archimedean t−norm, then it has zero divisors if and only if $g(0) < +\infty$. A 'canonical' example of a t−norm with zero divisors is given by the Lukasiewicz t−norm

$$W(x,y) = \max\{x + y - 1, 0\}.$$

We call a strictly increasing function ϕ form the unit interval onto itself an *automorphism* of the unit interval. Any automorphism of the unit interval is a continuous function satisfying boundary conditions $\phi(0) = 0$ and $\phi(1) = 1$.

Theorem 2.1. A t−norm T is a continuous Archimedean t−norm with zero divisors if and only if there exists an automorphism ϕ of the unit interval [0,1] , such that

$$T(x,y) = W^{\phi}(x,y) = \phi^{-1}(W(\phi(x), \phi(y))).$$

The automorphism ϕ from Theorem 2.1 is a *generator* of T.

A *negation* N is defined as a strictly decreasing function N: $[0,1] \rightarrow [0,1]$ satisfying

$$N(N(x)) = x, \text{ for all } x \in [0,1].$$

Thus defined negation is a continuous function satisfying boundary conditions $N(0) = 1$ and $N(1) = 0$.

A standard example of a negation is given by $N(x) = 1 - x$.

Theorem 2.2. N is a negation function if and only if there exists an automorphism ϕ of the unit interval, such that

$$N(x) = N^{\phi}(x) = \phi^{-1}(1 - \phi(x)).$$

The automorphism ϕ from Theorem 2.2 is a *generator* of N.

Suppose a t−norm T and a negation function N are given. Then the function S given by

$$S(x, y) = N(T(N(x), N(y)))$$

is the *t−conorm* of T. For example, if N is the standard negation $N(x) = 1 - x$, the t−conorm of W is given by

$$W^*(x, y) = \min\{x + y, 1\}.$$

A triple <T, S, N> is called a De Morgan triple in fuzzy set theory [3]. In this paper we assume that T is an Archimedean t−norm with zero divisors and T and N are generated by the same automorphism ϕ. Therefore, the automorphism ϕ may be regarded as a 'parameter' in our models. For a given ϕ, elements of the De Morgan triple have the following representations

$$T(x,y) = \phi^{-1}(\max\{\phi(x) + \phi(y) - 1, 0\}),$$

$$S(x,y) = \phi^{-1}(\min\{\phi(x) + \phi(y), 1\}),$$

$$N(x) = \phi^{-1}(1 - \phi(x)).$$

3. MODELS FOR VALUED PREFERENCE AND INDIFFERENCE RELATIONS

Let L be the usual first order language with identity based on the following symbols:

a) variables − x, y, z, u, w, ... ;

b) n-ary function symbols and the n-ary predicate symbols;

c) propositional connectives $\sim$, $\wedge$;
d) quantifiers $\forall$ and $\exists$.

The only *logical* predicate symbol is $=$. All other function and predicate symbols are *nonlogical* symbols. Formulas in L are constructed in the usual way. We write $\mathbb{F}(x, y, \ldots, w)$ to indicate that the free variables of the formula $\mathbb{F}$ are among the variables x, y, ... , w. Propositional connectives $\vee$ and $\rightarrow$ are defined by means of $\sim$ and $\wedge$ as follows:

$\mathbb{A} \vee \mathbb{B}$ is an abbrevation of $\sim((\sim\mathbb{A}) \wedge (\sim\mathbb{B}))$;

$\mathbb{A} \rightarrow \mathbb{B}$ is an abbrevation of $(\sim\mathbb{A}) \vee \mathbb{B}$.

We also introduce a nonlogical predicate symbol $\mathbb{R}$ satisfying the following nonlogical axioms:
i) $(\forall x)(\forall y)\ \mathbb{R}(x, y) \vee \mathbb{R}(y, x)$ *completeness*
ii) $(\forall x)(\forall y)(\forall z)\ (x\mathbb{R}y \wedge y\mathbb{R}z) \rightarrow x\mathbb{R}z$ *transitivity*

Suppose that a De Morgan triple $<T, S, N>$ generated by ϕ is given. We use the following interpretations for propositional connectives:

$$\sim\alpha = N(\alpha) \quad \text{and}$$

$$\alpha \wedge \beta = T(\alpha, \beta),$$

for all α, $\beta \in [0, 1]$.
Then, in accordance with our definitions,

$$\alpha \vee \beta = N(T(N(\alpha), N(\beta))) = S(\alpha, \beta)$$

and

$$\alpha \rightarrow \beta = S(N(\alpha), \beta) = N(T(\alpha, N(\beta))),$$

for all α, $\beta \in [0, 1]$.

Since t–norms and conorms are associative functions, we will not use parenthesis in expressions like $\alpha \wedge \beta \wedge \gamma$ or $\alpha \vee \beta \vee \gamma$.

The function $\forall$ is defined for any non-empty set $X \subseteq [0, 1]$ by $\forall X = \mathit{inf} X$.

The following lemma establishes an important property of functions $\wedge$ and $\sim$.

Lemma 3.1. $\alpha \wedge (\sim\beta) = 0$ iff $\alpha \leq \beta$.

Proof. By definition,

$$\alpha \wedge (\sim\beta) = T(\alpha, N(\beta)) = \phi^{-1}(\max\{\phi(\alpha) + \phi(N(\beta)) - 1, 0\}) =$$
$$\phi^{-1}(\max\{\phi(\alpha) + 1 - \phi(\beta) - 1, 0\}) = \phi^{-1}(\max\{\phi(\alpha) - \phi(\beta), 0\}).$$

The assertion of the lemma follows immediately.

□

In particular, $\alpha \wedge (\sim\alpha) = 0$, for all $\alpha \in [0, 1]$, or, equivalently, $\alpha \vee (\sim\alpha) = 1$, for all $\alpha \in [0, 1]$, which is the Law of Excluded Middle.

A *model* for L is a triple $<A, Id, R>$, where A is a non-empty set, Id is a characteristic function of the identity relation on A, and R is a function $A \times A \rightarrow [0, 1]$.

For each formula $\mathbb{F}(x, y, \ldots, w)$ and interpretation of x, y, ... , w as elements a, b, ... , m in A, we define a value function $F(a, b, \ldots, m)$ with codomain $[0, 1]$ by induction on the formulas using interpretations for propositional connectives introduced above. For instance, $R(a, b)$ is the function from the triple introduced above. If, say, $F(x, y)$ is a formula such that $F(a, b)$ is defined for all a and b in A, then define

$(\sim F)(a,b) = \sim(F(a,b)),$

$(F \wedge G)(a,b) = F(a,b) \wedge G(a,b),$

$(\forall y\ F)(a) = \forall\{F(a,b) \mid b \in A\} = \inf\{F(a,b) \mid b \in A\}$, etc.

We say that $F(a, b, \ldots, m)$ is *true* if $F(a, b, \ldots, m) = 1$ (designated value is 1, cf. [6]); in particular, our axioms and properties are assumed to be true formulas.

We call $R(a,b)$ introduced above a *valued weak preference relation* on the set A. We assume that this relation satisfies the following conditions which are interpretations of the nonlogical axioms i) and ii).

$\forall\{R(a,b) \vee R(b,a) \mid a, b \in A\} = 1$, and

$\forall\{(R(a,b) \wedge R(b,c)) \rightarrow R(a,c) \mid a,b,c \in A\} = 1,$

or, equivalently,

$R(a,b) \vee R(b,a) = 1$ *completeness*

$[(R(a,b) \wedge R(b,c)) \rightarrow R(a,c)] = 1$ *transitivity*

for all a, b, and c in A.

We define a valued *indifference relation* I on A by

$$I(a,b) = R(a,b) \wedge R(b,a),$$

for all a and b in A, and a valued *strict preference relation* P on A by

$$P(a,b) = R(a,b) \wedge (\sim R(b,a)),$$

for all a and b in A (compare with the definitions for binary relations in Introduction).

In the next section we establish basic properties of I and P. These properties are introduced here. In classical theory a binary predicate $\mathbb{R}$ is called symmetric if

$(\forall x)(\forall y)\ \mathbb{R}(x,y) \rightarrow \mathbb{R}(y,x)$

and antisymmetric if

$(\forall x)(\forall y)\ \mathbb{R}(x,y) \rightarrow (\sim\mathbb{R}(y,x)).$

Accordingly, we define properties of a valued binary relation R:

$[R(a,b) \rightarrow R(b,a)] = 1$ *symmetry*

$[R(a,b) \rightarrow (\sim R(b,a))] = 1$ *antisymmetry*

We will need different representations for the basic properties of valued binary relations. They are established in the following lemma.

<u>Lemma 3.2</u>. Listed below are equivalents of the respective properties

$R(a,b) = R(b,a)$ *symmetry*

$R(a,b) \wedge R(b,a) = 0$ *antisymmetry*

$R(a,b) \wedge R(b,c) \leq R(a,c)$ *transitivity*

for all a, b, and c in A.

<u>Proof</u>. i) $1 = R(a,b) \rightarrow R(b,a) = (\sim R(a,b)) \vee R(b,a) = \sim(R(a,b) \wedge (\sim R(b,a)))$ is equivalent to $R(a,b) \wedge (\sim R(b,a)) = 0$. By Lemma 2.2, it is equivalent to $R(a,b) \leq R(b,a)$. Similarly, $R(a,b) \leq R(b,a)$.

ii) $1 = R(a,b) \rightarrow (\sim R(b,a)) = (\sim R(a,b)) \vee (\sim R(b,a)) = \sim(R(a,b) \wedge R(b,a))$ is equivalent to $R(a,b) \wedge R(b,a) = 0$.

iii) $1 = [(R(a,b) \wedge R(b,c)) \rightarrow R(a,c)] = \sim(R(a,b) \wedge R(b,c) \wedge (\sim R(a,c)))$

is equivalent to $R(a,b) \wedge R(b,c) \wedge (\sim R(a,c)) = 0$. By Lemma 2.2, it is equivalent to $R(a,b) \wedge R(b,c) \leq R(a,c)$.

□

4. MAIN THEOREMS

Suppose that R is a valued weak preference relation on A, and I and P are valued indifference and strict preference relations associated with R. In this section we prove basic properties of these relations.

Theorem 4.1. I is a symmetric and transitive valued binary relation.

Proof. i) $I(a,b) = R(a,b) \wedge R(b,a) = R(b,a) \wedge R(a,b) = I(a,b)$. By Lemma 3.2, I is symmetric.

ii) By transitivity of R,
$I(a,b) \wedge I(b,c) = R(a,b) \wedge R(b,a) \wedge R(b,c) \wedge R(c,b) = [R(a,b) \wedge R(b,c)] \wedge [R(c,b) \wedge R(b,a)] \leq R(a,c) \wedge R(c,a) = I(a,b)$. By Lemma 3.2, I is transitive. □

Theorem 4.2. P is antisymmetric and transitive valued binary relation.

Proof. i) $P(a,b) \wedge P(b,a) = R(a,b) \wedge (\sim R(b,a)) \wedge R(b,a) \wedge (\sim R(a,b)) = 0$, by Lemma 3.1.

ii) $P(a,b) = R(a,b) \wedge (\sim R(b,a)) = \phi^{-1}(\max\{\phi(R(a,b)) + 1 - \phi(R(b,a)) - 1, 0\}) =$

$= \phi^{-1}(\max\{\phi(R(a,b)) - \phi(R(b,a)), 0\})$.
If $P(a,b) > 0$, then $\phi(P(a,b)) = \phi(R(a,b)) - \phi(R(b,a))$.

To prove transitivity we need to prove that $P(a,b) \wedge P(b,c) \leq P(a,c)$. It suffices to consider the case when $P(a,b) > 0$ and $P(b,c) > 0$. Then we have to prove the following:

$$\phi^{-1}(\max\{\phi(P(a,b)) + \phi(P(b,c)) - 1, 0\}) \leq P(a,c),$$

or, equivalently,

$$\phi(R(a,b)) - \phi(R(b,a)) + \phi(R(b,c)) - \phi(R(c,b)) \leq 1 + \phi(R(a,c)) - \phi(R(c,a)). \quad (4.1)$$

By transitivity of R, $R(b,c) \wedge R(c,a) \leq R(b,a)$, or, equivalently,

$$\phi^{-1}(\max\{\phi(R(b,c)) + \phi(R(c,a)) - 1, 0\}) \leq R(b,a),$$

that is

$$\phi(R(b,c)) + \phi(R(c,a)) \leq 1 + \phi(R(b,a)),$$

or

$$\phi(R(b,c)) \leq 1 + \phi(R(b,a)) - \phi(R(c,a). \quad (4.2)$$

Since R is a complete valued relation, $R(b,c) \vee R(c,b) = 1$, or, equivalently,

$$\phi(R(b,c)) + \phi(R(c,b)) \geq 1.$$

Substituting (4.2) into the last inequality, we obtain

$$1 \leq \phi(R(b,c)) + \phi(R(c,b)) \leq 1 + \phi(R(b,a) - \phi(R(c,a)) + \phi(R(c,b),$$

implying

$$\phi(R(b,a)) + \phi(R(c,b)) \geq \phi(R(c,a)). \quad (4.3)$$

By transitivity of R

$$\phi(R(a,b)) + \phi(R(b,c)) \leq 1 + \phi(R(a,c)).$$

Subtracting (4.3) from the last inequality, we obtain (4.1).

□

Actually, P satisfies a stronger property than antisymmetry:

$\min\{P(a,b), P(b,a)\} = 0$ *strong antisymmetry*

for all a and b in A.

Indeed, if $R(a,b) \leq R(b,a)$, than $P(a,b) = R(a,b) \wedge (\sim R(b,a)) = 0$, by Lemma 3.1. Otherwise, $P(b,a) = 0$, by the same argument.

5. CONCLUSION

The model presented in the paper is based on Lukasiewicz-like multiple valued logic; it uses triangular norms with zero divisors. A similar approach works in the case of t−norms without zero divisers. We intend to present a comprehensive study of valued preference relations in a future publication.

REFERENCES

[1] Dubois and H. Prade, Fuzzy Sets and Systems: Theory and Applications (Academic Press, New York, 1980)
[2] P.C. Fishburn, Binary choice probabilities: on the varieties of stochastic transitivity, J. Math. Psychol., 10 (1973) 329-352.
[3] E.P. Klement, Operations on fuzzy sets and fuzzy numbers related to triangular norms, in: Proc. of the 11th ISMVL, University of Oklahoma, 1981, 218-225.
[4] D.H. Krantz, R.D. Luce, P.Suppes, and A. Tversky, Foundations of Measurement (Academic Press New York, 1971).
[5] S. Ovchinnikov and M. Roubens, On strict preference relations, Fuzzy Sets and Systems, to appear.
[6] N. Recher, Many-Valued Logic (McGraw Hill, New York, 1969).
[7] B. Schweizer and A. Sclar, Probabilistic Metric Spaces (North-Holland, Amsterdam, 1983).

REVEALED FUZZY PREFERENCES

Bondareva O.N.
Institute of Mathematics and Mechanics
Leningrad University
198904 Leningrad, USSR

Abstract: It is proved that there is a one-to-one mapping between the class of all antireflexive binary nonfuzzy relations and a class of choice functions with special properties. The direct mapping is the rule of choice, the inverse mapping is the revealed preference. A fuzzy extension of these notions are given showing this result also holds for fuzzy choice on nonfuzzy sets i.e. for fuzzy preferences. Because the properties of choice functions allow for a trivial fuzzy extension, it is convinient to describe properties of fuzzy relations in terms of choice functions, i.e. to consider the revealed fuzzy preferences. The above principle of fuzzy extension is applied to a fuzzy multiperson choice problem.

Keywords: choice, revealed preference, core, von Neumann-Morgenstern solution, multiperson choice.

1 INTRODUCTION

Consider a choice on an arbitrary nonfuzzy set A, finite or infinite. Let $R(x,y)$ be an arbitrary relation on A, R may be fuzzy, i.e. $0 \leq R(x,y) \leq 1$. In this paper fuzzy choice on nonfuzzy sets is considered. There are two standard rules of choice with respect to a nonfuzzy relation R: (1) select the best alternatives, (2) select undominated alternatives. These rules are equivalent in the following sense: if $C(X)$ $(X \subset A)$ is the set selected from X by the first rule under R then $C(X) = C'(X)$ where $C'(X)$ is obtained by the second rule but under the following relation R': $xR'y$ iff xRy is not true. In this paper the second rule is prefered called, as in game theory, the core.

The rule of choice maps a set of relations into a set of choice functions. If this is a one-to-one mapping the relation is named the revealed preference. In other

J. Kacprzyk and M. Fedrizzi (eds.), Multiperson Decision Making Using Fuzzy Sets and Possibility Theory, 71–79.

words, the revealed preference has the following properties: each choice function C from the given class allows to reconstruct a relation R such that this C is obtained by this choice rule with R. Such a mapping exists for the set of all antireflexive relations and a set of choice functions fulfiling the well-known axioms of choice. The fuzzy extension of these axioms is trivial because they are expressed in term of operations $\cap$, $\cup$ and $\subset$ only.

The idea of revealed preferences may be used both for a fuzzy extension of choice with a relation and to state an equivalence of some results in term of relations and choice functions in the nonfuzzy case.

The von Neumann-Morgenstern solution is a standard notion in game theory. Inspite of this it is a direct extension of the first rule to a wider class of relations though it is less known in the theory of choice. The von Neumann-Morgenstern solution is a set V such that all its elements must be independent and each element x outside of V must be 'worse' than some element $y \in V$, i.e. yRx. For example, if R is a weak order, the set of all maximal elements is the von Neumann-Morgenstern solution.

The result about revealed preferences for this rule is proved. It is given a fuzzy extension of the von Neumann-Morgenstern solution.

The principle of revealed preferences is applied to multiperson choice. A fuzzy multiperson choice is defined.

2 Some preliminaries

Let A be an arbitrary set, finite or infinite, of alternatives for choice. Nonfuzzy subsets of A are denoted by capital letters $X, X', \ldots$; fuzzy subsets are denoted both by special letters $\mu, \mu', \ldots$ and the same letters $X, X', \ldots$. So $X(x)$ is the membership function of X, if X is nonfuzzy then $X(\cdot)$ is a characteristic function of the set X, i.e. $X(x) = 1$ if $x \in X$ and $X(x) = 0$ otherwise. Similarly, $R \subset A \times A$ is both a nonfuzzy binary relation and a fuzzy relation $R(x, y)$. So a relation, fuzzy or nonfuzzy, is denoted by $R(x, y)$, $R \in [0, 1]^{A \times A}$. The only restriction for relations is the antireflexivity (irreflexivity), i.e. $R(x, x) = 0$.

The relation R need not be transitive. There are two reasons for transitivity to be innatural. The first is explained by the next example. Let two persons have preferences R' and R'', linear orders, and a collective relation is $R = R' \cup R''$, i.e. the couple prefers x to y if at least one person prefers x to y. This very natural relation is not transitive. And second: only the antyreflexivity has an evident extension to the fuzzy case.

A nonfuzzy choice function is $C: \; 2^A \to 2^A, \quad C(X) \subset X$.

A fuzzy choice is considered here on the nonfuzzy sets only, so the choice function is here $C: \; 2^A \to [0, 1]^A, \quad C(X) \subset X$.

Notice that $C(X) \subset X$ means $C(X)(x) \leq X(x)$. Because $X(x) = 0$ for $x \notin X$, then $C(X)(x) = 0$ for $x \notin X$; this condition will be later omitted but always implied.

Denote (1) $\mathcal{R}$ — a set of antireflexive relations and $\mathcal{R}_f$ — its fuzzy extension, (2) $\mathcal{C}$ — a set of choice functionsand $\mathcal{C}_f$ — its fuzzy extension,(3) $\mathcal{R}_1$ — the set of all weak orders, asymmetric and negatively transitive relations.

Recall some well-known properties of choice functions. Each of them has many names (and many authors), it is given only one here.

H (*heritage condition*): $C(X) \cap X' \subset C(X')$, $X' \subset X$.

C (*Condorcet condition*): $\bigcup_{y \in X} C(\{x, y\}) \subset C(X)$, $x \in X$.

S (*choice stability*): if $X' \subset C(X) \subset X$ then $C(X') = X$.

I (*independence of irrelevent alternatives*): if $C(X) \subset X' \subset X$ then $C(X') = C(X)$.

K (*Chernoff condition*): if $X' \subset X$ and $X' \cap C(X) \neq \emptyset$ then $C(X') = C(X) \cap X'$.

Because all this conditions are expressed with the operations $\cap$, $\cup$ and $\subset$ only, their fuzzy extension are trivial. So all this conditions are considered to be written for fuzzy sets.

3 Rules of Choice and Revealed Preferences

The rule of choice F allows to find an 'optimal' set (may be fuzzy) for each $X \subset A$ if a preference relation is known. Formally: $F : \mathcal{R} \times 2^A \to [0,1]^A$, $F(R, X) \subset X$, i.e. $F(R, \cdot)$ is a choice function for a fixed R. Hence the rule F defines the mapping from the set of relations $\mathcal{R}$ to a set of choice functions.

There are two standard rules: first (F_1) — select all $x \in X$ each 'better' that all other, and second (F_2) — select all $x \in X$ such that there is no $y \in X$ 'better' than x. Thus: $F_1(R, X) = \{x \in X : xRy, y \in X\}$, $F_2(X, R) = \{x \in X : \nexists y \in X : yRx\}$.

These rules are eqivalent in the following sense: let $R' = A \times A - R$, i.e. $xR'y$ iff xRy is not true, then $F_1(R, X) == F_2(R', X)$ (see, for example, Aizerman and Malishevski(1981)).

Note that F_1 is correct for the reflexive relations and F_2 — for the antireflexive ones. F_2 is more preferable because $F_2(X)$ is nonempty for a larger class of relations.

F_2 is called in game theory the core and denoted C. Therefore

$$C_R(X) = X - R(X), \tag{1}$$

where $R(X) = \bigcup_{x \in X} R(x)$. $R(x) = \{y : xRy\}$. $R(X)$ being considered as a fuzzy set represented by the membership function $R(x)(y) = R(x, y)$, then $R(X)(y) = \sup_{x \in X} R(x, y)$ if $\cup$ is interpreted as **sup**. If the negation is defined by $(X - B)(x) = X(x) - B(x)$, then

$$C_R(X)(y) = 1 - \sup_{x \in X} R(x, y), \quad y \in X, \tag{2}$$

Definition 3.1. If rule $F : \mathcal{R} \to \mathcal{C}$ has an inverse function $R(C) : F(R(C), X) = C(X)$, then $R(C)$ is called the *revealed preference.*

Therefore F determines the one-to-one mapping between the sets $\mathcal{R}$ and $\mathcal{C}$ in this case.

4 A THEOREM ABOUT REVEALED PREFERENCES

The following theorem was proved by many authors for the nonfuzzy case(may be in another form and for a finite A) as, e.g., in Aizerman and Malishevski (1981). Its fuzzy analogon was proved independent in Aizerman and Litvakov (1988) for a special case and in Bondareva(1988). This theorem describes the class $\mathcal{C}$ corresponding to the class $\mathcal{R}$ of all antireflexive relations with respect to the core.

Lemma 4.1.(of binarity). If a choice function $C : 2^A \to [0,1]^A$ satisfies the conditions H and C, then $C(X)(x) = \inf_{y \in X} C(\{x,y\})(x),\ x \in X$.

This result is extended to the choice from fuzzy sets in Bondareva(1988).

Denote $\mathcal{C}(H, C)$ the set of all choice functions satisfying H and C.

Suppose that $C(\{x,y\})$ is nonempty for all $x, y \in A$. Only such classes of choice functions are considered.

Theorem 4.2. The fuzzy core, the rule given by (2), defines a one-to-one correspondence between the set $\mathcal{R}_f$ of all antireflexive relations and the set $\mathcal{C}_f(H, C)$, fuzzy choice functions satisfying H and C; the revealed preference is

$$R(C)(x,y) = 1 - C(\{x,y\})(y). \tag{3}$$

Proof is following from lemma 4.1. See Bondareva(1988).

Corollary 4.3. The conditions H and C are necessary conditions for any class of relations to have revealed preferences via the core.

5 PROPERTIES OF REVEALED PREFERENCES

Because of theorem 4.2 it is possible to express some properties of relations by properties of corresponding choice functions.

Recall some known properties of nonfuzzy prefernces.

Proposition 5.1. The antireflexivity of a relation is equivalent to the condition $C(\{x\}) = \{x\},\ x \in A$.

Proposition 5.2. The nonemptiness of a choice from each finite set is equivalent to the acyclicity of the revealed preference.

Proposition 5.3. Each revealed preference R defined by $C \in \mathcal{C}(H, C, I, D)$ is transitive where D: $C(\{x,y,z\}) \neq \emptyset$ for each triple $\{x,y,z\} \subset A$. If A is finite, the inverse is also true.

Notice that the last is not true for an infinite A: there are transitive relations having the choice functions not belonging to $\mathcal{C}(H, C, I, D)$. For instance, let R be a relation such that xRy iff $|x| > |y|$ and $xy > 0$, on $A = (-\infty, \infty)$. Put $X = [-5, \infty)$ then for $X' = [-5, 5]$ the condition I does not hold because $C(X') = \{-5, 5\}$ but $C(X) = \{-5\}$, and $C(X) \subset X'$ but $C(X) \neq C(X')$.

Proposition 5.4. The class $\mathcal{R}_1$ of weak orders is equivalent to $\mathcal{C}(K, D)$.

It implies that each revealed preferences $R(C) \in C(K, D)$ is a weak order.

6 THE POSTULATE OF REVEALED PREFERENCES. STRICT PREFERENCES

It is possible to perform some transformations on the relation or the choice function in a choice problem with a relation R. If the final choice is such that it is possible to reveal the initial relation R, then any information is not lost (if information is received, this is another problem). This condition is here called the principle of revealed preferences. Let a nonfuzzy $\mathcal{R}$ be the set of revealed preferences for $\mathcal{C}$ with respect to a rule F; $\mathcal{R}_f, \mathcal{C}_f$ and F_f are their fuzzy extensions, and then F_f must be a one-to-one mapping $\mathcal{R}_f \rightarrow \mathcal{C}_f$. So Theorem 4.2 means that the extension of the core (2) satisfies the postulate of revealed preferences.

Consider from this point of viev the reduction of R to the strong preference P: xPy iff xRy and not yRx. Only if R is a reflexive and symmetric relation it is possible to reconstruct R from P. Hence only in this case it is possible to reveal both R and P from the choice function. The extension of this reduction to fuzzy relations does not satisfy the principle of revealed preferences. For example, Orlovsky (1978) (see also Zhukovin, Burstein and Korelov, 1987) defines a relation $R^S(x, y) = R(x, y) - R(y, x)$ if $R(x, y) > R(y, x)$ and $R^S(x, y) = 0$ otherwise, and a choice function $C(X)(y) = 1 - \sup_{x \in X} R^S(x, y)$. As it follows from Theorem 4.2 R^S is a revealed preference but it is impossible to receive R from C (only R^S).

The second example is about the transitivity of a fuzzy relation. Many authors (see,e.g., Batirshin,1979; and Ovchinnicov and Roubens,1989) define the transitivity as: $R(x, y) \geq \inf_{z \in A}(R(x, z), R(z, y))$. According to the principle of revealed preferences this condition must follows from the conditions C, H, I and D (see Proposition 5.3). It is difficult to verify this conditions because the fuzzy D is undefined. Maybe the notions of transitivity, the strict relation, ets. are specifically nonfuzzy ones. Let as show that it is true for the equivalence relation. Let R be an equivalence relation on A; then $C(X) = X$ or $C(X)(y) = 1$, $y \in X$, if X is nonfuzzy. Because C is the core for the negation relation R' (R is reflexive), the condition $C(\{x, y\})(y) = 1$ implies $R'(x, y) = 0$. It means that $R(x, y) = 1$ for all negation functions (see Ovchinnikov and Roubenc, 1989). Hence, the equivalence relation on nonfuzzy set must be nonfuzzy only.

7 MULTIVALENT CHOICE. VON NEUMANN-MORGENSTERN SOLUTIONS

Return to rule F_1. Consider an example of a two-person choice: $A = \{a, b, c, d\}$ and two persons 1 and 2 have their preference relations R' and R'', the linear orderings expressed with profiles: $R' : abcd$, $R'' : cbad$. Put $R = R' \cup R''$ i.e. x is 'better' than y for the couple if it is 'better' at least for one of them. Hence $R = \{(a,c),(c,a),(a,b),(b,a),(b,c),(c,b),(a,d),(b,d),(c,d)\}$. Note that aRc, aRb, aRd, similar, b and c is 'better' than any other object of choice. Note also that the core is empty. The one-point sets $\{a\}$, $\{b\}$, $\{c\}$ are von Neumann-Morgenstern solutions in terms of game theory. It is therefore necessary to define a multivalent choice.

Definition 7.1. A multivalent choice function is $\overline{C} : 2^A \to [0,1]^A$, so that $\overline{C} = \{C\}$ where each $C : 2^A \to [0,1]^A$ is an ordinary choice function $C(X) \subset X$.

If there are no $C(X)$ for X, it means that $\overline{C}(X) = \emptyset$.

For the above example $\overline{C}(A) = \{C(A), C'(A), C''(A)\}$ where $C(A) = \{a\}$, $C'(A) = \{b\}$ and $C''(A) = \{c\}$.

Definition 7.2. A von Neumann-Morgenstern solution (NMS) for a nonfuzzy relation R on the set $X \subset A$ is

$$V_R(X) = X - R(V_R(X)), \tag{4}$$

A set $V_R(X)$ is the set with the properties: 1) internal stability: xRy, $x, y \in V_R(X)$, is impossible 2) external stability: for any $y \in X - V_R(X)$ there is an $x \in V_R(X)$ such that xRy.

Denote $\overline{V}_R(X) = \{V_R(X)\}$ so $\overline{V}_R(X)$ is a multivalent choice function.

The multivalent choice and NMS as its example was investigated in Kitainik(1988) and in Bondareva(1988).

Each $C(A)$, $C'(A)$, $C''(A)$ is NMS in the above example.

Define the revealed preference *via* NMS as follows:

$$xRy \text{ iff } \{x\} \in \overline{C}(\{x,y\}), \tag{5}$$

There is an obvious extension $\overline{C}$, $\overline{S}$, $\overline{I}$ of properties C, S, I of choice functions. For example, $\overline{C}$ is: if x belongs to each $C(\{x,y\}) \in \overline{C}(\{x,y\})$ for each pair $\{x,y\}$, $y \in X$, then x belongs to each $C(X) \in \overline{C}(X)$, that is $\overline{C} : C(X)(y) \geq \inf_{C \in \overline{C}} inf_{x \in X} C(x,y)(y)$, $y \in X$, $C(X) \in \overline{C}(X)$.

Similarly: $\overline{S}$: if X' is contained in some $C(X) \in \overline{C}(X)$ then each $C(X') = X'$.

$\overline{I}$: if some $C(X) \subset X' \subset X$ then $C(X) \in \overline{C}(X')$.

Suppose, as above, that $\overline{C}(\{x,y\}) \neq \emptyset$ for each $\{x,y\} \subset A$.

Theorem 7.3. The set $\mathcal{R}$ of all antireflexive relations corresponds to the set $\mathcal{C}(\overline{C}, \overline{S}, \overline{I})$ of multivalent choice functions $\overline{C}(X) = \{V_{R(C)}(X)\}$, conversely each

$\overline{C} \in \mathcal{C}(\overline{C}, \overline{S}, \overline{I})$ defines the revealed preference $R(\overline{C})$ by (5) such that $\overline{C}(X) = \{V_{R(C)}(X)\}$.

This theorem is proved in Bondareva (1988).

Notice that the ordinary choice function is also a multivalent choice function with $\overline{C}(X) = \{C(X)\}$ consisting of a unique $C(X)$. So it is possible to find the intersection of the sets $C(C, H) \cap C(\overline{C}, \overline{S}, \overline{I}) = \mathcal{C}(C, H, I)$ because $H \subset S$.

Corollary 7.4. For $\mathcal{C}(C, H, I)$ both Theorems 7.3 and 4.2 are true and conditions (3) and (5) define the same relation.

Indeed, $\{x\}$ is a unique NMS iff y does not belong to $C(\{x, y\})$.

The notion of a NMS is extended to the fuzzy choice. A direct fuzzy extension of condition (4) is rather impossible, but the conditions $\overline{C}, \overline{S}, \overline{I}$ have trivial extension to the fuzzy choice on arbitrary sets.

Rewrite the conditions $\overline{S}$ and $\overline{I}$ for fuzy sets in another form. Condition $\overline{S}$ is: if $\mu' \leq C(\mu) \leq \mu$ then $C(\mu') = \mu'$. If $\delta\mu$ denotes the function $\delta\mu(x) = \delta(x)\mu(x)$, then each $\mu' \leq C(\mu)$ is expressed as $\mu' = \delta C(\mu)$ where $0 \leq \delta(x) \leq 1$, $x \in A$. So $\overline{S}$ is equivalent to $C(\delta C(\mu)) = \delta C(\mu)$. Similarly, $\overline{I}$ is expressed by $C(\delta C(\mu) - (1-\delta)\mu) = C(\mu)$ where $(1 - \delta)(x) = 1 - \delta(x)$.

Hence a fuzzy NMS satisfies the equations

$$V(\delta V(\mu)) = \delta V(\mu))$$

$$V(\delta V(\mu) - (1 - \delta)\mu) = V(\mu)$$

with a fixed $\mu \in [0, 1]^A$ and an arbitrary $\delta \in [0, 1]^A$.

However, how to solve these equations remains unknown.

8 MULTIPERSON CHOICE AND REVEALED PREFERENCES

Consider a choice on A with n relations $R_1, R_2, \ldots, R_n$. Let the choice be multiperson (or multicriterial).

There are some forms of this problem: 1) for given $R_1, \ldots, R_n$ construct a relation $R = f(R_1, \ldots, R_n)$, a social preference, or the group preference; 2) for given $R_1, \ldots, R_n$ construct a choice function $C(R_1, \ldots, R_n)$, a group choice function; 3) for given choice functions $C_1, \ldots, C_n$ of n persons construct a group choice function $C(C_1, \ldots, C_n)$.

Notice that some results in form 2 or 3 are not new because of their equivalence to known ones in form 1. All further conditions will be given in form 1, though it is not difficult to construct their equivalents in forms 2 and 3.

There are two basic conditions of multiperson choice:

PO (*Pareto optimality*): if $R_1 = \cdots = R_n$ then $f(R_1, \ldots, R_n) = R_1$.

AC (*Arrow condition*): $f(R_1, \ldots, R_n)(x, y) = f(R_1(x, y), \ldots, R_n(x, y))$.

The latter is often called the condition of independence of irrelevant alternatives, but it is not equivalent to condition I for the choice function with the same name. For example, let $A = \{a, b, c, d, e\}$, $R' = \{(a, c), (a, d), (b, e)\}$, $R'' = \{(a, c), (d, b)\}$. Note that both R' and R'' are transitive. The group preferense $R = R' \cup R'' = \{(a, c), (a, d), (d, b), (b, e)\}$. Here $C_R(A) = \{a\}$. In $X = \{a, c, d, e\}$ the element e is not dominated, hence $C_R(X) = \{a, e\}$, and condition I does not hold, although R satisfies AC.

The Arrow problem in our language is to find existence conditions for a one-to-one mapping from $\mathcal{R}^n$ to $\mathcal{R}$. The Arrow theorem is: if $\mathcal{R}$ is the set of all weak order, then each PO and AC function $f(R_1, \ldots, R_n)$ is equal to some R_i. This result does not seemed to be a paradox beacause the social preferences must not be generally transitive if all individual preferences are transitive. Recall the function $R_0 = \cup_{i=1}^{n} R_i$. In the choice function language $C_{R_0}(X) = \cap_{i=1}^{n} C_{R_i}(X)$, $X \subset A$. For weak orders $C_{R_0}(X)$ is the total maximum if it exists. $C_{R_0}(X)$ is often empty. Another collective preference $R_I = \cap R_i$ leads to $C_{R_I}(X) = \cup_{i=1}^{n} C_{R_i}(X)$. $C_{R_I}(X)$ is the strong Pareto optimum: for $x \in C_{R_I}(X)$ there is no y: yR_ix for all i. The relation R_I is transitive for transitive R_i, $i = 1, \ldots, n$, but it is not a weak order if all R_i are weak orders.

Both R_0 and R_I satisfy the conditions AC and PO. All the functions satisfying AC and PO are described as follows (see Aizerman and Aleskerov, 1983). A coalition is $S \subset \{1, \ldots, n\} = I$. Let $\sigma(x, y)$ be a system of coalitions for $\{x, y\} \subset A$. Each relation R satisfying AC is expressed in form

$$R(x, y) = \max_{S \in \sigma(x,y)} \min_{i \in S} R_i(x, y), \quad x, y \in A, \tag{6}$$

This is true for an infinite A also.

It is not difficult to verify the following fact.

Proposition 8.1. The nonfuzzy condition (6) describes the class of all nonfuzzy relations $R = f(R_1, \ldots, R_n)$ satisfying AC and PO iff $\sigma(x, y) \neq \emptyset$, for each $(x, y) \in A \times A$.

If all R_i are antireflexive and $\sigma(x, x) \neq \emptyset$, $x \in A$, then R defined (6) is also antireflexive.

9 A FUZZY MULTIPERSON CHOICE

The conditions PO and AC have a trivial fuzzy extension. Condition (6) is the same for the fuzzy case. Proposition 8.1 is not true in the fuzzy case because not all relations satisfied PO and AC satisfy (6). For example, the relation

$$R(x, y) = \sum_{i \in I} \alpha_i R_i(x, y), \quad \alpha_i \geq 0, \quad \sum_{i \in I} \alpha_i = 1,$$

satisfy PO and AC but not (6).

Since there is no obvious fuzzy extension of the weak order, it does not seem to exist an analogon of the Arrow imposibility theorem.

As it has been mentioned, the transitive individual relations generate an intransitive group relation. Similar, it seems natural for nonfuzzy individual relations to generate a fuzzy group relation. For example, the relation $R(x,y) = 1/n \sum R_i(x,y)$ for nonfuzzy $R_i, i \in I$, is fuzzy. This relation generates all nonfuzzy majority relations in the following manner. Let $Q = \{(x,y) \in A \times A : R(x,y) > k/n\}$, $k < n$, then $Q \subset A \times A$ is the relation: xQy iff at least k persons prefer x to y.

There are other expressions of majority rules:

$$R(x,y) = \max_{|S| \geq k} \min_{i \in S} R_i(x,y), \tag{7}$$

It is a special form of (6) with $\sigma(x,y)$ independent of x, y. Define $a_i' = k \max_{1 \leq i \leq n} a_i$ if there are $k-1$ elements $a_i \geq a_i'$. Then (7) may be rewritten as $R(x,y) = k \max_{1 \leq i \leq n} R_i(x,y)$, this is fuzzy k-majority relation.

REFERENCES

Aizerman M.A.,Aleskerov F.T.(1983) Arrow problem in the theory of group choice. Avtomatika i Telemekhanika 9, 127–151.

Aizerman M.A., Litvakov B.M.(1988) On some extencions the theory of variants choice. Avtomatika i Telemekhanika 3, 92–104.

Aizerman M.A., Malishevski A.V.(1981) General theory of best variants chioce: some aspects. IEEE Trans.Automat. Control 26, 1030–1040.

Batirshin I.Z.(1979) On transitivity of fuzzy orders. In: Issledovanie Operazij i Analiticheskoe Proektirovanie v Tekhnike.Kasanski Avia. Inst., Kazan, 67–73.

Bondareva O.N.(1987) A finite approximation of choice on infinite set. Isv. AN SSSR. Tekhn. Kibern. 1, 18–23. Engl.transl. in: Engineering Kybern. Soviet Journal of Comp. and Syst. Sciences.

Bondareva O.N.(1988) An axiomatization of the core and von Neumann-Morgenstern solutions as functions of non-fuzzy and fuzzy choice. Vestnik LGU 8, 3–7. Engl. transl in: Vestnik Leningrad Univ.Math. 21, N2, 1–7.

Kitainik L.M.(1988) Fuzzy binary rlations and decisions procedures. Isv.AN SSSR. Tekhn. Kibernetika 6, 12–30.

Orlovsky S.A.(1978) Decision-making with a fuzzy preference relation. Fuzzy Sets and Systems 1(3), 155–167.

Ovchinnikov S. and Roubens M.(1989) Submitted to Fuzzy Sets and Systems.

Zhukovin V.E., Burstein F.V. and Korelov E.S.(1987) A decision making model with vector fuzzy preference relation. Fuzzy Sets and Systems 22, 71–79.

CATEGORIES OF FUZZY RELATIONS IN DECISION MAKING

Vladimir B. Gisin
Department of mathematics
Pedagogical Institute
Leningradskoe shosse 112/1-1-56
125445 Moscow, USSR

Abstract: in this paper we consider categories of fuzzy relations from the point of view of a general theory of correspondence categories. Several constructions of decision making are defined and studied by methods of the theory of categories.
Keywords: correspondence category, lattice, fuzzy set, choice function, preference relation.

1. INTRODUCTION

Many important results of decision making analysis are based on algebraic properties of fuzzy relations which form ordered categories with involution, or shortly OI-categories. Methods of the theory of categories are especially usefull in the investigation of properties and constructions related with several objects. A categorical approach allows to clarify some problems. For example, existence of factorizations with respect to fuzzy equivalences, functional decompositions, etc. A general theory of correspondence categories is accounted in Calenko, Gisin and Raikov (1984), Gisin and Calenko (1984, 1985).

The language of correspondence categories allows to obtain fuzzy analogues of "crisp" notions and form techniques of the translation from "crisp" into "fuzzy". The language of correspondence categories is richer than the language of categories. In correspondence categories we have an associative composition, units, an order and an involution. In spite of abstractness, the correspondence categories reflect

J. Kacprzyk and M. Fedrizzi (eds.), Multiperson Decision Making Using Fuzzy Sets and Possibility Theory, 80–89.

properties of correspondences of sets rather well: a relation between correspondences being formulated in the language of OI-categories is true iff it is true for correspondences over sets. This statement is a matter of the fundamental theorem of the theory of correspondence categories. It shows a natural framework of general categorical methods in the theory of fuzzy correspondences.

In correspondence categories may be interpreted several logical operations: existential quantifiers, conjunctions. In categories of fuzzy relations may be interpreted in addition other propositional connectives, universal quantifiers. Using such interpretations we obtain abstract algebraic analogues of operators and formulas of decision making analysis.

Using an abstract approach we can come into new problems. For example, given a preference relation r, we can construct an associated domination relation (strict preference relation) s. It was shown in Kuzmin (1982) that if r is transitive then s is also transitive (in the case of a linear scale of fuzziness). In general this statement does not hold. In the case of L-fuzzy relations it is true iff L satisfies the dual analogue of an important property of intuitionistic disjunction, see Rasiowa and Sikorski (1970).

2. CORRESPONDENCE CATEGORIES

2.1. Basic notions

A category C consists of a class of objects and disjoint sets of morfisms C(X,Y) for all pairs of objects X, Y. If $f \in C(X,Y)$ then we also write also $f: X \twoheadrightarrow Y$. In C we have an associative composition of morphisms (if $f: X \twoheadrightarrow Y$, $g: Y \twoheadrightarrow Z$ then $fg: X \twoheadrightarrow Z$) and unit morphisims $\Delta_X: X \twoheadrightarrow X$ (an index may be omitted) for all objects. In OI-categories all sets C(X,Y) are ordered, and for every morphism $f: X \twoheadrightarrow Y$ there exists its involution $f^*: Y \twoheadrightarrow X$. Involution and composition are monotone, and in addition $f^{**}=f$, $(fg)^*=g^*f^*$ for all morphisms f, g.

Morphisms of OI-categories will be called correspondences. In OI-categories may be defined many important classes of correspondences. A correspondence $r: X \twoheadrightarrow Y$ is called: D-regular, if $rr^* \geq \Delta_X$; I-regular, if $r^*r \leq \Delta_Y$; B-regular, if $r^*r \geq \Delta_Y$; K-regular, if $rr^* \leq \Delta_X$. A correspondence is called BK-regular, if it is both B- and K-regular, and similarly in other cases. A correspondence is called functional, projection, an injection if it is DI-regular, DIB-regular, DIK-regular, respectively ; r is difunctional if $r = rr^*r$. Obviously, every functional correspondence is difunctional. A correspondence r is called reflexive, symmetric, transitive if $r \geq \Delta$, $r=r^*$, $r^2 \leq r$, respectively . A correspondence is an equivalence if it is reflexive, symmetric, and transitive; a preorder if it is reflexive, and transitive; r is an order if r is a preorder, and Δ is the greatest symmetric morphism such that $\Delta \leq r$. Dually (with respect to $\leq$ and *) are defined co-notions, for example, c is a coequivalence if $c \leq \Delta$, $c = c^*$, and $c^2 \geq c$.

Given correspondences $r: X \to Y$, $s: Z \to Y$ we denote by r/s the greatest correspondence $t: X \to Z$ such that $ts \leq r$. It can be readily checked that r/r is a preorder.

An important class of OI-categories is formed by the so-called modular categories. An OI-category C is modular if it satisfies the following axioms.

I1. Ordered sets $C(X,Y)$ are inf-semilattices.

I2. For all $r: X \to Y$, $s: X \to Z$, $t: Y \to Z$ it holds

$$(r \cap st^*) \geq rt \cap s. \tag{1}$$

We say that in an OI-category C all correspondences have functional decompositions if C satisfies the following axiom

I∅. Every morphism $r: X \to Y$ is representable in the form $r = f^* g$, where f and g are functional morphisms:

$$X \overset{f}{\leftarrow} U \overset{g}{\rightarrow} Y \tag{2}$$

An OI-category is said to be a correspondence category if it satifies axioms I∅ - I2. A correspondence category is uniquely (up to an isomorphism) determined by its subcategory of functional morphisms. The following theorem is one of the main structure theorems about correspondence categories.

2.1.2. Theorem. Any small correspondence category can be embedded (with respect to OI-structure) into the category of (crisp) correspondences of sets. (A category is small if the class of objects is a set).

2.1.3. Remark. A formula of the form $R(x,y)$, where R is a predicate, may be interpreted in an OI-category C by a morphism $r \in C(X,Y)$. If C is modular we can interprete an existesial quantifier: $\exists y R(x,y)$ may be interpreted by $\Delta x \cap rr^*$. Suppose in addition that in C all sets $C(X,Y)$ contain the greatest correspondences Ωxy and for all pairs of morphisms r, s which have a common codomain there exists r/s. Then a formula of the form $\forall y R(x,y)$ may be interpreted by $\Delta x \cap r/\Omega xy$. Note that $R(x,x)$ may be interpreted by $(\Delta x \cap r)\Omega xx$ (here $X=Y$). The interpetations mentioned above can be explained with the point of view of the topos theory. They will be used in constructions related with choice operators in 4.1.

2.2. Completions of correspondence categories by symmetric idempotents

An equivalence in an OI-category is called realizable if it is of the form ff^* where f is a functional morphism. If there exists a projection $p: X \to Y$ such that $e = pp^*$ then equivalence e is called strictly realizable. In this case Y is a quotient object of X by e.

We say that an OI-category satisfies axiom R if every equivalence is strictly realizable. Dually, an OI-category satisfies axiom R^* if every coequivalence is strictly realizable (i. e. every coequivalence is of the form m^*m where m is an injection).

Note that if C is a correspondence category then every coequivalence c in C is realizable. If C is modular then c is a couquivalence iff $c \leq \Delta$. If C satisfies R, or R^* then every functional morphism f is representable in the form f=pm where p is a projection an m is an injection. If C is a correspondence category such that all equivalences are realizable in C, then C satisfies R if and only if C satisfies R^*. If C is a correspondence category satisfying R then for any object A coequivalences c such that $c \leq \Delta_A$ are in one-to-one correspondence with subobjects of A.

Now we shall describe a completion of an OI-category C. Denote by SI(C) the category of symmetric idempotents. The class of objects of SI(C) is formed by such morphisms a that $a = a^*$ and $a = a^2$. If a, b are objects in SI(C) then SI(C)(a,b) is formed by all correspondences f from C (more precisely, by all triples (f,a,b)) such that af = f = fb. The structure of an OI-category in SI(C) is defined in a natural way. OI-category C can be fully embedded in SI(C) by $X \mapsto \Delta_X$, $C(X,Y) \ni r \mapsto (r, \Delta_X, \Delta_Y)$.

Consider full subcategories Eq(C) and Coeq(C) in SI(C). Objects of Eq(C) are equivalences from C, Objects of Coeq(C) are coequivalences from C. Categories Eq(C) and Coeq(C) are minimal extensions of C satisfying R and R^*, respectively . If C is modular then SI(C) is modular and SI(C)= Eq(Coeq(C)) is a minimal extension of C satisfying both R and R^*. Let C satisfy the following axiom:

IØ'. Every correspondence r in C is representable in the form $r = f^*cg$ where f, g are functional and c is a coequivalence.

Then SI(C) satisfies IØ. So if C is a modular category satisfying IØ' then SI(C) is a correspondence category.

3. CATEGORIES OF FUZZY RELATIONS

3.1. <u>Basic properties</u>

Let L be a complete lattice with a monotone binary operation $*$ such that $(L,*)$ is a commutative semigroup with a zero and a unit. The following two cases are the most interesting for us: 1) $* = \cap$, 2) L=[Ø;1] and $*$ is a t-norm. Let L-Fur be the category of L-fuzzy relations. Objects of L-Fur are sets. Morphisms from X to Y are defined by mappings $X \times U \to L$, composition is the sup-$*$-composition: given $r: X \nrightarrow Y$, $s: Y \nrightarrow Z$ then rs is defined by

$$(rs)(x,z) = \sup(r(x,y) * s(y,z)). \tag{3}$$

3.1.1. Proposition. The composition defined by (3) is associative if and only if $*$ is distributive with respect to sup, i. e.

$$a * \sup A = \sup(a*A) \tag{4}$$

for all $a \in L$, $A \subseteq L$.

Later on we suppose that (4) holds in L. The following operation may be defined in L:

$$a|b = \sup\{\, x \in L \mid x*b \leq a \,\}.$$

If $*=\cap$ then L is a complete Heyting algebra and $a|b = b \twoheadrightarrow a$. It is well-known that $b \cap b \twoheadrightarrow a = b \cap a$. If $*$ is a continuous t-norm then $(a|b)*b = a \cap b$. Using $|$ we obtain the following negation: $\neg a = \emptyset | a$, i. e.

$$\neg a = \sup\{\, x \in L \mid x*a = \emptyset \,\}.$$

Operations sup, $|$, $\neg$ may be pointwise extended to $C(X,Y)$. Note that sup is distributive with respect to sup-$*$-composition. Hence, given $r:X \twoheadrightarrow Y$, $s:Z \twoheadrightarrow Y$ there exists $r/s:X \twoheadrightarrow Z$, $r/s = \sup\{\, t \mid ts \leq r \,\}$. It can be readily checked that

$$(r/s)(x,z) = \inf(r(x,y)|s(z,y)). \tag{5}$$

Every set L-Fur(X,Y) has the greatest element Ω_{XY} and the least element ω_{XY} (as a rule index will be omitted): $\Omega_{XY} \equiv 1$, $\omega_{XY} \equiv \emptyset$.

3.1.2. Proposition. L-Fur is a modular category if and only if $*=\cap$.

3.2. Functional correspondences

Here and later on $* = \cap$.

3.2.1. Proposition. An L-fuzzy correspondence $r:X \twoheadrightarrow Y$ is functional iff for every $x \in X$ nonzero elements $r(x,y) \in L$ form a partition of 1.

Proof is immediate: by definition,

$$\sup(r(x,y) \cap r(x',y)) \geq \Delta_X(x,x'),$$
$$\sup(r(x,y) \cap r(x,y')) \leq \Delta_Y(y,y').$$

Consequently, for all $x \in X$ we have $\sup r(x,y) = 1$ and $r(x,y) \cap r(x,y') = \emptyset$ if $y \neq y'$.

If an L-fuzzy correspondence $r:X \twoheadrightarrow Y$ is functional then $r(x,y) \cup \neg r(x,y) = 1$ for all $x \in X$, $y \in Y$. Denote by B(L) the set of all elements $a \in L$ such that $a \cup \neg a = 1$. B(L) is a Boolean algebra and any functional L-fuzzy correspondence is B(L)-fuzzy correspondence. In particular, all functional $[\emptyset;1]$-fuzzy correspondences are crisp. Hence all realizable equivalences and coequivalences are crisp. Therefore, $[\emptyset,1]$-Fur does not satisfy R, R^*.

3.2.2. Proposition. Let K be a Boolean algebra. Every K-fuzzy correspondence $r:X \twoheadrightarrow Y$ is representable in the form $r = p^{\#}fg^{\#}q$, where p, f, g, q are functional K-fuzzy correspondences.

Proof. Let p, q be projections of $X \times Y$ onto X and Y, respectively. Let U be disjoint union $(X\times Y)_1 \sqcup (X\times Y)_2$. Further, let g be injection $(X\times Y)_1 \twoheadrightarrow U$ and let $f:(X\times Y)_2 \twoheadrightarrow U$ be defined in the following way: $f((x,y),(x',y')_1) = \emptyset$ if $x \neq x'$ or $y \neq y'$; $f((x,y),(x,y)_1) = r(x,y)$, $f((x,y),(x,y)_2) = \neg r(x,y)$. Obviously p, g, q are functional; f is functional by 3.2.1. For $f^{\#}g$ we have $(f^{\#}g)((x,y),(x',y'))$ is $r(x,y)$ if $(x,y) = (x',y')$, else it is $\emptyset$. Hence it follows that $p^{\#}fg^{\#}q = r$.

3.2.3. Proposition. L-Fur satisfies I∅ if and only if $L = \{\emptyset,1\}$.

Proof. "If" is well-known. Conversely, let L-Fur satisfy I∅. Put $X = \{x\}$. Given a set Y and a functional L-fuzzy correspondence $f:Y \rightarrow\!\!\!\!\!\!+ X$, we have $f(y,x) = 1$ for all $y \in Y$. So if f, $g:Y \rightarrow\!\!\!\!\!\!+ X$ are functional and $Y \neq \emptyset$ then $(f^{\#}g)(x,x) = 1$. Hence $r:X \rightarrow\!\!\!\!\!\!+ X$ (defined by arbitrary element $r(x,x) \in L$) has a functional decomposition iff $r(x,x) = \emptyset$ or $r(x,x) = 1$. Since L-Fur satisfies I∅ we have $L = \{\emptyset,1\}$.

3.2.4. Proposition. L-Fur satisfies I∅'.

Proof. Let $r:X \rightarrow\!\!\!\!\!\!+ Y$ be an L-fuzzy correspondence. Let p, q be projections of $X \times Y$ resp. on X, Y, and let $c:X \times Y \rightarrow\!\!\!\!\!\!+ X \times Y$ be an L-fuzzy relation defined by: $c((x,y),(x',y')) = \emptyset$ if $x \neq x'$ or $y \neq y'$; $c((x,y),(x,y)) = r(x,y)$. It can be easily seen that p, q are functional, $c \leq \Delta_{X \times Y}$ is a coequivalence, and $r = p^{\#}cq$.

3.3. <u>Sets with fuzzy equalities</u>

In this subsection we consider categories Coeq(L-Fur) and SI(L-Fur). Objects of Coeq(L-Fur) are L-fuzzy subsets; Coeq(L-Fur) is a correspondence category, but in general it does not satisfy R. Objects of SI(L-Fur) are sets with L-fuzzy equalities. Any object of SI(L-Fur) can be represented in the form (X,a) where X is a set, and a is an L-fuzzy relation on X such that $a(x,x') = a(x',x)$ and $a(x,x') \cap a(x',x'') \leq a(x,x'')$ for all $x, x', x'' \in X$. Morphisms from (X,a) to (Y,b) are L-fuzzy correspondences r such that $a(x,x') \cap r(x',y) \leq r(x,y) \leq a(x,x)$ and $b(y',y) \cap r(x,y') \leq r(x,y) \leq b(y,y)$ for all $x, x' \in X$, $y, y' \in Y$. A morphism r is functional if in addition $\sup r(x,y) = a(x,x)$ for all $x \in X$ and $r(x,y) \cap r(x,y') \leq b(y,y')$ for all $x \in X$, $y, y' \in Y$. The full subcategory of functional morphisms of SI(L-Fur) is isomorphic to the topos of L-sets, and SI(L-Fur) is isomorphic to the correspondence category over this topos.

3.4. <u>Additional operations</u>

Let C be an OI-category such that all sets C(X,Y) are Heyting algebras (see 3.1). If C is modular then for all r, s, t (such that all compositions below are defined) the following inclusion holds

$$r^{\#}(rt \rightarrow\!\!\!\!\!\!+ s) \leq t \rightarrow\!\!\!\!\!\!+ r_{\#}s. \qquad (6)$$

In virtue of modularity, $t \cap r^{\#}(rt \rightarrow\!\!\!\!\!\!+ s) \leq r^{\#}(rt \cap rt \rightarrow\!\!\!\!\!\!+ s)$. Since $rt \cap rt \rightarrow\!\!\!\!\!\!+ s \leq s$, we have $t \cap r^{\#}(rt \rightarrow\!\!\!\!\!\!+ s) \leq r_{\#}s$. Hence it follows (6). In particular, if $s = \omega$ we obtain $r^{\#}(\neg(rt)) \leq$

$\neg t$. The latter inclusion plays an important role in relational algebra (see Jonsson (1982)).

If L is a complete lattice then the following operation may be defined in L:

$$a \div b = \inf\{ x \in L \mid x \cup b \geq a\}. \qquad (7)$$

If inf is distributive with respect to U, i. e. if L satisfies the dual of (4), then L becomes a Brouwerian algebra. We have in L the following negation $\ulcorner a = 1 \div a$.

Further we shall suppose that L is a complete lattice endowed with $\twoheadrightarrow$ and $\div$ which is both Heyting and Brouwerian algebra, or shortly L is a complete BH-algebra.

Now we describe one construction which is usefull in the theory of paraconsistent logics (see Smirnov, 1984, 1987). Let K be a BH-algebra and let K^* be its dual. Put $L = K \times K^*$. Obviously L is a BH-algebra, so we have two negations $\neg(a,b) = (\neg a, \ulcorner b)$, $\ulcorner(a,b) = (\ulcorner a, \neg b)$. In addition in L may be defined the following negation: $\hat{}(a,b) = (b,a)$. It holds $\hat{}\hat{}(a,b) = (a,b)$, $\hat{}\neg\hat{} = \ulcorner$, $\hat{}\ulcorner\hat{} = \neg$.

Valuations in $K \times K^*$ may be treated as follows: The first component is the degree of "yes", the second one is the degree of "no". Given a set A, a $K \times K^*$-fuzzy subset X of A is determined by two functions $y_x, n_x : X \twoheadrightarrow K$, such that $\mu_x(x) = (y_x(x), n_x(x))$.

OI-categories of $K \times K^*$-fuzzy correspondences have rather interesting construction. This is a subject of further investigations.

4. PREFERENCE RELATIONS AND CHOICE IN CATEGORIES OF FUZZY RELATIONS

In this section we shall show applications of the language of correspondence categories. In this language we describe choice functions, and constructions related with preferences.

4.1. Choice operators

Let C be a correspondence category endowed with additional operations /, U, $\div$ and morphisms ω, Ω, and satisfying axiom R. For example C = SI(L-Fur) where L is a complete BH-algebra. For any object A the set of all subobjects of A is in one-to-one correspondence with the set S(A) of all coequivalences c such that $c \leq \Delta_A$ (see 2.2). Taking it into account we mean by choice operator Φ a mapping $\Phi : S(A) \twoheadrightarrow S(A)$. Important properties of choice operators (see Aizerman (1977), Aizerman and Litvakov (1988)) may be formulated in the category language :

M: if $c' \leq c$ then $\Phi(c') \leq \Phi(c')$;

H: if $c' \leq c$ then $\Phi(c') = c' \cap \Phi(c)$;

C: if $c = c' \cup c''$ then $\Phi(c') \cap \Phi(c'') \leq \Phi(c)$.

Let $r \in C(X,X)$. It can be viewed various choice operators related with r. We consider three such operators. Together with every operator we consider an example. In all examples A is a crisp set, $X \leq A$, is a fuzzy subset, c is a coequivalence on A such that $c(x,x) = \mu_X(x)$, $f_X(x)$ is a corresponding choice function (instead of $f_A(x)$ we write $f(x)$).

4.4.1. $\Phi(c) = c \cap rc\Omega$ (maximal result). In this case $f_X(x) = \sup(r(x,y) \cap \mu_X(y)) \cap \mu_X(x)$. We have $f(x) = \sup r(x,y)$. Obviously Φ satifies M. Using techniques of modular categories we shall show that operator Φ is equivalent to operator Φ', where $\Phi'(c) = c \cap rcr^*$. It is clear that $\Phi(c) \leq \Phi'(c)$ because $cr^* \leq \Omega$. Conversely, $\Phi'(c) = c \cap rcr^* = c \cap rc(cr^* \cap \Omega) \geq c \cap \Delta \cap rc\Omega$ (by I2) $= c \cap rc\Omega = \Phi(c)$.

4.1.2. $\Phi(c) = c \cap (\Delta \div \Omega cr)$ (graph-dominant choice). If there exists y such that $\mu_X(y) = 1$ and $r(y,x) = 1$ then $f_X(x) = \emptyset$, otherwise $f_X(x) = \mu_X(x)$. In this case Φ satisfies H. If $c' \leq c$ then $\Phi(c') = c' \cap (\Delta \div \Omega c'r) \geq c' \cap (\Delta \div \Omega cr) = c' \cap c \cap (\Delta \div \Omega cr) = c' \cap \Phi(c)$. Suppose in addition that in $C(A,A)$ the following identity holds

$$w \div (u \cup v) = (w \div u) \cap (w \div v) \tag{8}$$

(in the case when $C \leq SI(L\text{-}Fur)$ we can replace this conditiion by the requirement that (8) holds in L). Then Φ satisfies C (and, hence, M). Really, $\Phi(c') \cap \Phi(c'') = (c' \cap (\Delta \div \Omega cr)) \cap (c'' \cap (\Delta \div \Omega c''r)) = c' \cap c'' \cap (\Delta \div (\Omega c'r \cup \Omega c''r)) = c' \cap c'' \cap (\Delta \div \Omega(c' \cup c'')r) \leq (c' \cup c'') \cap (\Delta \div \Omega(c' \cup c'')r) = \Phi(c' \cup c'')$. There are possible another generalizations, for example $\Phi(c) = c \cap (c \div \Omega cr)$. But the latter does not satisfy M.

4.1.3. $\Phi(c) = c \cap \neg(cr^*c \,/\, \Omega)$ (the chice of nondominated elements). Here we have $f_X(x) = \emptyset$ if there exists y such that $\mu_X(y) > \emptyset$ and $r(y,x) > \emptyset$, otherwise $f_X(x) = \mu_X(x)$. If $X = A$ then $f(x)$ is the characteristic function of the subset of nondominated elements. It can be easily seen that Φ satisfies H (but does not satisfy M even in a "crisp" case).

Examples mentioned above show that various choice operators can be described in the language of correspondence categories. Now we shall consider another example.

4.1.4. Two following relations play an important role in investigation of graph-dominant choice (see Aizerman and Litvakov (1988)): maximal relation r_1, and minimal relation $r_\emptyset$ such that $r_\emptyset$, r_1, r give the same result being applied to A. In the language of correspondence categories $r_\emptyset$ and r_1 may be defined as follows:

$$r_\emptyset = (r \div \Omega(r \cap \Delta)) \cup (r \cap \Delta), \quad r_1 = r \cup \Omega(r \cap \Delta).$$

We have for r_1: $\Delta \div \Omega r_1 = \Delta \div \Omega(r \cup \Omega(r \cap \Delta)) = \Delta \div (\Omega r \cup \Omega^2(r \cap \Delta)) = \Delta \div (\Omega r \cup \Omega(r \cap \Delta)) = \Delta \div \Omega r$. For $r_\emptyset$: since $r_\emptyset \leq r$ and $\Omega r_\emptyset = \Omega((r \div \Omega(r \cap \Delta)) \cup (r \cap \Delta)) = \Omega(r \div \Omega(r \cap \Delta)) \cup \Omega(r \cap \Delta)) = \Omega(r \div \Omega(r \cap \Delta)) \cup \Omega^2(r \cap \Delta)) = \Omega((r \div \Omega(r \cap \Delta) \cup \Omega(r \cap \Delta)) \geq \Omega r$, we have $\Omega r_\emptyset = \Omega r$. Hence it follows that $\Delta \div \Omega r = \Delta \div \Omega r_\emptyset$ (see 4.1.2).

4.2. Transitivity of domination relation

Let $r \in C(A,A)$ be treated as a preference relation. We can define associated indifference relation $i = r \cap r^{*}$, and domination relation (strict preference relation) $s = r \div i$. Note that $s = r \div r^{*}$. If r is reflexive and transitive then it can be readily checked that i is an equivalence. But in general s is not transitive. Later on we suppose that $C \leq SI(L\text{-}Fur)$. In Kuzmin (1982) it was shown that with $L = [0;1]$ if r is transitive then s is also transitive.

4.2.1. Example. Let $L = \{0,1,a,b,c\}$ be the lattice ordered by

$$0 < c < a,\ b < 1 \tag{9}$$

Let $A = \{x,y,z\}$ and $r(x,x) = r(y,y) = r(z,z) = 1$, $r(x,z) = r(y,x) = r(z,x) = r(z,y) = c$, $r(x,y) = a$, $r(y,z) = b$.
It can be easily seen that r is transitive. Since $a \div c = a$ and $b \div c = b$ we have $s(x,y) = a$, $s(y,z) = b$, and $s(u,v) = 0$ in other cases.
Then $s^{2}(x,z) = a \cap b = c$. So $s^{2} \leq s$ does not hold.

L-fuzzy relation $r \in C(A,A)$ is called regular if $r(x,y) \geq r(y,x)$ or $r(y,x) \geq r(x,y)$ for all $x,y \in A$. Regularity of r means that $(r \div r^{*}) \cap (r^{*} \div r) = \omega$ (dually, $(r \nrightarrow r^{*}) \cup (r^{*} \nrightarrow r) = \Omega$). We say that L-Fur (or just L) preserves transitivity if $s = r \div r^{*}$ is transitive for all regular and transitive r.

4.2.2. Proposition (Gisin (1988)). L preserves transitivity if and only if in L the following identity holds

$$(u \cap v) \div w = (u \div w) \cap (v \div w). \tag{10}$$

Let T2 be the lattice of interval numbers, i. e. T2 consists of such pairs of numbers (a_1,a_2) that $a_1 \leq a_2$. Let $0 < \alpha < 1$, $b = (0,1)$, $a = (\alpha,\alpha)$, $c = (0,\alpha)$. Then we have in T2 $a \cap b = c$, $a \div c = a$, $b \div c = b$. Hence (10) fails in T2. Therefore T2 does not preserve transitivity. Further, let T3 be the lattice of triangular numbers, i. e. T3 consists of such triples of numbers (a_1,a_2,a_3) that $a_1 \leq a_2 \leq a_3$. Consider embedding $T2 \to T3$ defined by $(a_1,a_2) \mapsto (a_1,a_1,a_2)$. It can be easily seen that T2 is closed in T3 with respect to $\div$. Hence it follows that (10) fails in T3. Therefore T3 does not preserve transitivity.

One of possible definitions of asymmetric relation is related with operation $r \div r^{*}$: s is asymmetric if $s = s \div s^{*}$. By this definition an asymmetric relation s associated with arbitrary relation r may be defined by $s = r \div r^{*}$. Really, since $(u \div v) \div v = u \div v$ in any BH-algebra L, and $s \leq r$, we have $s \div s^{*} = (r \div r^{*}) \div s^{*} \geq (r \div r^{*}) \div r^{*} = r \div r^{*} = s$. On the other hand, $s \leq s \div s^{*}$. Hence $s = s \div s^{*}$.

REFERENCES

Aizerman M.A., N.V.Zavalishin, and E.S.Piatnicki (1977) Global set functions in the theory of choice. Automatica and Telemechanica N 3, 111 - 125; N 5, 103 - 113 (in Russian).

Aizerman M.A., and B.M.Litvakov (1988) On some optimization problems in the general theory of choice. Automatica and Telemechanica N 3, 92 - 105 (in Russian).

Calenko M.S., V.B.Gisin, and D.A.Raikov (1984) Ordered categories with involution. Diss. Math., v. 227, Warsaw.

Gisin V.B., and M.S.Calenko (1984) An algebraic theory of systems and its applications I, in: Systems studies (Yearbook), Nauka, Moscow, 130 - 151 (in Russian).

Gisin V.B., and M.S.Calenko (1984) An algebraic theory of systems and its applications II, in: Systems studies (Yearbook), Nauka, Moscow, 113 - 135 (in Russian).

Gisin V.B. (1988) On transitivity of domination relations, in: The choice problem on crisp and fuzzy data, Institute for systems studies, Moscow, 6, 15 - 22 (in Russian).

Jonsson B. (1982) The theory of binary relations. Preprint.

Kuzmin V.B. (1982) Multiperson decision making in spaces of crisp and fuzzy binary relations. Nauka, Moscow (in Russian).

Rasiowa H., and R.Sikorski (1970) The mathematics of metamathematics. Monografie Matematyczne, v. 41, PWN, Warsaw.

Smirnov V.A. (1984) On one system of paraconsistent logics, in Smirnov V.A.(Ed.): Multivalued and paraconsistent logics, Institute of Philosophy, Moscow, 129 - 133 (in Russian).

Smirnov V.A. (1987) Logical methods of science. Nauka, Moscow, (in Russian).

DETERMINATION AND INTERPRETATION OF THE FUZZY UTILITY OF AN ACT IN AN UNCERTAIN ENVIRONMENT

Bernadette MATHIEU-NICOT
Faculté de Sciences Economiques et de Gestion
Université de Dijon
4 Boulevard Gabriel
21000 Dijon
France

Abstract : The aim of this paper is to draw up the series of axioms of utility applicable when imprecision and uncertainty are to be taken into account. FUZZY POSSIBILISTIC UTILITY is based on the subjective but nonetheless rigorous characteristics of possibility. It has a certain affinity not only with the exactness of Bernoulli's theory but also with the imprecision of the human theory of Shackle. In its applications, under certain circumstances, fuzzy utility comes down to determining a membership function of a fuzzy subset. Calculating the mathematical expectation of fuzzy utility may be compared to calculating a fuzzy interval.

Keywords : Fuzzy utility, possibility, weakened rationality, fuzzy intervals.

1. INTRODUCTION

Decision making in the face of uncertainty is born of the fact that the consequences of chosen actions hinge on events which may or may not come about. These events do not lend themselves to any ideal standard computation. Their origins are not well-known and we cannot compare them with or liken them to past investigations. They are by their very nature uncertain and inaccurate. The theory of probability proves inadequate because of the lack of experimental information. We must approach such an uncertain world with suitable mathematical tools. By virtue of its non-additive and non-distributive properties, Zadeh's theory of possibility enables such situations to be measured and explained.

The selection procedure in an uncertain and inaccurate world makes use of certain axioms (such as fuzzy complete preorder, a fuzzy mixture operation, continuity and independence of axioms) and a density of possibilty in order to determine a function of fuzzy utility. This function is a numerical translation of the decision maker's imprecise preferences. So we shall see, to start with, that the series of axioms of the fuzzy utility of a decision can be given by :

$$W(H_j) = \bigvee_{\mathbf{s}_k} [U(H_j, \mathbf{s}_k) \wedge \Pi(\mathbf{s}_k)] \quad \text{with } \max_{\mathbf{s}_k} \Pi(\mathbf{s}_k) = 1$$

Only ordinary (non fuzzy) mathematical concepts are printed bold face. However, when there can be no possible ambiguity, this convention will be dropped.

J. Kacprzyk and M. Fedrizzi (eds.), Multiperson Decision Making Using Fuzzy Sets and Possibility Theory, 90–97.

$U(H_j, \mathbf{S_k})$ is the utility of a fuzzy decision H_j, knowing the event $\mathbf{S_k}$ and $\Pi(\mathbf{S_k})$ is the possibility of the event $\mathbf{S_k}$.

We shall go on to draft an interpretation and outline the different consequences of such a function of choice in the sphere of economics. Subsequently we shall give the algorithms for calculating or comparing fuzzy possibilistic utility. The decision is reached then by comparing the densities of possibility of the different acts. Moreover, the inaccurate quantities which are translated by densities of possibility can be defined by fuzzy intervals. The fuzzy utility of a decision is denoted by specific measures expressing the behaviour of indifference, preference or aversion of the decision maker towards uncertainty.

2. FUZZY POSSIBILISTIC UTILITY

2.1 Series of axioms of fuzzy possibilistic utility

2.1.1 The 'problem'

A fuzzy decision problem in the face of uncertainty is defined by using the possibility of an event $\mathbf{S_k}$. Consider the following four elements :

$< \mathbf{S}, \mathbf{D}, \pi(\mathbf{S_k}), U(H_j, \mathbf{S_k}) >$

The series of axioms of the fuzzy possibilistic utility rests on a weak Borel field $\mathcal{S}$.

$\mathbf{S}$ represents the set of states of nature.

$\mathbf{D}$ represents the set of all a priori possible actions.

$\mathcal{S} \subset L^{\mathbf{D}}$, $L = [0,1]$.

$\mathcal{S} = \{H_1, H_2, ..., H_j ... H_r\}$ the fuzzy subset H_j.

$H_j \subset \mathbf{D}$ is called a fuzzy decision.

$\pi(\mathbf{S_k})$ is the possibility of $\mathbf{S_k}$ events.

As in the usual theory, possibility is defined in a Borel field β.

$U(H_j, \mathbf{S_k})$ is the fuzzy expected utility function. The fuzzy possibilistic utility of a fuzzy decision H_j is :

$$W(H_j) = \Pi(X_j) = \bigvee_{\mathbf{S_k}} [\mu_{H_j}(\mathbf{S_k}) \wedge \pi(\mathbf{S_k})]$$

$$= \bigvee_{\mathbf{S_k}} [U(H_j, \mathbf{S_k}) \wedge \pi(\mathbf{S_k})]$$

$$= \bigvee_{\mathbf{S_k}} [U(X_{jk}) \wedge \pi(\mathbf{S_k})] \quad \text{with } \underset{\mathbf{S_k}}{\text{Max}}\, \pi(\mathbf{S_k}) = 1$$

$\Pi(X_{jk})$ is the possibility of a fuzzy decision H_j, knowing the event $\mathbf{S_k}$, the latter is expressed by a fuzzy utility. It also corresponds to the possibility of a fuzzy subset X.

$$\forall X \in \mathcal{S}, \forall \mathbf{S_k} \in \mathbf{S}, \Pi(X_{jk}) = \bigvee_{\mathbf{S_k}} (\mu_{X_{jk}}(\mathbf{S_k}) \wedge \pi(\mathbf{S_k}))$$

Just as with the possibility of a fuzzy subset which is translated by an uncertain mathematical expectation, the possibility of the $\vee\wedge$ contingency can be associated with a fuzzy subset X.

The decision maker will make the choice which gives the maximum $\vee\wedge$ contingency of the fuzzy utility of a fuzzy act.

2.1.2 Axiom-system, existence and continuity of fuzzy possibilistic utility.

The decision maker's selection criterion is materialized by the notion of preference relative to the various acts. This preference is expressed by a binary relation written $\lesssim$ defined on $\mathbf{D}$ such that for action $d_1 \in \mathbf{D}$, $d_2 \in \mathbf{D}$, $d_1 < d_2$ is read "d_2 is preferred to d_1"; $d_1 \sim d_2$ is read "Action d_1 is indifferent to action d_2". $\mathbf{H}$ is the set of acts made technically possible.

Axiom 1. A fuzzy complete preorder exists on the sets of actions $\mathbf{H} \subset \mathbf{D}$. ($\mathbf{H}$ has a fuzzy complete preordered structure if the fuzzy relation $\mathfrak{R}$ is reflexive and Max - Min transitive).
Axiom 1'. The possibility of the event defined by Zadeh (1978) exists.
Axiom 2. The set $\mathbf{H} \subset \mathbf{D}$ is called $\vee\!\wedge$ mixture operation towards $\mathcal{S}$. Given $\mathbf{H} \subset \mathbf{D}$ we call fuzzy $\vee\!\wedge$ mixture operation towards $\mathcal{S}$ the following convex arrangement. Consider three fuzzy subsets H_1, H_2, Ω.

$$\forall \mathbf{S}_k \in \mathbf{S},\ \mu_{(H_1, H_2, \Omega)}(\mathbf{S}_k) = \mu_\Omega(\mathbf{S}_k) \bullet \mu_{H_1}(\mathbf{S}_k) + [1-\mu_\Omega(\mathbf{S}_k) \bullet \mu_{H_2}(\mathbf{S}_k)]$$

The latter is also denoted by $(H_1, H_2, \Omega) = \Omega H_1 + \bar{\Omega} H_2$.

The series of axioms of fuzzy possibilistic utility generalizes not only neo-Bernoullian axiomatics but also the fuzzy stochastic utility of Mathieu-Nicot (1985). We note more particularly that the property of substitution (even the fuzzy one) is not imperative since event processing in uncertainty is accomplished by other operations (Max, Min) than addition.

Suppose the Max - Min contingency of the utility of the fuzzy decision H_j :

$$W(H_j) = \Pi(X_{jk}) = \bigvee_{\mathbf{S}_k} [U(H_j, \mathbf{S}_k) \wedge \pi(\mathbf{S}_k)] \quad \text{(I) with } \operatorname*{Max}_{\mathbf{S}_k} \pi(\mathbf{S}_k) = 1$$

It squares with the possibility of a fuzzy event also called the possibility of a sensation by Kaufmann (1977). This sensation represents the conditional utility of a fuzzy decision.

We also write expression (I) as :

$$W(H_j) = \operatorname*{Sup}_{\mathbf{S}_k \subset \mathbf{S}} [\operatorname{Min} U(H_j, \mathbf{S}_k), \pi(\mathbf{S}_k)] = \Pi(X_{jk})$$

We prove the existence and continuity of such a function with possibility subsets. Indeed

$$\Pi(X_{jk}) = \operatorname*{Sup}_{\mathbf{S}_k \in X_{jk}} \pi(\mathbf{S}_k).$$

As $\Pi(X_{jk})$ is a fuzzy measure, a fuzzy integral in Sugeno's sense exists :

$$\Pi(X_{jk}) = \int_{\mathbf{s}} X_{jk}\, d\Pi$$

X_{jk} is the utility of the fuzzy decision H_j. Knowing the event $\mathbf{S}_k$ we can interpret it as a fuzzy variable denoted X'.

So we have :

$X' : \mathbf{S} \rightarrow \mathbb{R}$ is a fuzzy variable.
Its density of possibility is $\varphi : \mathbb{R} \rightarrow [0, 1]$ and $\varphi(x) = \text{Poss}\,\{x' = x\}$.
Consider the fuzzy subset F of $\mathbb{R}$; we have:

$$\text{Poss}\,\{X' = F\} = \operatorname*{Sup}_{x \in \mathbb{R}} \{\operatorname{Min} F(x), \varphi(x)\} \text{ and } \text{Poss}\,[X'=F] = \operatorname*{Sup}_{F} \varphi(x)$$

More precisely, this latter expression means that we have maximized φ on the fuzzy constraint F. Moreover, for any $A \subset \mathbb{R}$, if $\Pi_x(A) = \operatorname*{Sup}_{x \in A} \varphi(x)$ is the possibility measure of φ, we have :

$$\text{Poss}\,\{X = F\} = \int_{\mathbb{R}} F d\Pi_x \quad \text{where} \quad \int_{\mathbb{R}} F d\,\Pi = \operatorname*{Sup}_{\alpha \geq 0} [\operatorname{Min}(\alpha, \Pi\{F \geq \alpha\}]$$

This result is proved with Sugeno's theorem (1974). If φ is continuous, F is superlatively half continuous, strictly converse; and sup F is compact.
Then $\text{Poss}\,\{X' = F\} = \text{Poss}\, X' \in A$ where $\forall \alpha_0 \in [0,1]\ A = \{x \in \mathbb{R} : F(x) \geq \alpha_0\}$

2.2 Interpretation of decision theory as regards uncertainty

2.2.1 Measure, preference and value in uncertainty

The decision theory as regards uncertainty hinges on two ideas : firstly, the stochastic theory does not give a good interpretation of real behaviour; secondly, it does not include in all cases the basic time dimensions. The theory of probability does not set up an exact representation of actual decisions because its dimensions are inadequate to this use, and because the general decision procedure is not always realistic. The theories of uncertainty reject determinism in order to include imagination and human creative intuition. Uncertainty would not be probabilized owing to the fact that time is not reversible. Just as with the potential surprise of Shackle (1969, 1979), possibility and necessity measures can be considered as uncertain measures. They express the degree of belief of the decision maker, thanks to his intuition and his inspiration. He can not only discriminate and arrange the degree of real or imaginary feelings in uncertainty, but he can also assign a cardinal measure thanks to the interval concept. Both the consideration of the imagined and possible events and the assessment of acts, which is more or less inaccurate, mean that the fuzzy preference of an act is not only restricted to extreme cases, to neutrality or to a simple average of possible acts, but that it also sorts out the more or less preferable solutions.

So fuzzy possibilistic utility depends not only on the inaccurate preference of an act but also on its possibility. For all degrees of possibility, fuzzy utility is characterized by a fuzzy measure which is expressed by a single specific number. Fuzzy possibilistic utility translates the overall behaviour of the decision maker as regards the different imagined acts.

2.2.2 Behaviour in uncertainty

Classical decision theories revolve around the idea of rationality and their purpose is to pose the normative problem of selecting the guidelines for choice. The postulates of the traditional theory of expected utility are derived analytically from the notion of rationality, this theory is however empirical in that the axioms only impose formal constraints on behaviour. Such is not the case with the theses which take uncertainty into account. In fact Shackle's conception which uses potential surprise to represent uncertainty, is transposed essentially by abandoning the systematic search for rationality and is characterized by its concern with explaining the process of concrete decisions. Time, for Shackle, appears as a succession of "moments in being" which all possess their own individuality. Let's suppose that at a given moment a businessman decides to invest a certain amount. With hindsight his decision may turn out to have been a mistake. Should he misread the potential of the market, for example, it may even be that at a certain moment he regrets his choice. Does this change of attitude translate irrationality? In reality the state of mind of the decision maker at the moment of taking the decision and the state of mind of the same decision maker two years on are two different things. It is meaningless to say that one appears more rational than the other. In a word, the fact that the expectations of a person making a decision prove exact or not has no bearing on the rationality or irrationality of the decision. At the moment it was taken, the decision maker opted for the act which seemed the more rational to him. Rationality, if it has any meaning, is intimately tied in with the conditions which characterize the problem of choice when it arises. Shackle, as it were, trusts the decision maker. Any decision stems from a specific act selection mechanism, the rationality of which cannot be denied. Our analysis confines itself essentially to measuring possibility, while noting in passing that a series of axioms of decision theory in uncertainty can be built on the basis of necessity. This theory introduces new means, its aim is not to renew existing theories but to complete them, it sets out simply to give a different way of seeing, a different outlook on concepts and models of uncertainty and imprecision. The theory of "fuzzy possibilistic expected utility" is derived from the axiom of fuzzy total preorder. It follows from it that in such a context the decision maker's reasoning is weakened.

Taking uncertainty into consideration serves to strengthen the relative nature of an approach which purports to be strictly rational. Total rationality is in practice unworkable in any uncertain situation. Total maximization is impossible. Such an optimum will not then necessarily be sought, but a reasonable and satisfactory degree will do the job.

"Fuzzy possibilistic expected utility" which is founded logically on the subjective, but also rigorous nature of possibility has affinities with the accuracy of Bernoullian theories as well as the inaccuracy of human theories taken up by that of Shackle. It forges a link between a normative conception on the one hand and a rather descriptive and explicative one on the other. In other words "fuzzy possibilistic utility" enables the transfer to be made from theory to practice by reducing the gap between actions and concrete behaviour and those forecast by the theory.

This conception may also be likened to the principle of "limited rationality" of Simon.

According to Simon, people seldom try to find the optimal act in a problem of decision. They choose instead the most satisfactory consequences of the acts open to them. Two characteristics of the choice appear fundamental to him : choice is exercised first of all in relation to a definition of the situation which will always be a simplified approximate representation of reality. Next, the elements of this definition are not given objectively but stem from the subjective appreciation of the decision maker. The economic actor remains a rational being, and is happy to be so, but he is aware that his rationality is limited. The limits do not appear by chance, they develop and are fixed on the basis of the interplay of human relations within complex systems with their own characteristics and regulations. It is within the framework of such interplay that people define acceptable solutions. To understand them, references have to be changed. The theory of possibilistic fuzzy utility takes into account all the acts made possible for technical reasons. They characteristically take account of the imprecise preferences of the decision maker - these preferences are measured by a membership function - note besides that Simon's (1957) rule of satisfaction differs from that of optimization in that there is no need to explore the entire set of possible acts. The difference is expressed by the fact that the optimization separates the moment of evaluation from that of choice. (All possible elements of the space considered are evaluated and those which are thought to be possible and maximum are retained.) The rule of satisfaction consists in a succession of pairs of evaluation and choice. (Each act is in turn assessed then subjected to choice i.e. rejected or accepted.) If the sequential aspect is ignored the principle of satisfaction appears as a special case of classical optimization. Satisfactory solutions result from maximizing a function with two values : zero (non-satisfaction) and one (satisfaction). Which means that the purpose of the sequential aspect is to take into account the range of intermediate situations. However it may exclude extreme situations as the process stops when the first satisfactory state is reached.

Thus when the decision maker is aware of the deformation or imperfection of the information available to him, he no longer seeks to maximize in the full sense of the word. He will not necessarily look for the perfect solution, but rather the ideal approximation of the optimal solution, given his level of aspiration and satisfaction.

In fact the nature of rationality does not consist in a necessary movement between cause and effect, but in a necessary agreement between the laws of nature and the law of thought which enables certain characters of experience to be forecast.

By imagining an infinite number of situations, what is possible serves as the starting point of any analysis of uncertainty or imprecision. The problem is what evidence will fill the gap between the contingency of what is possible as conceived by imagination and what is necessary as shown by experience. In order to reconcile what is imagined possible and what is perceived as real, the existence of something outside the conscious mind must be posited. The possibility of an event enables the equivalence of uncertainty and reality to be extended to a random and no longer determining reason.

3. UTILITY FUNCTION AND FUZZY SUBSET

3.1 Representation of possibilistic fuzzy utility by a specific membership function

In general, the fuzzy utility of a decision can be expressed by a fuzzy number which translates the quality of an act from the point of view of the imprecise preferences of the economic agent. It is an element of information linked to the act. Fuzzy utility makes it possible to draw up an order of imprecise preference from the consequences of various possible acts for the decision maker, taking into account the behaviour of the decision maker in the face of uncertainty.

The resulting fuzzy utility is characterised as a cardinal measure.

It can be shown that the membership function of a fuzzy subset plays the role of a normalized fuzzy utility function. The purpose of the membership function is to translate numerically the intensity of the decision maker's preferences between the different measures of consequences X. The assessment can be made exactly. The shape of the curve means the decision maker's behaviour can be expressed, just as with fuzzy utility.

Measuring this behaviour is done with the help of a continous scale from[0, 1]. Some five to seven levels between 0 and 1 are determined to translate the decision maker's threshold of perception. Experimental studies show that above and beyond these seven levels the shades of meaning between two consecutive levels become imperceptible. The solution consists in expressing in words the levels of compatibility between objective and consequence (such as : fully, fairly average, weakly compatible, incompatible) and assigning a value between zero and one to each level.

Two approaches are suggested by Dubois (1980) to identify a membership function.

The first consists in making set X and set X' discrete and then asking the decision maker to assess (f (A, B, C, D, E)) for each consequence $x' \in X$, and then to smooth over the result.

The second approach suggests making an L. R. type fuzzy number. It is up to the decision maker to provide the parameters directly (core and support of the fuzzy number) as well as the shape of the function (it may be linear, concave, convex or have both concave and convex segments). Utility as a cardinal representation of preferences is very close to the concept of fuzzy subsets according to the hypothesis whereby the degree of compatibility of each state of nature is defined in relation to the subset $Hj \subset \mathbf{D}$. Where **D** is the set of acts to be classified according to the evaluation criteria of their consequences.

This membership function corresponds closely to a representative value function of the utility of a decision X_j knowing the state of nature $\mathbf{S_k}$ to which levels of possiblity are subsequently attached. All the same, a fuzzy subset is not always a disguised utility function in the strict sense of the term. The utility function such as it was introduced into the classical theory presents some special features.

The utility function has the property of continuity whereas the membership function may well be built on a finite system of reference. Moreover, the specific nature of the axioms (of Von Neumann and Morgenstern, Savage etc.) which have characterized the utilty function so far, when a decision is to be taken in the face of a risky or uncertain future, leads naturally enough to the use of the theory of probability. The theory of choice is in fact based on a relation of preference over a set of probability distributions, so that choice in a risky universe boils down to selecting a probability distribution among a set of distributions characterizing the consequences of the actions.

The definition of a membership function does not rely necessarily on the existence of probability, but rather on the existence of relative preference between elements of the system of reference.

Finally it should be noted that the utility function, in contrast to fuzzy subsets, generally refers to an amount of money; it can be evaluated on a cost scale. The utility expresses the interest presented by an action compared with a certain monetary equivalent.

Thus by characterizing a membership function, the concept of fuzzy utility comes up again. Norwich and Turksen (1982) show to this end that in order to establish the uniqueness of a membership function, to within one transformation class, it is possible to provide the Cartesian product X.X (identifying the set of consequences) with a fully pre-ordinated binary preference relation. The advantage of this more operational approach of fuzzy utility lies in showing that if the relative preferences can be compared over a continuum X, and if this preference relation satisfies certain axioms like those of the fuzzy expected utility, then there is an evalutation function, unique to within one fine transformation, over an interval which enables the objects of reference system X to be classed.

3.2 Representation of a fuzzy expected utility function by a fuzzy interval

We have seen that $\Pi(X_{jk})$ is the possibility of the fuzzy decision H_j, knowing event $\mathbf{S}_k$. The latter is interpreted by a fuzzy utility function. $\Pi(X_{jk})$ is also interpreted in terms of fuzzy "possibilistic" subsets. When consequence X_j of an action j is expressed by a specific fuzzy subset such as a fuzzy number, Moore (1979), or a fuzzy interval, the resulting "fuzzy possibilistic utility" is comparable to such subsets.

More precisely when the utility of consequence X_{jk} is represented by a fuzzy number, and the membership function characterizing the fuzzy number is a measure of possibility, $U(X_{11})$ ...$u(X_{nm})$ form in this case fuzzy intervals written :

$$[U(\underline{X}_{jk}),\ U(\bar{X}_{jk})]$$

Thus the mathematical expectation of a fuzzy decision corresponds to the fuzzy interval :

$$\Pi\ (X_j) \in [\Pi(\underline{X}_{jk}),\ \Pi(\bar{X}_{jk})]$$

The purpose of this method of evaluating the mathematical expectation of fuzzy utility in practice is to use all the results of fuzzy arithmetic applied to fuzzy intervals.

4. CONCLUSIONS

The association : possibility and fuzzy utility serves to give a better account of human psychology without drawing too heavily on introspection.

The idea that the behaviour of man is dictated by the search for greater satisfaction (in the widest sense of the word) seems right. This statement is only acceptable in as much as it remains vague. It cannot hold up a rigid superstructure linked to binary logic alone. "Fuzzy possibilistic utility" is a step towards bringing together the exactness of Bernouillian theories and the imprecision of human theories adopted in the theory of Shackle. The selection behaviour which arises in a fuzzy and uncertain context links in with that of the "behaviourist theory" of Simon. The accuracy which we find in Bernoullian theories comes out in the construction of the series of axioms of "fuzzy possibilistic utility", through recourse respectively to expectancy or contingency $\vee\!\wedge$, and by taking into account the dispersion of uncertain values.

The imprecision of human theories is expressed by the somewhat vague preferences of decision makers and by taking into consideration the possible imagined events.

Just as with Simon's model, the decision maker's behaviour in an imprecise and uncertain universe is characterized by "limited rationality" which is also "weakened". It is "weakened" by the decision maker's imprecise behaviour which is made concrete by a supple but rigorous series of axioms (total fuzzy preorder).

Rationality is finally limited in that the decision maker does not necessarily choose the maximal solution (with extreme values) but the one which gives him maximum satisfaction (intermediate values) without having recourse for all that to a sequential procedure.

The criteria of rationality of the decision maker is weakened in comparison with that of the neo-Bernoullian theory. This new dimension does not affect the decision maker's behaviour fundamentally as he always tries to optimize his decision according to his environment while being aware that the optimum he assigns is not necessarily the possible economic optimum of the classical economic theory.

The algorithms offered in an imprecise and uncertain context are based primarily on the theory of fuzzy subsets or on the calculation of fuzzy intervals. The latter presents the advantage of being suitable for replacing probabilist methods without increasing the necessary calculations. For choice to be possible, there must be a common unit enabling the whole range of consequences to be evaluated. The evaluation of this unit is performed by the fuzzy utility function. It is through this function (or particularized membership function) that the theory of fuzzy subsets assumes a meaning.

REFERENCES

Dubois D. (1980) G.L.S. Shackle : Economist and Forerunner of the Theory of Possibility. (Un economiste précurseur de la théorie des possibilités : G.L.S. Shackle). Bulletin Busefal, Spring 1980, 70-78.

Dubois D. and Prade M. (1985) The Theory of Possbility (Théorie des possibilités). Masson, Paris.

Kaufmann A. and Gupta M. (1985) Introduction to Fuzzy Arithmetic. Van Nostrand Reinhold, The Netherlands.

Mathieu-Nicot B. (1985) Mathematical Expectation of Fuzzy Utility (Espérance mathématique de l'utilité floue) Coll. I.M.E. N° 29, Dijon.

Moore R.E. (1979) Methods and Applications of Interval Analysis, in SIAM, vol.2.

Neumann Von J. and Morgenstern O. (1953) Theory of Games and Economic Behaviour. Princeton University Press.

Norwich A.M. and Turksen I.B. (1982) The Fundamental Measurement of Fuzziness in Fuzzy Set and Probability Theory : Recent Developments. R.Yager (Ed.), Pergamon Press, 49-60.

Ponsard C. (1982) Partial Spatial Equilibria with Fuzzy Constraints. Journal of Regional Science 22, 159, 175.

Savage L.J. (1972) The Foundations of Statistics. Dover Publications Inc., New York.

Shackle G.L.S. (1967) Decision, Determinism and Time (Décision, déterminisme et temps). Dunod, Paris.

Shackle G.L.S. (1979) Imagination and the Nature of Choice. Edinburgh University Press.

Simon H.A. (1957) Models of Man. Wiley, New York.

Simon H.A. (1966) Theories of Decision Making in Economics and Behavioural Science, in Survey of Economic Theory, III.

Sugeno M. (1974) Theory of Fuzzy Integrals and its Applications. Tokyo Institute of Technology.

Zadeh L.A. (1978) Fuzzy Sets as a Basis for a Theory of Possibility. F.S.S, 1, 3-28.

EXTENDING AGGREGATION OPERATORS FOR MULTICRITERIA DECISION MAKING

Livia D'Apuzzo, Massimo Squillante and Aldo G.S. Ventre

Istituto di Matematica della Facolta' di Architettura
dell'Universita' degli Studi di Napoli "Federico II".
Via Monteoliveto 3. I-80134 Napoli, Italy.

Abstract. A decision making model in a continuum is considered in order to present the settings and some results by Yager [7], Aczél and Saaty [1] in a unique framework.

Keywords. Aggregation, mean operators, rearrangements.

1. INTRODUCTION

This paper is motivated by the works by J. Aczél and T.L. Saaty [1] and R.R.Yager [7] concerning functions that synthesize, i.e. aggregate, quantifiable judgements a(1),...,a(n) on a set X of alternatives.Such aggregation functions act either over the judgements expressed by a finite number n of individuals on a given alternative x in X, or over the degrees to which a single alternative satisfies a finite number of criteria.In both cases the problem of defining aggregation functions can be formulated as: given the quantities a(1),...,a(n), determine a function F whose value F(a(1),...,a(n)) synthesizes the quantities so as to satisfy suitable reasonable conditions. In both mentioned papers [1,7] the considered functions assume the form of mean operators.

In some phenomena that occur in social sciences and in economics the number of individuals that participate either in a decision making process or in the formation of an opinion is very large and often the contribution of a single individual is neglegible. Furthermore sometimes both judgements and behaviors depend on criteria that evolve in time, possibly with suitable regularity characters. Therefore it seems a good way, in order to deal with these situations, to consider decision making models with a continuum of criteria or individuals on the analogy of the work by R. Aumann [2] (see also [3] and [6]).In doing so we shall exhibit a unique framework into which most of the results in [1 and 7] can be recovered.

J. Kacprzyk and M. Fedrizzi (eds.), Multiperson Decision Making Using Fuzzy Sets and Possibility Theory, 98–104.

2. MEAN OPERATORS

Let X be a set of alternatives and T={t(1),t(2),...,t(n)} a set of criteria,or individuals. For every x in X and t(i) in T,let a(x;t(i)) the degree to which x satisfies the criterion t(i). To ease the notations, we shall obmit the variable x, whenever possible. So a(x;t(i)) = a(t(i)).The following minimal conditions have to be satisfied [1 and 7] by an aggregation functional F(a(t(1)),...,a(t(n)) :

1. Commutativity over the arguments.

2. Monotonicity.

3. Idempotency: F(a,...,a) = a.

Let now suppose that the set T of the criteria is a continuum, and define for every x in X a function A(t) with t in T, that expresses the degree to which x satisfies the criterion t. The aggregation operator F acts on A(t) in such a way that the conditions 1. to 3. are satisfied. We will indicate a general class of aggregation operators in terms of suitable means,by fixing T as the real interval I=[0,1]. If f is a bounded function in T, the quasi-arithmetic mean is defined [5] by

$$M(\Phi;f) = \Phi'\left(\int_T w(t)\Phi(f(t))dt \right),$$

where Φ is a continuous and strictly monotonic function,Φ' is its inverse,and $\int_T w(t)dt = 1$.

It is worth recalling that quasi-arithmetic means have been investigated in a different context by B. de Finetti [4].
Let B(t) denote the decreasing rearrangement [5] of A(t).

Proposition 1. The functional

$$F(A) = M(\Phi;B(t)) \qquad (1)$$

enjoies the following properties:
1. if A(t) and A'(t) have the same decreasing rearrangemen , then F(A(t)) = F(A'(t)) (commutativity) ;

2. if $A(t) \leq A'(t)$, then $F(A(t)) \leq F(A'(t)$ (monotonicity) ;

3. if $A(t) \equiv a$, then F(A(t)) = a (idempotency, or consensus property).

The proofs of 1. and 3. are staightforward. Proof of 2. follows from strict monotonicity of Φ. Thus the functional

F(A) defined in (1) is an aggregation operator.

3.OPERATORS FOR THE EXTREME AGGREGATION CASES

It is worth observing that from monotonicity and idempotency of operators F we have

$$e' \leq F(A) \leq e''$$

with e' = essinf A(t) and e' = esssup A(t).

Let now consider the functionals

$$F'(A(x)) = e'(x) \quad \text{and} \quad F''(A(x)) = e''(x)$$

that will be considered as aggregation operators in the following extreme situations. When choosing the alternative x we can be led by the only exigence that x satisfies almost all the criteria of T, with the exception of a subset of measure zero; therefore we choose F'(A(x)) = e'(x) as the aggregation operator. Of course we shall select the alternative x corresponding with sup e'(x), where the supremum is taken over X. On the other hand if the alternative must be selected on the basis of a good satisfaction f some subset of T with positive measure,-let say "a few criteria"-, the aggregation is given by F"(A(x)) = e"(x).
Following Yager [7] let us denote these extreme decision making situations as "pure anding" and "pure oring",respectevely, and F'(A) and F"(A) pure anding and pure oring operators.

4. SLIDING TOWARDS PURE ANDING AND PURE ORING

The behavior of the operator F depends on functions Φ and w. Once function Φ is fixed a suitable choice of w makes F more or less close to the pure anding or pure oring operator. Let s be an element of the open interval (0,1),

$$w(s;t) = \begin{cases} 0 & \text{for } t \in [0,1-s) \\ 1/s & \text{otherwise} \end{cases}$$

$$w'(s;t) = \begin{cases} 1/s & \text{for } t \in [0,s] \\ 0 & \text{otherwise} \end{cases}$$

If B(x) is supposed to be continuous, then

$$F(A(x))=F(w;A(x)= \Phi'((1/s) \int_{[1-s,1]} \Phi(B(x;t))dt)=B(x;t')$$

$$F(A(x))=F(w';A(x)=\Phi'((1/s) \int_{[0,s]} \Phi(B(x;t))dt)=B(x;t'') ,$$

where t' and t'' are suitable points in $[1-s,1]$, and $B(x;t')$ and $B(x;t'')$ tend to $e'(x)$ and $e''(x)$, as s tends to 0. In conclusion $F(w(s;t))\to F'$ and $F(w'(s;t))\to F''$, as $s\to 0$.

Let now consider, together with Yager [7], a functional

$$\text{orness}:w \dashrightarrow s\in I$$

in order to evaluate the closeness of F to the operator of pure oring. such a functional shuold satisfy the conditions:

1. $\text{orness}(w)=1/2$, for $w=1$
2. $\text{orness}(w(s;t))\to 0$, as $s\to 0$
3. $\text{orness}(w'(s;t))\to 1$, as $s\to 0$.

to this aim, we easily check that the definition

$$\text{orness}(w)= \int_I (1-t)w(t)dt$$

fulfils the conditions above.

From a mean value property we have the following proposition that extends a result due to Yager [7].

Proposition 2. Let w and w' be continuous weights. If

$$\int_{[0,x]} w(t)dt \geq \int_{[0,x]} w'(t)dt,$$

fo every x, then $\text{orness}(w)\geq\text{orness}(w')$.

5. DISPERSION OF THE WEIGHT

The following example shows that different weighting functions can have the same orness. Let us set for x in (0,1) and $s \le 2\min(x,1-x)$,

$$w(x,s;t) = \begin{cases} 1/s & \text{for } t \in [x-s/2,\ x+s/2] \\ 0 & \text{otherwise} \end{cases}$$

These functions exhibit different dispersions of the weight; the concentration near x increases as $s \to 0$. Anyway we have $\text{orness}(w(x,s)) = 1-x$, for every s; in particular for $x=.5$, $w(x,s)$ has the same orness that the constant maximum weighting function $w' \equiv 1$.We introduce the following measure of dispersion in order to describe the behavior of weights with the same orness and different concentrations:

$$\text{dispersion}(w) = D(w) = \exp\left(-\int_T g(w,t)\,dt\right),$$

where

$$g(w,t) = \begin{cases} 0 & \text{if } w(t)=0 \\ w(t)\log w(t) & \text{if } w(t)>0 \end{cases}$$

For instance, $D(w(.5,s)) = s$ according to the decrease of the concentration when s tends to 1.

6. MEAN POWER OPERATORS

Let now consider the special class $\{F(r)\}$, r real, in the set of all aggregation operators obtained by setting

$$\Phi(y) = \begin{cases} y^r & \text{for } r \neq 0 \\ \log y & \text{for } r = 0 \end{cases}$$

$$F(r;A(x)) = M(r;B(x)) = \Phi'\left(\int_T w(t)\Phi(B(x;t)\,dt\right), \qquad r \neq 0$$

$$F(0;A(x)) = M(0;B(x)) = \exp\left(\int_T w(t)\log B(x;t)\,dt\right).$$

It is easy to see that the class $\{F(r)\}$ is increasing w.r. to r, and

$$F(r;A(x)) \to F(0;A(x)), \text{ as } r \to 0$$

$$F(r;A(x)) \to e'(x) = F'(A(x)), \text{ as } r \to -\infty$$

$$F(r;A(x)) \to e''(x) = F''(A(x)), \text{ as } r \to +\infty$$

Let now remark that the synthesis of the judgements depends on the choice of the index r.The functionals $F(-\infty)$ and $F(+\infty)$ correspond with the extreme situations of pure anding and pure oring, respectevely. Indeed $F(-\infty;A(x)) = e'(x)$. Then the synthesis operator ranges from pure anding to pure oring situations. So we can define a measure of the orness in terms of the index r. The following conditions should be met by such a measure H :

$$H(-\infty) = 0, \; H(0) = 1/2, \; H(+\infty) = 1,$$

and H is isotone with r.
Let us show, as an example of the function H,the following

$$H(r) = \exp(2r)/\exp(2r)+1.$$

Finally let us list some properties of mean power aggregation operators.

1. Reciprocity. For every $r \in R$

$$F(-r;1/A) = 1/F(r;A),$$

in particular

$$F(0;1/A) = 1/F(0;A)$$

2. Associativity holds for $r = 0, +1, -1, +\infty, -\infty$.

3. Traslativity of F(1),

$$F(1;A(x,t)+h) = F(1;A(x,t))+h,$$

i.e. an increase of h in the judgements implies the same increase in the synthesis.

4. Decomposability of F(r) holds for every r.

CONCLUSION

We have defined aggregation operators for multicriteria decision making on a continuum as generalized means operators. They fulfils at least commutativity, monotonicity and idempotency requirements, what meets Yager's assumptions. Furthermore, besides these properties some more other are recovered common to the handlings by [1] and [7], namely the forms of pure anding and pure oring operators in terms of extrema, the action of the weight function in sliding the functional towards the pure anding and the pure oring, the properties related to the measure of orness and to dispersion, the positive homogeneity of F(r), the reciprocity of F(0).

REFERENCES

1. J.ACZÉL and T.L.SAATY, Procedures for synthesizing ratio judgements,J.Math.Psychology 27, 93-102 (1983).

2.R.AUMANN, Markets with a continuum of traders,Econometrica 32, 39-50 (1964).

3.P.DUBEY and L.S.SHAPLEY, Non cooperative general exchange with a continuum of traders, Rep. no. 51/79, Hebrew Univ. Jerusalem.

4.B.deFinetti, Sul concetto di media, G.I.I.A., vol.II, 369-396 (1931).

5.G.H.HARDY, J.E.LITTLEWOOD and G.POLYA, Inequalities, Cambridge University Press, 1952.

6.H.SKALA, Some problems of measure theory which are related to economic theory, Stochastica 6, n.3, 305-320 (1982).

7.R.R.YAGER, On ordered weighted averaging aggregation operators in multicriteria decision making, IEEE Trans. Sys. Man Cyb. 18, 1, 183-190 (1988).

RANKING ALTERNATIVES BY WEAK TRANSITIVITY RELATIONS

Luciano Basile
Istituto di Matematica, Facoltà di Architettura.
Università degli Studi di Napoli, via Monteoliveto, 3. 80134, Napoli, Italy.

Abstract

A ranking of alternatives is constructed in the framework of the fuzzy preference relations that are complementary to the unit (CU-relations).

It is proved that the further assumption of weak transitivity is a sufficient condition for the ranking of the alternative set.

Keywords: Fuzzy preference relations, CU-relation, weak transitivity, weak coherence, ranking.

1. INTRODUCTION *

We present here a model for suitable preference relations into which the ranking of the alternatives is obtained in simple structure (lattice). In this framework the assumption of complementarity to the unit is reasonable. Furthermore we would like stress that the request of weak transitivity is common both in many practical problems of selection the best alternative, and ranking.

Let $X = \{x_1, x_2, ..., x_n\}$ be a given set of alternatives, and R a fuzzy binary relation on X, i.e. a fuzzy subset of XxX. The value $r(x_i, x_j)$ of R indicates the degree to which x_i dominates x_j.

Definition. A fuzzy relation R_{cu} is said to be a CU (complementary to the unit) preference relation if [3]:

(1a) $r(x_i, x_j) + r(x_j, x_i) = 1$, for every $x_i, x_j \in X$

(1b) $r(x_i, x_i) = 0.5$, for every $x_i \in X$.

* I am very grateful to A.G.S. Ventre, who provided helpful comments on earlier drafts of this paper.

J. Kacprzyk and M. Fedrizzi (eds.), Multiperson Decision Making Using Fuzzy Sets and Possibility Theory, 105–112.

It is worth stressing that the definition above is slightly different from the following due to Orlovsky [4].

Definition. A fuzzy preference relation R is defined as follows:

$$r(x_i, x_j) = \begin{cases} r(x_i, x_j) - r(x_j, x_i), & \text{when } r(x_i, x_j) \geq r(x_j, x_i) \\ 0, & \text{otherwise.} \end{cases}$$

J. M. Blin and A. B. Whinston [1] and J. M. Blin [2] analyze a *fuzzy* preference relation, induced by individual opinions of experts, which set up a ranking of alternatives rather than a preference on any pair of alternatives (x_i, x_j). They obtain a kind of fuzzy preference relations which can be easily reduced to CU or Orlowsky's relations and suggest a procedure to rank the alternatives. In particular, the ranking is obtained only if R_{cu} is transitive; otherwise, it is necessary to remove the element r_{ij} which originates the intransitivity.

We will show that the transitivity of the relations R_{cu} is not requested as a necessary condition in order to rank the set of alternatives; indeed, the same ranking may be obtained under a weaker condition.

The paper is structured as follows. In Section 2 the concept of *weak transitivity* (or *coherence*) for a relation R_{cu} is defined. The sufficiency of the weak transitivity condition to give rise to a ranking over the set X of the alternatives is proved in Section 3 and an example is illustrated in Section 4. Finally, an outline of the results is provided in order to exhibit the efficiency of the suggested algorithm in order to recognize the coherence of R_{cu} and find the ranking over the set X.

2. WEAKLY TRANSITIVE (OR COHERENT) RELATIONS R_{cu}.

Definition. A CU fuzzy preference relation R is said to be weakly transitive (or coherent) if

(1) $r_{ij} > 0.5$ and $r_{jk} > 0.5 \Rightarrow r_{ik} > 0.5$, for every i, j, k $\in$ {1,2, ...,n}

In fact the above definition is weaker than the transitivity one

$$r_{ik} \geq \max_j [\min (r_{ij}, r_{jk})], \quad \text{for every i, k.}$$

One can point out the difference between the two definitions by observing that:

if R is transitive, then $R^2 \leq R$, on the other hand

if R_{cu} is weakly transitive, then the position of the entries less than 0.5 in R^2_{cu} is the same as the ones in R_{cu}.

Indeed we get in R_{cu}:

$$r_{ik}^{(2)} > 0.5 \Rightarrow \quad \text{there is at least one } j : r_{ij} > 0.5 \quad \text{and} \quad r_{jk} > 0.5 \Rightarrow r_{ik} > 0.5$$

$$r_{ik}^{(2)} = 0.5 \quad \Rightarrow \quad \text{there is exactly one } k : r_{ik} > 0.5 \quad \text{and} \quad r_{kk} = 0.5 \Rightarrow r_{ik} > 0.5$$

$$r_{ik}^{(2)} < 0.5 \Rightarrow \forall\, j, \; r_{ij} < 0.5 \text{ or } r_{jk} < 0.5 \Rightarrow \text{for } j = k \;\; r_{ik} < 0.5, \text{ being } r_{kk} = 0.5.$$

3. RANKING UNDER WEAK TRANSITIVITY

Theorem 1. If R_{cu} is weakly transitive relation on X, then there exists one and only one row whose elements r_{ij} are greater than 0.5, for $i \neq j$.

Proof: Let us proceed to the proof by induction on n.

For $n = 2$, the implication follows from the definition of R_{cu}.

Now, given R_{cu}, let us suppose that the implication hold for every relation R' on $[X - \{x_n\}]$, and let us prove that the implication is also true for R_{cu}.

Let us assume, by induction hypothesis,

$$\bigwedge_{j\neq i} r_{ij} > 0.5 \quad , \quad j = \{1, 2, ..., n - 1\}.$$

The element r_{in} may be greater or less than 0.5: if $r_{in} > 0.5$, the implication is proved; if $r_{in} < 0.5$, then $r_{ni} > 0.5$. Furthermore, if all the elements r_{nj}, with $j \neq i \neq n$, of the n-th row are greater than 0.5, then the implication is true. If at least one element r_{nj} of the n-th row is less than 0.5, then, by the weak transitivity of R', setting $j \neq i \neq n$, we get

$$r_{jn} > 0.5 \quad \text{and} \quad r_{ni} > 0.5 \quad \Rightarrow \quad r_{ji} > 0.5,$$

i.e. $r_{ij} < 0.5$, which contradicts the hypothesis $\bigwedge_{j\neq i} r_{ij} > 0.5$.

Therefore R_{cu} includes at least an i-th row such that $\bigwedge_{j\neq i} r_{ij} > 0.5$.

Finally, the unicity can be demonstrated observing that if there were two rows, the i-th and the k-th, such that $\bigwedge_{j\neq i} r_{ij} > 0.5$ and $\bigwedge_{j\neq k} r_{kj} > 0.5$ then $r_{ik} > 0.5$ and $r_{ki} > 0.5$.

Q.E.D.

Let us remark that the inverse of th. 1 is not true.

Theorem 2. Let X' be any subset having at least 2 elements in the alternative set X. The following propositions are equivalent:

a) R is weakly transitive.

b) Any restriction R' of R to X' is weakly transitive.

c) There is one and only one row, both in R and every restriction R' of R to X', such that $\bigwedge_{j \neq i} r_{ij} > 0.5$.

Proof.

a) $\Rightarrow$ b). Obvious.

b) $\Rightarrow$ c). By theorem 1.

c) $\Rightarrow$ a). Suppose that there exist indices, i, j, k such that

$r_{ij} > 0.5, \quad r_{jk} > 0.5 \quad \text{and} \quad r_{ik} < 0.5.$

Let us consider the restriction R' of R to X' = $\{x_i, x_j, x_k\}$. Hence we get:

R'	x_i	x_j	x_k
x_i	0.5	> 0.5	< 0.5
x_j		0.5	> 0.5
x_k			0.5

and every row has a minimum less than 0.5, which contradicts the hypothesis.

4. RANKING THE ALTERNATIVES

We are now able to show that, in view of Th. 2, the condition of weak transitivity of R_{cu} allows to construct a ranking of X.

Let x_i be the first alternative, corresponding to $\bigwedge_{j \neq i} r_{ij} > 0.5$, in symbols,

$$x_i > \{x_1, x_2, ..., x_{i-1}, x_{i+1}, ..., x_n\}.$$

Let us now consider the restriction R' of R to the set $[X - \{x_i\}]$, which also is weakly transitive, and take x_k such that

$$x_i > x_k > \{x_1, ..., x_{i-1}, x_{i+1}, ..., x_{k-1}, x_{k+1}, ..., x_n\}.$$

We obtain the ranking by iterating the procedure.

Example:

R_{cu}	x_1	x_2	x_3	x_4	$\bigwedge_{j\neq i} r_{ij}$
x_1	0.5	0.6	0.7	1	0.6
x_2	0.4	0.5	0.8	0.1	0.1
x_3	0.3	0.2	0.5	0.3	0.2
x_4	0	0.9	0.7	0.5	0

$$\bigwedge_{j\neq 1} r_{1j} = 0.6 \Rightarrow x_1 > \{x_2, x_3, x_4\}$$

R'	x_2	x_3	x_4	$\bigwedge_{j\neq i} r_{ij}$
x_2	0.5	0.8	0.1	0.1
x_3	0.2	0.5	0.3	0.2
x_4	0.9	0.7	0.5	0.7

$$\bigwedge_{j\neq 3} r_{3j} = 0.7 \Rightarrow x_1 > x_4 > \{x_2, x_3\}$$

R''	x_2	x_3	$\bigwedge_{j\neq i} r_{ij}$
x_2	0.5	0.3	0.3
x_3	0.7	0.5	0.7

$$\bigwedge_{j\neq 2} r_{2j} = 0.7 \Rightarrow x_1 > x_4 > x_3 > x_2$$

The ranking of the alternatives $X = \{ x_1, x_2, x_3, x_4\}$ is the following:

$$x_1 > x_4 > x_3 > x_2.$$

5. *WEAKLY COHERENT* RELATIONS R_{cu}

The previous sections have been devoted to the concept of weak transitivity for a relation R_{cu} as a sufficient condition for a ranking of the alternatives.

We have examined situations in which there is no indifference in the preference among the alternatives. In other words, such an R_{cu} has, in particular, the value 0.5 only in the elements r_{ii}.

Now, if we introduce the possibility of *indifference* to preferring an alternative to another, then a ranking is possible in a subset C in the power set P(X) rather than in X.

To this purpose let us introduce the following

Definition. A relation R_{cu} is said to be *weakly coherent* if

$$r_{ij} \geq 0.5 \quad \text{and} \quad r_{jk} \geq 0.5 \quad \Rightarrow \quad r_{ik} \geq 0.5 \qquad \text{for every } i, j, k \in \{1, ..., n\}.$$

Obviously a weakly transitive relation is weakly coherent.

In this case, the existence of at least one row with a minimum greater than 0.5 is obtained from a reasoning analogous to that used in the case of weak transitivity, but the unicity is not ensured anymore, in fact it is possible to have at least one row with a minimum equal to 0.5.

In order to introduce a strict order (>) in a suitable partition of X, let us define the following *weak-preference* relation defined in X.

Definition. A weakly coherent relation R_{cu} is said to be a *weak preference* relation, denoted $x_i \gtrsim x_j$, if

$$\exists\, k \in \{1, ..., n\} : r_{ik} \geq 0.5 \text{ and } r_{kj} \geq 0.5.$$

$\gtrsim$ is a preorder.

a) (reflexivity) $x_i \gtrsim x_i$, $\forall\, i \in \{1, ..., n\}$, obviously.

b) (transitivity) $x_i \gtrsim x_j$ and $x_j \gtrsim x_k \;\Rightarrow\; x_i \gtrsim x_k$, $\forall\, i, j, k \in \{1, ..., n\}$.

Indeed

$$x_i \gtrsim x_j \;\Rightarrow\; \exists\, h \in \{1, ..., n\} : r_{ih} \geq 0.5 \text{ and } r_{hj} \geq 0.5 \;\Rightarrow\; r_{ij} \geq 0.5$$

and

$$x_j \gtrsim x_k \;\Rightarrow\; \exists\, u \in \{1, ..., n\} : r_{ju} \geq 0.5 \text{ and } r_{uk} \geq 0.5 \;\Rightarrow\; r_{jk} \geq 0.5.$$

Therefore $x_i \gtrsim x_k$.

Now, to get a strict preference of x_i on x_j ($x_i > x_j$) it is required a set in which it is not true simultaneously $x_i \gtrsim x_j$ and $x_j \gtrsim x_i$, i.e. it cannot be the case that $x_i \sim x_j$ (indifference between x_i and x_j).

Then, if in a set one considers the relation $\gtrsim$ and the no indifference $\not\sim$ between x_i and x_j, then a relation of strict order is obtained.

Indeed, in addition to a) and b) it is also verified:

c) (asymmetry): $x_i \gtrsim x_j$ and $x_i \not\sim x_j \;\Rightarrow\; x_j \not\gtrsim x_i$.

Indeed:

$$x_i \gtrsim x_j \;\Rightarrow\; \exists\, k \in \{1, ..., n\} : r_{ik} \geq 0.5 \text{ and } r_{kj} \geq 0.5 \;\Rightarrow\; r_{ij} \geq 0.5.$$

So $r_{ji} \leq 0.5$, and being $x_i \not\sim x_j \;\Rightarrow\; r_{ji} < 0.5$.

Finally, to get a ranking when a weak coherent R_{cu} is assigned, it is necessary to group in a single class all the alternatives having a row minimum equal to 0.5. In this way we get a set of classes of alternatives, among which there is a strict order, being verified both the preorder relation of weak preference and the $C_i \not\sim C_j$.

Example.

R_{cu}	x_1	x_2	x_3	x_4	x_5	x_6	$\bigwedge_j r_{ij}$
x_1	0.5	0.8	0.5	0.8	0.5	0.8	0.5
x_2	0.2	0.5	0.2	0.7	0.3	0.1	0.1
x_3	0.5	0.8	0.5	0.6	0.5	1	0.5
x_4	0.2	0.3	0.4	0.5	0.2	0.3	0.2
x_5	0.5	0.7	0.5	0.8	0.5	0.7	0.5
x_6	0.2	0.9	0	0.7	0.3	0.5	0

The relation R_{cu} is weakly transitive because it together with its restictions admits at least one row such that $\bigwedge_j r_{ij} = 0.5$. Investigating the relation R_{cu} we note that there are three rows, whose minimum is equal to 0.5, which implies that x_1, x_3 and x_5 are preferable to the others.

We can say then the alternatives $\{x_1, x_2, x_3\}$ form a class I, which dominates all the other alternatives.

The restriction R' of R_{cu}, obtained by eliminating x_1, x_3, x_5,

R'	x_2	x_4	x_6	$\bigwedge_j r_{ij}$
x_2	0.5	0.7	0.1	0.1
x_4	0.3	0.5	0.3	0.3
x_6	0.9	0.7	0.5	0.5

admits just one row, the 6th, whose minimum is equal to 0.5. This ensures that x_6 is preferable to the other alternatives and forms a class II.

The restriction R'' of R', obtained by eliminating x_6

R''	x_2	x_4
x_2	0.5	0.7
x_4	0.3	0.5

admits just one row x_2, with minimum 0.5.

The alternatives x_2 and x_4 form classes III and IV, respectively.

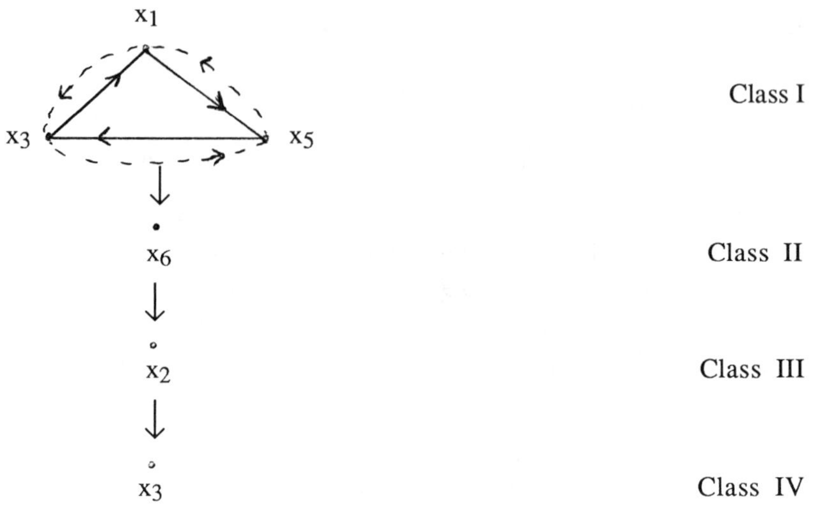

$$\{x_1, x_3, x_5\} > \{x_6\} > \{x_2\} > \{x_4\}.$$

CONCLUDING REMARKS

We have described a ranking procedure under weak transitivity (or coherence) and weak coherence relations.

REFERENCES

[1] J.M. Blin - A.B. Whinston, "Fuzzy Sets and Social Choice", *Journal of Cybernetics*, 1974, 3, 4, pp. 28-36.

[2] J.M. Blin, "Fuzzy Relations in Group Decision Theory, *Journal of Cybernetics*, 1974, 4, 2, pp. 17-22.

[3] A. Fadini - L. Basile, "Su certe relazioni sfocate di preferenza per processi decisori, *'Atti' dell'Accademia Pontaniana*, vol. XXVIII, 1979, pp. 267-284.

[4] S.A. Orlovsky, "Decision-making with a fuzzy preference relation", *Fuzzy Sets and System 1*, 1978, pp. 155-167.

CALCULATING THE MEAN KNOWLEDGE REPRESENTATION FROM MULTIPLE EXPERTS

Ricardo José Machado
IBM Rio Scientific Center
P.O Box 4624
20.071 - Rio de Janeiro, Brasil

Armando Freitas da Rocha
Departamento de Fisiologia e Biofísica
Instituto de Biologia - UNICAMP
13081 - Campinas, Brasil

Beatriz de Faria Leão
FUC - Instituto de Cardiologia
Av. Princesa Isabel, 395
90000 - Porto Alegre, Brasil

ABSTRACT: This paper presents a method to extract the consensus knowledge from multiple human experts on a specific problem domain. The method may be useful for the knowledge engineer, interested in building expert systems based on the knowledge from a population of experts, as well as for the epistemologist, interested in understanding the characteristics of the reasoning models used by experts.

KEYWORDS: multiexpert, consensus, knowledge

1. INTRODUCTION

The increasing interest on the expert system technology by the big business world puts stress on a new demand: to model the knowledge shared by a group of experts (**Corporative Knowledge**) instead to use the expertise of just one subject.

J. Kacprzyk and M. Fedrizzi (eds.), Multiperson Decision Making Using Fuzzy Sets and Possibility Theory, 113–127.

For instance envision a large corporation (eg. a bank) interested in building a corporative knowledge base using the expertise of its more prominent specialists in a particular problem domain (eg. credit approval). This approach requires **averaging** the knowledge of the target population.

Most of the Artificial Intelligence researchers points out the knowledge acquisition as the critical component in the development of expert systems and as the main obstacle to the widespread utilization of this technology. Naturally, if the knowledge acquisition from a single expert is by itself a hard task, the difficulty degree can appear multiplied when we deal with a population of experts. Hence it is fundamental to have a knowledge acquisition method that is simple, uniform, of fast application, suitable to several domains and able to provide data easy to transcribe into the different classical knowledge representation formalism's.

Our proposition is to follow the method proposed by Rocha et al for knowledge and cognition analysis in a population (Greco, Rocha, 1984, 1987; Leão and Rocha, 1987; Theoto et al., 1987). This method has showed to attend largely the requirements before mentioned. The method is based on the concept of jargon codification presented by Rocha and Rocha (1985) or Theoto and Koizumi (1987) to obtain an uniform and systematic description of the knowledge shared by the population and on the graph theory to express the structure of this knowledge. It has been successfully applied in the acquisition of heuristic information in medical diagnosis problems (Leão, 1987,1988) and in the analysis of text understanding (Greco and Rocha, 1987; Theoto et al, 1987). The method is reproduced in the Appendix 1. Greco and Rocha (1988) showed the soundness of this approach for the analysis and description of text understanding in a population, finding a significant correlation between this understanding and the electrical brain activity, measured through the electro-encephalogram. Theoto et al (1987) showed that this approach can be used to characterize the different understandings that a same text can have in populations with different knowledges about an specific subject. Leão and Rocha (1987), Leão (1988) using this technique demonstrated that the knowledges of experts and non-experts are expressed through quite well different structures. More recently this approach started to be used in the acquisition of knowledge in the area of arterial hypertension.

The calculus used in these early papers for obtaining the mean graphs representing the shared knowledge is based on the quartiles statistical analysis of the evidences occurrence. This is actually the weakest point of the methodology. The arcs of the mean graph in this approach are obtained by frequency from the quartiles, what causes arcs with few occurrences to be included in the mean graph in some instances. This paper presents a method to calculate the mean knowledge from multiple human experts on a specific problem domain avoiding the pointed out inconveniences existent in the previous approach. This paper also introduces several parameters useful for the characterization of the expert knowledge and for the measurement of the agreement level among the members of a population. The concepts and algorithms here introduced may be useful for the knowledge engineer, interested in extracting the consensus knowledge from a population for building an expert system, as well as to the epistemologist, interested in understanding the characteristics of the expert knowledge, the differences between the reasoning models of experts and non-experts, and how the heuristics developed by the human professionals through years of experience are stored and modified.

2. DESCRIPTION OF THE METHOD

2.1 Knowledge graphs

We define a **Knowledge Graph (KG)** as the minimal directed AND/OR acyclic graph representing the knowledge of an expert for a specific diagnostic hypothesis. These graphs are produced by the application of the knowledge acquisition method described in the Appendix 1. The graphs themselves may be accepted as a formal knowledge representation as well as they may be used for conversion into other classical knowledge representations, such as rules or frames.

Figure 1 exhibits a typical knowledge graph, where we can identify three types of nodes:

Hypothesis-node - is the highest level node in the graph and represents the diagnostic hypothesis taking into consideration.

Evidence-nodes - are the nodes situated in the lowest level of the graph. They represent the symptoms, signs and test results, in the case of medical problems, or data supporting decisions in other fields of application. These nodes are ranked from left to right in decreasing order of importance to support the final hypothesis.

Intermediate-nodes - represent different aggregations of evidences used by the expert in his reasoning. Often they can be conceptually identified as pathophysiological states, syndromes, etc.

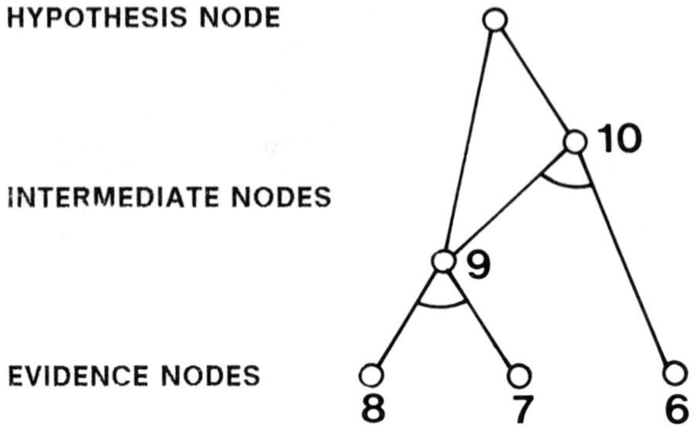

Figure 1 - Knowledge graph of one expert about one diagnostic hypothesis

The expert assigns to each node of the KG a note between 0 and 10 that represents the possibility of the hypothesis being supported by the information issued by the node (or the membership degree of this node to the hypothesis). The experts usually employ AND nodes as intermediate nodes, and one OR node for the hypothesis node. Because of this we can easily split the knowledge graph, through the separation of the arcs reaching the hypothesis node, in several subgraphs that can be interpreted as different pathways available to the expert to reach the hypothesis, departing from the evidences.

2.2 Construction of the knowledge graph of the experts population

The algorithm presented in this section allows us to build a graph that accumulates the overall knowledge of a population of experts, about a specific hypothesis. The aggregation criterion considers only the topology of structures, disregarding completely the labels of the nodes. For instance, a structure with the evidences (E1, E2 e E3) is considered equivalent to another formed by the evidences (E2, E1, E5), if the topology of both structures is equal and the evidences occupy equivalent positions in the lists of evidences of the KG's to which they belong. For instance ((E1,E2),E3) = ((E2,E5),E1). This approach is feasible since the evidence nodes are, by assumption, ranked by decreasing importance. It is based on the work of Yager (1987) for the aggregation of evidence in neural networks. The key point here is that the topological matching analysis searches for a reasoning structure independently of the actual pieces of information supporting the hypothesis. On the contrary, the intuitive approach criterion would try to match the structures not only by topology but also by their labels. It is not adopted in our method since very often it cannot be applied due to the high level of disagreement that we can find in a population of human experts, and due to the difficulty or even impossibility of encountering a large number of experts on a specific subject.

The algorithm also computes the occurrence frequency of each arc in the experts population graph. This measure is fundamental for the pruning process that is adopted to calculate the mean graph in the section 2.4.

ALGORITHM: POPULATION_KG_CALCULATION

* *Aggregation of knowledge graphs*

- Select arbitrarily one expert KG as the embryonic graph of the population, defining for each arc a counter with initial value 1, and for each node one accumulator with initial value equal to the membership of the node in the original graph.
- Aggregate sequentially the other KG's disregarding the labels of their evidence nodes.
- For each pathway in the graph being aggregated do
  - If there exists a topologically similar structure in the population KG being formed, occupying the same positions in the rank of evidences and using the same type (AND/OR) of corresponding intermediate nodes.

    Then - Add one to the arc counters of this structure
    - Add the memberships of the nodes of the pathway being aggregated to the corresponding membership accumulators.

    Else - Aggregate the new structure to the population graph, defining counters with initial value 1 to each arc.
    - Add the memberships of the nodes of the pathway being aggregated to the corresponding membership accumulators of the population graph.

* *Calculation of the nodes memberships*

- Divide the value of the membership accumulator of each node of the population graph by the sum of the counters of the arcs that leave this node.

* *Determination of the labels for the evidence nodes*

- For each position at the lower level of the population KG determine the frequency distribution of evidences for this position.
- Walking from left to right in the lower level of the population KG, select as the most representative label to the current position the most frequent evidence in the distribution, if it was not used before. (If it was used, select the next more frequent, obeying the same restriction).

2.3 Definition of parameters to characterize knowledge

The following defined parameters allow to characterize the knowledge graph for a specific diagnostic hypothesis.

Evidence Use Ratio (EU) - is obtained dividing the number of evidence nodes of the KG by the number of evidence nodes used in the population KG. The domain of this parameter is [0, 1]. It assumes the value 1 for the members of the population that uses the maximum number of evidences.

Fuzziness Degree (F) - is obtained dividing the number of arcs increased of 1 by the number of nodes of the KG. F belongs to the interval $[1, \infty]$. F is 1 for a tree, growing larger as additional arcs are aggregated to the graph.

Slimness Degree (S) - is obtained as the product of EU and F. The knowledge graph is as much slim as less fuzzy it is, and as less evidences are used in the reasoning. The domain of S is $[0, \infty]$. S assumes, for instance, the value 1 for an individual that employs the maximum number of evidences and organizes his knowledge as a tree.

Difficulty Degree (D) - is defined as the average value of the products between the fuzziness degrees and the number of evidences of the KG's of the individuals. It can be calculated by the product:

$S_{average}$ · Maximum number of used evidences.

This parameter expresses the difficulty degree that a group of people encounters when dealing with a specific hypothesis. More difficult problems will require the use of a larger number of evidences and more fuzzy graphs.

Agreement degree (A) - This parameter is able to measure the level of agreement among the members of a population, concerning a specific hypothesis. It is defined as:

$$A = S_{average} / S_{population}$$

where: $S_{average}$ is the average value of the slimness degrees of the KG's o the individual experts, $S_{population}$ is the slimness degree of the population KG. A belongs to the interval [0,1]. If A is small then there is low agreement among the members of the experts population. If A is equal to 1 then there is full agreement among the experts.

The KG of figure 1 presents EU= .37 , F= 1.16 and S= .43, assuming that the maximum number of evidences used by the experts in the population for that diagnostic hypothesis was 8.

2.4 Determination of the mean knowledge graph

We present next an algorithm to calculate the mean knowledge graph, based on a pruning procedure of the arcs of the population KG whose frequency is lower than a predefined threshold.

ALGORITHM: MEAN_GRAPH_CALCULATION

- For an integer threshold varying from 1 to the number of individuals of the population do
  - Prune the population KG according to the threshold, executing the algorithm POPULATION_KG_PRUNING.
  - Calculate the slimness degree (S_p) of the pruned graph.
- Select as the mean KG the pruned graph that
minimizes $|S_p - S_{avg}|$.
(S_{avg} = average value of the slimness degrees of the KG's of the individual experts)

ALGORITHM POPULATION_KG_PRUNING

- Remove from the population KG all the arcs whose frequency is lower than the threshold.
- Remove also all the arcs and nodes of the subgraphs that become disconnected from the hypothesis node in the pruning process.

3. EXPERIMENTAL RESULTS

We tested the proposed method with knowledge graphs about congenital heart diseases, elicited by Leão (1988) from experts and non-experts of the Institute of Cardiology of R. G. do Sul, the Heart Institute, the Institute D. Pazzaneze, and the Cardiology Division of EPM, located in Brazil.

	Diagn	Navg Evid	Navg Node	Navg Arcs	EU avg	F avg	S avg	Navg Paths	Clust Size	D	A
EXPERT	PDA	3.5	6.5	6.6	.70	1.15	.81	2.0	1.8	4.1	.46
	VSD	6.1	10.1	10.3	.61	1.10	.67	2.1	2.7	6.7	.27
	ASD	4.6	8.1	8.2	.65	1.13	.75	2.2	1.8	5.2	.29
	AVSD	5.5	8.9	8.8	.61	1.10	.67	1.9	3.0	6.0	.24
	AOCO	3.8	6.2	5.8	.49	1.08	.52	1.8	2.3	4.1	.20
NON EXPERT	PDA	7.4	11.8	11.6	.49	1.06	.52	1.6	2.8	7.8	.21
	VSD	9.0	13.3	13.0	.64	1.05	.64	1.7	3.3	9.1	.22
	ASD	7.9	11.9	11.9	.65	1.09	.71	1.9	3.5	8.5	.24
	AVSD	7.6	11.8	11.3	.54	1.05	.56	1.7	2.6	7.9	.22
	AOCO	7.5	11.5	11.1	.53	1.06	.56	1.6	2.9	7.9	.25

Table 1 - Comparison between experts and non-experts

We worked with 12,19,18,17 and 13 knowledge graphs from experts and 10,16,15,12 and 8 KG's of non-experts for the diseases: Patent Ductus Arteriosus (PDA), Ventricular Septal Defect (VSD), Atrial Septal Defect (ASD), Atrioventricular Septal Defect (AVSD), and Aortic Coarctation (AOCO), respectively. Figure 2 illustrates some experts and non-experts knowledge graphs for the diagnostic hypothesis PDA. To simplify the figure we did not present the membership degrees. Figure 3 shows the knowledge graphs for the population of experts and of non-experts for the same diagnostic hypothesis PDA. The frequencies of arcs are indicated through numbers in the graph corresponding to experts, and through the line thickness in the graph of non-experts. Figure 4 presents the experts and for non-experts mean graphs for

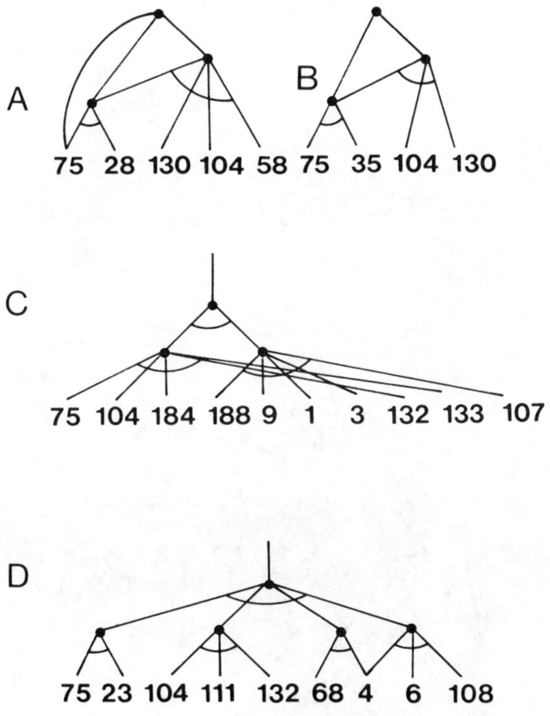

Figure 2 - Examples of knowledge graphs about the hypothesis PDA, taken from experts (A,B) and non-experts (C,D)

pathologies: PDA, VSD, ASD, AVSD and AOCO. Again to simplify the figure we did not present the membership degrees. Table 1 presents several statistics and measurements made on the populations of experts and of non-experts, for comparison. The columns of the table 1 represent sequentially from left to right: pathology code, average numbers of: evidence nodes, general nodes, and arcs in the KG's of the professionals; average values of the parameters: evidence usage ratio, fuzziness degree, slimness degree; average number of diagnostic pathways, average size of evidence clusters, difficulty degree presented by the problem and the agreement degree among members of the population.

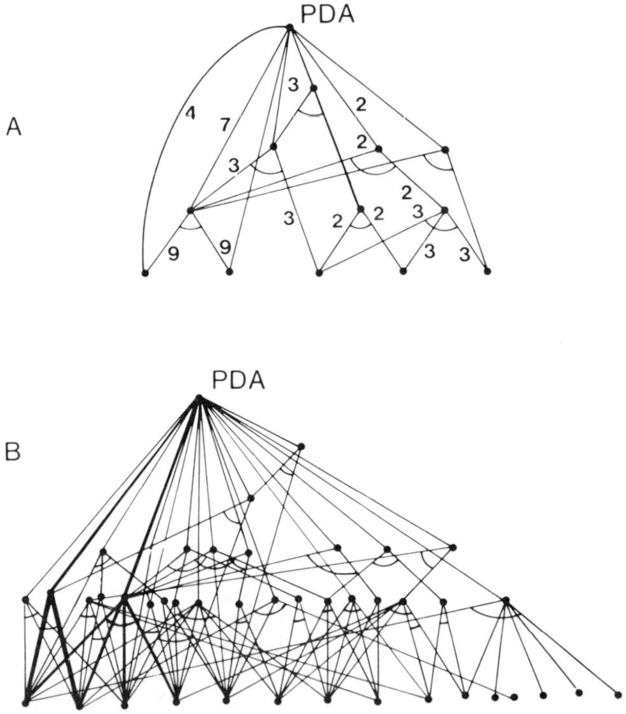

Figure 3 - Knowledge graphs for the population of experts (A) and of non-experts (B), for the hypothesis PDA

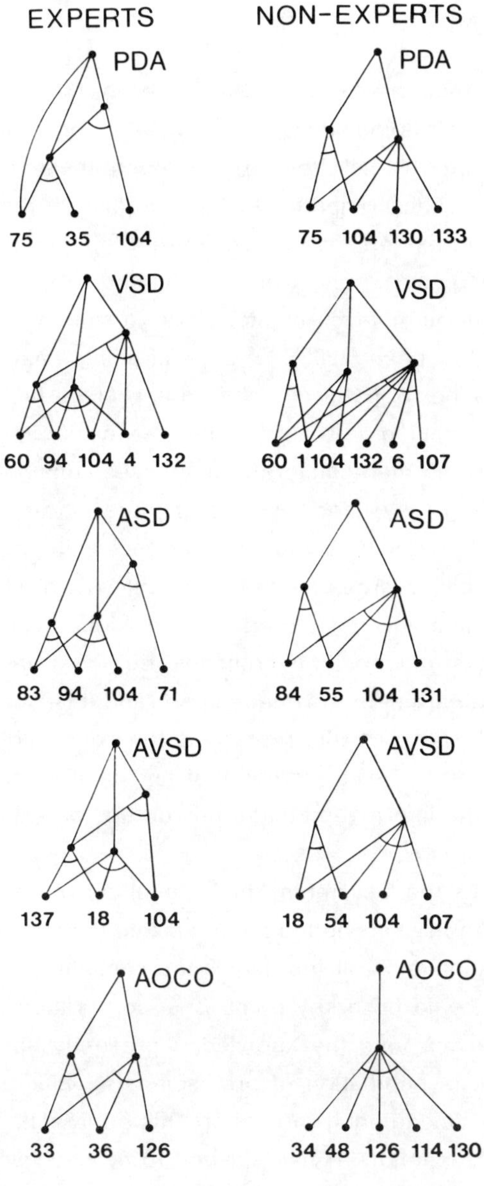

Figure 4 - Mean graphs for several diagnostic hypotheses in the population of experts and of non-experts

4. CONCLUSIONS

The application of the tools presented in this paper for the analysis of the reasoning models employed by cardiologists specialized and not specialized in congenital heart diseases, showed distinct behavior between these two groups of professionals. The experts when compared to the non-experts employed smaller and less complex reasoning models, as demonstrated by the construction of knowledge graphs with smaller structure, concerning the number of evidences, the number of nodes and arcs and the number of levels. These results are in agreement with the conclusions presented by Leão (1988). In spite of using smaller graphs, the experts displayed a larger number of pathways able to associate the evidences to the diagnostic hypotheses, evidencing a greater semantic capacity in comparison to non-experts. Also the experts formed clusters of evidences considerably smaller than the non-experts, even when they used clusters of large cardinality. The non-experts were frequently unable to include the key evidences employed by the experts in their reasoning, even using large clusters. The calculated difficulty degree (D) was smaller for experts than for non-experts. The highest agreement degrees (A) appeared for the easiest diagnostic hypotheses (eg: PDA) and for the experts. The same occurred for the parameters evidence use ratio (EU), fuzziness degree (F) and slimness degree (S). These results reproduce the conclusion of Leão (1988): "To be an expert means to think with models that are smaller, slimmer, more compact and, mainly, to value the information differently of the non-experts".

In the construction of expert systems, the acquisition of knowledge from multiple experts tends to enrich the knowledge base, making the system more complete and validating the information, since it represents the consensus of a group and not only one individual. This could be a key point for a corporation that needs to capture, consolidate, distribute and use the knowledge presently spread throughout some prominent experts in its population of professionals. Efficient methods for knowledge acquisition are fundamental to support this approach. The method of jargon codification of Rocha (appendix 1) has showed to be effective to elicit the heuristic knowledge employed by experts. An important obstacle to the multiple experts approach comes from the high divergence degree frequently encountered in the selection of evidences to support their decisions.

The method introduced in this paper allows to resolve easily these opinion conflicts, extracting the consensus knowledge of a population of experts. The algorithm revealed to be able to produce a mean graph, even in situations where there was a very high level of discordance in the use of the evidences, as occurs with difficult diagnostic hypotheses or in populations of non-experts. The proposed algorithm presents the following main advantages:

- easy computer implementation,
- consistency with the human intuition, preserving the structures more commonly used by the experts,
- the mean graphs may be converted into another knowledge representation formalism allowing to build expert systems that incorporate coherently the knowledge from multiple experts.

The analysis of the mean graphs allowed to reach the same results obtained with the statistical analysis of the individual graphs, demonstrating that these mean graphs express well the information contained in the reasoning models of the experts.

The mean graphs of the experts differ largely from the mean graphs of the non-experts. However it is possible to identify in the last ones some embryonic structures that typically belong to the graphs of experts. This suggests that, through a learning process, the non-experts could evolve, as in a neural network, toward the reasoning models of experts.

APPENDIX 1 - KNOWLEDGE ACQUISITION METHOD

We reproduce here the method for knowledge acquisition used by Rocha et al in his learning experiment and modified by Leão for application in the area of medical diagnosis.

Phase 0- Definition

1) To define the project scope (problem domain).
2) To define the diagnostic hypotheses that will be part of the project.
3) To define a list of symptoms, signs and tests results, with the assistance of a medical expert or a group of experts.

Phase 1- Knowledge acquisition

1) To select, without previous knowledge of the expert, one of the diagnostic hypotheses, if possible in the decreasing order of its frequency of occurrence.

2) To ask to the expert to indicate in the symptoms list the items necessary for the formulation of the diagnostic hypothesis, defining by this way a work subset.

3) To ask the expert to rank the work sublist according to the importance of the items for the diagnostic.

4) To ask the expert to assume the items of the ordered list as the evidence nodes of a knowledge graph and to associate them in the form that he judges necessary to conclude about the diagnostic hypothesis (generating intermediate nodes that converge to the diagnostic hypothesis).

5) To ask the expert to ascribe a membership degree between 0 and 10 for the information represented in each node of the graph in relation to the diagnostic.

6) To ask the expert to define the logical operators (AND, OR, NOT) associated to the nodes of the graph

7) To repeat the process for the other diagnostic hypotheses forming a family of graphs.

REFERENCES

- Greco G., A.F. Rocha, and M.T. Rocha (1984) Fuzzy Logical Structure of Text Decoding. Proceedings of the 6th International Congress of Cybernetics and Systems, Paris, V.1, pp.193-198.

- Greco G., and A.F. Rocha (1987) The Fuzzy Logic of Text Understanding. Fuzzy Sets and Systems, 23, pp. 347-360

- Greco G. and A,F. Rocha (1988) Brain Activity and Fuzzy Belief in Tamáa Zédémyi (Ed.): Recent Advances in Psychology n. 56, North Holland pp. 287-319

- Leão B.F., F.A. Lucchesi, and A. F. Rocha (1987) A Methodology Proposal for Knowledge Acquisition. Proceedings of the Seventh International Congress of Medical Informatics pp. 1042-1050

- Leão B.L. (1988) Construction of the Knowledge Base of an Expert System for Supporting Diagnosis of Congenital Heart Diseases. PhD thesis - Escola Paulista de Medicina, São Paulo

- Rocha A.F., and M.T. Rocha (1985) Specialized Speech: A First Prose for Language Expert Systems. Inf. Sci. 35 pp. 1-8

- Theoto M.R. et al. (1987) The Fuzzy Decodings of Educative Texts. Fuzzy Sets and Systems 23, pp. 331-345

- Theoto M.T. and M.S. Koizumi (1987) The Expert Environment - A Case Study. Preprints of the Second International Fuzzy Systems Association Congress, Tokio, V. 1 pp. 380-383

- Yager R.R. (1987) On the Aggregation of Processing Units in Neural Networks. I International Conference on Neural Nets, San Diego, V. 2 pp. 327-341

An approach to customized end-user views in multi-user information retrieval systems

Henrik L. Larsen*
Department of Computer Science,
Roskilde University, Postbox 260,
DK-4000 Roskilde, Denmark

Ronald R. Yager†
Machine Intelligence Institute,
Iona College, New Rochelle,
NY 10801, USA

Abstract

We present a method for representing and applying local views in multi-user information retrieval systems. Such views allow end users to express their information needs in queries using their own preferred vocabulary of subject terms. Central to this approach is a similarity matrix which is used to relate the local user vocabulary to the primary vocabulary used by the host system. The similarity matrix determines the overall degree to which a document represented at the host matches the information needs as represented by a query. The computations involved are performed in low order polynomial time and space. We briefly discuss the knowledge combination operators applied and some computational improvements of the method. Finally, we suggest an efficient way for interfacing the method to traditional information retrieval systems.

1 Introduction

Although decision makers and other information users today are offered online access from a number of external information base hosts, this possibility is not widely utilized. A major reason for this situation is that it is difficult for ordinary end users to query the bases in a way that will provide a satisfactory answer to their information needs. Especially is difficult for users that need information on some topic or subject to express their needs in a query using the terms applied by the host for indexing of information objects (documents); local user environments tend to develop their own specialized terms and concepts. Further, in query formulation, as well as in document indexing, the selection of terms is based on human judgements and may therefore not yield a precise characterization of, respectively, the information needs and the document. Finally, such users often have problems in expressing their needs in Boolean logic, as typically required by query languages, and in formulating the query correctly in the specific query language.

As a result of these difficulties, the set of documents that are found to match a query does often provide an unsatisfactory answer to the information needs. In such a situation, the user will typically try again using alternative query terms and possibly other logical bindings.

*Supported in part by Teknologirådet (the Danish Technology Council) under ATV grant No. TS-9.
†Supported in part the Airforce Office of Scientific Research.

J. Kacprzyk and M. Fedrizzi (eds.), Multiperson Decision Making Using Fuzzy Sets and Possibility Theory, 128–139.

In this paper, we propose as a solution to the information retrieval problem to provide each end-user group with a local view. A local view may be seen as a knowledge base that relates the locally preferred vocabulary to the primary view represented by the "official" vocabulary (thesaurus) used for indexing at the host. The knowledge base is created by a domain expert in the user group by locally refining the official vocabulary to the local needs. In the refinement, the expert supply the terms and the term similarities that represent the local conceptual model of the subject domain. The knowledge base is described by a similarity relation which is represented by a similarity matrix. The end user query is constructed simply by selecting a set of appropriate terms from the local vocabulary — thus, no formulation in Boolean logic or in a specific query language are required. The degree to which a document matches a query is based on the inferred effective similarity between document and query terms.

Related approaches to a solution include the following. Bates [1] describes the use of a common end-user thesaurus which in a sense enriches the indexer thesaurus with the terms that end users tend to apply. The users may browse the thesaurus in the search for appropriate query terms. Wang *et al.* [11] argue for the use of relational thesauri containing semantic relations. In a sense such a thesaurus provides a conceptual domain model. A relational thesaurus is used to augment each query term with a disjunction of semantically close terms. The authors report an improvement in retrieval performance by this approach. Bezdek *et al.* [2] describe the use of the transitive closure of fuzzy thesauri in information retrieval systems. Mili and Rada [9] describe a method for merging two thesauri to obtain a "better" thesaurus. They include a strategy for resolution of conflicts between thesauri.

An advantage of our solution is that multiple local views provide a better tailoring to the needs in the local environments than it is possible with one common user thesaurus. A major problem with such a thesaurus is that the sense in which a term is used may vary from one environment to another, and this cannot be represented consistently by a common thesaurus.

Our solution contains some elements from the approaches mentioned: The similarity matrix contains, in a sense, the kind of information that was utilized from end-user thesauri and relational thesauri in the approaches mentioned above. For the utilization in query handling, we compute the effective term similarities which represent the transitive closure. Further, the creation of the similarity matrix representing the local view, may be seen as the result of a kind of merge of the local "thesaurus knowledge" with the indexer thesaurus.

The way we understand the information retrieval problem, and the idea of applying a measure of the strength of term relationships were inspired by earlier work in the KIWI[1] project on user-friendly systems for querying information bases, see *e.g.* [10,7].

This paper is structured as follows. In section 2, we describe the informations retrieval problem considered. In section 3, we outline the idea in our solution by local views in multi-user information retrieval systems. In section 4, we describe the basic matching method in the solution, including computation of the effective similarities in a local view, and querying through such a view. In section 5, we demonstrate central characteristics of the method. In section 6, we deal with issues of importance for an efficient implementation of the method.

[1] KIWI: Knowledge-based User-friendly System for the Utilization of Information Bases (ESPRIT project 1117).

2 The information retrieval problem

The information retrieval problem may be expressed as the problem of identifying those documents in a collection that satisfy the user's information needs.

The document collection is represented by one or more information bases maintained at one or more information base hosts. In the informations bases the documents are represented by so-called surrogates. A surrogate may be seen as a record which contains fields for bibliographic information and for characterization of the contents of the document. A typical characterization field, often referred to as "controlled terms", is a field that contains a list of terms characterizing the subject of the document. These terms are selected from an official vocabulary, called a thesaurus, for the subject area of the information base. The selection is done by a human indexer, or an indexing system, by scanning the document text and choosing a set of appropriate official terms.

On the other hand, users translate their information needs to requests for information. A request may specify bibliographic characteristics and/or a subject that the user needs information about. The user represents the request as a query to the information retrieval system that provides the retrieval services from the information base host.

The retrieval system works, essentially, by comparing document surrogates to the query. If a surrogate matches the query, it —or some specified part of it— is retrieved. In this paper, we shall deal with the common problem of satisfying subject needs. Thus, we shall simply consider queries and surrogates as subject characterizations; when no confusion can arise, we shall use the term "document" in stead of the term "surrogate" for the representation of a document.

Since the answer to a user's information needs from a document collection is determined by comparing the selected representations of the needs and the documents, the answer that the user really obtains to the query may be seen as an "approximation" to the ideal answer to the needs, as illustrated by figure 1. The ideal answer may be seen as the set of those documents that would

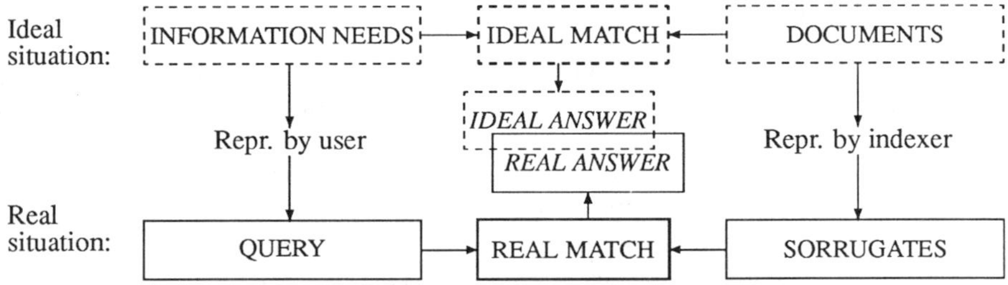

Figure 1: The matching problem in information retrieval

satisfy the information needs in the judgement of the user domain expert. Ideally, the real answer should by equal to the ideal answer. The degree of equality between the two answers is measured by the quality performance measures of information retrieval: recall and precision. The recall is the portion of the ideal answer obtained in the real answer, and the precision is the portion of the real answer that belongs to the ideal.

2.1 The problem of mismatching vocabularies

A major cause of poor end-user performance in information retrieval systems is that the query terms selected by end users often are at odds with those used for indexing of the documents in the collection queried [4]. To improve this situation, thesauri have been provided for the major subject domains. A thesaurus provide a sort of standard terminology, characterized by a vocabulary of terms, and some term relations, such as 'broader term' and 'narrower term'. Thesauri are typically used by indexers, and may be consulted by users for query formulation. Ideally, the indexer selects the terms from the thesaurus such that they conjunctively describe a "sufficiently" specialized subject classification of the document, and the users select from the same thesaurus the query terms that in conjunction precisely describe the class of documents that satisfy their information needs. However, reality is far from this ideal situation, mainly because of two kinds of problem: the representation problem, and the vocabulary mismatch problem.

First, it is difficult for indexers to describe the document contents sufficiently precise in the indexing; similarly, it is difficult for end users to characterize their precise information needs explicitly in a query. These difficulties are the major causes of the representation problem. Second, the terminology of a subject domain is neither universal, nor static [5]; especially, each end-user environment tends to develop its own specialized terminology for the subject domain of interest. These are the major causes of the vocabulary mismatch problem.

In dealing with these problems, the end user typically attempts to approach a satisfactory answer by reformulating the query by use of alternative terms. In the following, we propose a solution to the vocabulary mismatch problem. Central to the solution is the representation of a local view for each end-user group. A local view is represented by a similarity matrix which essentially maps the local user vocabulary to the primary vocabulary used by the host system.

3 Connecting the vocabularies by local views

We introduce the idea in our solution of the vocabulary mismatch problem in information retrieval is developed in the following.

Initially we start with a *primary vocabulary* of subject terms, maintained in a thesaurus and used by the host for indexing of documents. In the proposed approach, each local user group can provide its own *local vocabulary* through which the users in this group query the system. In particular, we envision a local domain expert to be responsible for this task. The domain expert refines the primary vocabulary to the local vocabulary by locally supplying the descriptor terms preferred in the local environment to the first vocabulary. We shall refer to the set of terms supplied as the *supplied vocabulary*. Thus, the local vocabulary is the union of the primary vocabulary and the supplied vocabulary.

In supplying a term, the expert essentially also supplies information about the degree of similarity to existing related terms in the primary vocabulary. These similarities are represented by a matrix $S = \{s_{ij}\}_{n \times n}$ where n is the size of the local vocabulary, and s_{ij} is the locally determined similarity between the i'th and the j'th term in the vocabulary. The matrix can be seen as a local knowledge base which refines the primary vocabulary to the needs in the local environment, or the local user group. The knowledge base provides the information needed to determine the effective similarities between the subject terms in a document and those in the query; the determined similarities are then used in computation of the overall degree to which the document matches the query. Since the knowledge applied is refined to the local needs, this degree is likely to be a close approximation to the degree to which the document satisfies the information needs represented by the query.

The knowledge base provides in a sense a user view that is customized for the local user group. We shall call such a view a *local view*, and the view provided by the host the *primary view*. The idea of providing local views to a primary view is illustrated by figure 2. We notice the apparent analogy

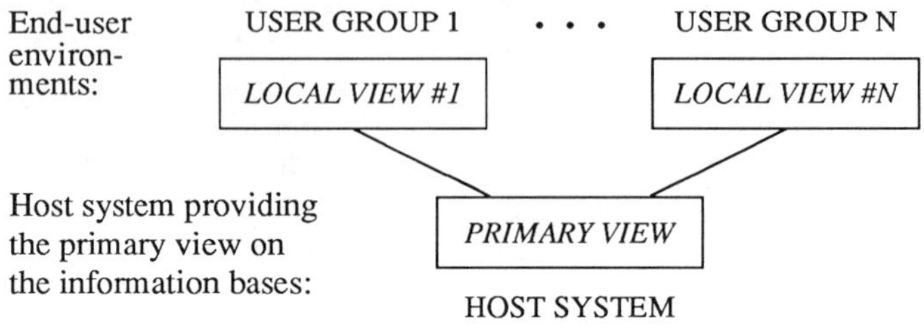

Figure 2: Local views to a primary view

between, on the one side, a local view for information retrieval and the local expert responsible for maintaining it, and, on the other side, a user view in database management systems and the local database administrator.

The user queries the document collection by characterizing his information needs by a set of terms selected from the local vocabulary. Thus, the resulting query is represented by a subset of the local vocabulary. The advantage of this approach is that it frees the user from the learning of a specific query language. Further, the user is not required to add a weight for the importance of a query term. This way to query through a local view is in particular friendly to the causal end user, since the user avoids the following problems in traditional query formulation: (1) learning the specific query languages required by different information retrieval systems, (2) guessing what the appropriate official subject terms may be, (3) representing needs correctly in the Boolean logic form as typically required by specific query languages, and (4) the requirement to assert importance weights, often serving to confuse ordinary users. We may assume that document terms are weighted. For documents, as for queries, the unweighted case is seen as just special case of the weighted case. Since our solution works for the latter case (queries and documents using weights), it also work in the first case (queries and/or documents not using weights). We shall in the description of the solution assume that queries are unweighted, while documents are weighted.

4 The matching method

Central to the matching procedure is the computation of the effective similarity matrix $E = \{e_{ij}\}$ from the initial similarity matrix $S = \{s_{ij}\}$, created by the expert. The value e_{ij} is the effective similarity between the terms t_i and t_j, given the initial, and possibly incomplete, similarities obtained from the expert and represented by the matrix S.

The matrix S is diagrammed by a weighted undirected graph G_S, such that the vertices represent the terms in the local vocabulary, and the edges represents the similarities supplied by the expert. A simple path between the vertices (terms) t_i and t_j in G_S is a list of vertices where no vertex is repeated, and successive vertices are connected by an edge. The strength of a simple path is the "product" of the similarities represented by its edges, thus the strength of the path $[t_i, t_j, t_k]$ is $s_{ij} * s_{jk}$[2]. The value e_{ij} in E is the strength of the strongest simple path between the vertices t_i and t_j; the value is zero if there is no path between the two vertices.

The degree to which a document D matches (satisfies) a query Q under the similarity matrix S, is $f_S(Q, D)$ defined as the minimum over the query terms of the strengths of the strongest path from a document term to the query term:

$$f_S(Q, D) = \min_{q \in Q} \max_{d \in D} e(q, d)$$

where $e(q, d) = e_{ij}$ for $(q, d) = (t_i, t_j)$.

4.1 The basic computational method

The local vocabulary is represented by the set $T = \{t_1, \ldots, t_n\}$ such that $A = \{t_1, \ldots, t_h\}$ is the primary vocabulary (used by the host for indexing), and $B = \{t_{h+1}, \ldots, t_n\}$ is the vocabulary supplied by the local expert.

The initial similarities given by the expert are represented by the matrix $S = \{s_{ij}\}_{n \times n}$, where $s_{ij} (\in [0, 1])$ is the similarity between the terms t_i and t_j. The matrix is symmetric and reflexive, *i.e.* $s_{ij} = s_{ji}$ for all pair (i, j), and $s_{ij} = 1$ for $i = j$.

The effective similarity matrix $E = \{e_{ij}\}_{n \times n}$ is computed by

$$E = S^{n-1} \tag{1}$$

such that

$$s^k_{ij} = \max_{u \in N_n} (s^{k-1}_{iu} * s_{uj}), \; k = 2, \ldots, n-1$$

where N_n denotes the set $\{1, \ldots, n\}$, and s^k_{ij} denotes the s_{ij} value in S^k.[3]

A user query is represented by a set Q of terms, called query terms, selected from the local vocabulary T. The query is represented by a tuple $(q_1, \ldots, q_n)$, where $q_i = 1$ if $t_i \in Q$, and $q_i = 0$ otherwise.

A document is represented by a set D of terms, called document terms, from the primary vocabulary A. The document terms may be associated a weight expressing the degree to which the document belongs to the class characterized by the term. Thus, a document is represented by a tuple of weights $(d_1, \ldots, d_n)$, $d_i \in [0, 1]$, where $d_i > 0$ if $d_i \in D$, and $d_i = 0$ otherwise; especially $d_i = 0$ for $i > h$.

In the situation where we want to determine the degree to which a given document matches a given query, our knowledge is represented by the following matrices:

$$\begin{aligned} E &= \{e_{ij}\}_{n \times n} && \text{(the effective similarity matrix)} \\ D &= \{d_i\}_{1 \times n} && \text{(the document)} \\ Q &= \{q_j\}_{n \times 1} && \text{(the query)} \end{aligned}$$

[2] The operator "*" may be the arithmetic multiplication operator. We shall return to a discussion of the operator in section 5.2.

[3] Viewed as fuzzy relations, the matrix E as defined by formula 1 represents the transitive max$-*$ closure of the relation represented by the matrix S.

In determining the degree of match, we first compute the matrix $R = \{r_{ij}\}_{1\times n}$ as the max$-*$ product of the matrices D and E, *i.e.*

$$R = DE \tag{2}$$

such that

$$r_j = \max_{i\in N_n}(d_i * e_{ij})$$

The value r_j is the strength of the strongest path from a document term to the term t_j; it may be seen as the degree to which the document belongs to the class represented by the term t_j.

Finally, we find the degree to which D matches Q, $f_S(Q,D)$, by the following conjunctive query evaluation:

$$f_S(Q,D) = \min_{j\in N_n} \max(1-q_j, r_j) \tag{3}$$

Since $q_j \in \{0,1\}$, $f_S(Q,D)$ is the minimum (over the query terms) of the degrees to which D belongs to the class characterized by a query term.[4]

The degree of match, as computed by the formulas 2 and 3, determines a ranking of the document collection queried. The answer to the query is the subset of the documents whose degree of match is at least a user specified threshold $\alpha\,(\in [0,1])$. Thus, the answer to the query Q posed to the document collection Ω, given the local view specified by the matrix S (developed into the matrix E), and the threshold α, is the set

$$A_S(Q,\Omega,\alpha) = \{D\in\Omega \mid f_S(Q,D)\geq\alpha\} \tag{4}$$

5 Central characteristics of the method

In this section, we discuss central characteristics of the matching method described in the preceding section. In section 5.1, we show that formula 1 computes the matrix E as expected. In section 5.2, we discuss the operator "$*$" (used in formula 1 and 2) for which we initially suggested "arithmetic multiplication". In section 5.3, we discuss relaxations of the conjunctive query interpretation represented by formula 3. In section 5.4, we suggest extensions of the representations to importance weighting of queries, and asymmetric similarities.

5.1 The effective similarity matrix

Formula 1 computes the effective similarity matrix E such that for all (i,j), e_{ij} is the maximum strength over all simple paths between t_i and t_j in the graph diagramming the initial similarity matrix S. We shall show this in the following. First, we define the length of a path as the number of (distinct) edges in the path, and by the predicate $\mathcal{M}_\ell(M)$ we express that the matrix M represents, for all (i,j), the maximum strength over all paths of length $\leq \ell$ between t_i and t_j. Thus, the task is to show that $\mathcal{M}_\infty(E)$ holds for $E = S^{n-1}$. The proof proceeds as follows.

Assume that $\mathcal{M}_k(S^k)$ holds for $k\geq 1$. Then $\mathcal{M}_{k+1}(S^{k+1})$ holds since $s_{ij}^{k+1} = \max_{u\in N_n}(s_{iu}^k * s_{uj})$ yields the maximum of s_{ij}^k and the strength of all possible extensions of the path with one edge. Since $S^1 = S$, and $\mathcal{M}_k(S^k)$ clearly holds for $k=1$, then (by induction on k) $\mathcal{M}_k(S^k)$ holds for all $k\geq 1$. Since no non-cyclic path can be longer than $n-1$, and the sequence $s_{ij}^1, s_{ij}^2, \ldots$ clearly

[4] The value of $f_S(Q,D)$ is also called the retrieval status value.

is non-decreasing, we have $\mathcal{M}_\infty(E)$ for $E = S^k$ $(k \geq n-1)$, in particular for $E = S^{n-1}$. We notice that the strength of a path can never be increased by adding a cycle to the path. Assume s is the strength of the non-cyclic path, and s' is the strength of the cycle; the strength of the first path extended with the cycle is then $s * s'$. Since $s' \leq 1$, we have $s \geq s * s'$.

5.2 The operator $*$

The operator $*$, which we initially assumed to be arithmetic multiplication, could in fact be any t-norm [3] which is bounded as follows:

$$\max(0, a+b-1) \leq a * b \leq \min(a, b) \tag{5}$$

It is easily verified, that the arithmetic multiplication operator satisfies this requirement. The bounds express, as illustrated by figure 3, the minimum and the maximum possible degree of overlap of

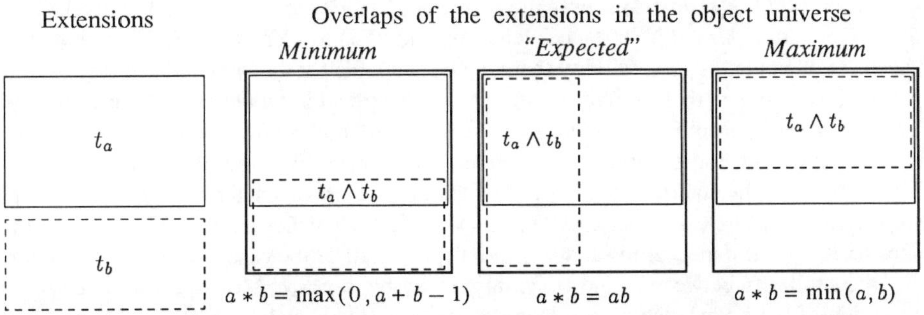

From left to right the figure illustrates the relative size of the extensions of t_a and t_b, and their minimum, "expected", and maximum overlaps in the object universe represented by the unit squares (framed by thick lines) .

Figure 3: Illustration of the bounds for the operator $*$

the extensions of two classes, characterized by the terms t_a and t_b; the arithmetic product ab is the expected degree of overlap, assuming probabilistic independence of t_a and t_b.

A family of parameterized operators, $*_p$, where p is the parameter, spanning the space bounded by formula 5, is defined by[12]

$$a *_p b = 1 - \min(1, ((1-a)^p + (1-b)^p)^{1/p}) \quad (p \geq 1) \tag{6}$$

which for $p = 1$ equals the lower limit for the $*$ operator, and for $p \to \infty$ gives the upper limit. We notice, that arithmetic multiplication is not in this family.

5.3 The conjunctive query interpretation

By formula 3, the user query is interpreted conjunctively, *i.e.* the degree to which a document matches a query is the degree to which the document satisfies the least satisfied query term. This is indeed the most "restrictive" query interpretation. The least restrictive case is represented by the disjunctive query interpretation obtained by

$$f'_S(Q, D) = \max_{j \in N_n} \min(q_j, r_j)$$

which computes the degree of match as the maximum (over the query terms) of the degrees to which the document belongs to the class characterized by a query term. By averaging the satisfactions of the query terms we may represent an interpretation between the most restrictive and the least restrictive cases. An informative averaging measure for the degree of match is [6]

$$f''_S(Q,D) = \min(U, 1 - U + L),$$

where L and U are, respectively, the lower bound $f_S(Q,D)$ and the upper bound $f'_S(Q,D)$ for the degree of match.

5.4 Representation of queries and term similarities

We have above assumed that the users do not employ importance weighting of query terms. However, the formulas for computation of the information retrieval status value, as either $f_S(Q,D)$, $f'_S(Q,D)$, or $f''_S(Q,D)$, do also satisfy the case where queries are represented by the importance weights of the terms. Use of importance weighting allows a more precise specification of information needs by the query — a possibility that may be desirable for the more "advanced" user.

Further, we assumed that similarities are symmetric strengths between terms, such that the matrix S (and therefore also E) is symmetric. However, often asymmetric strengths will provide a more correct measure for the purpose. Consider, for instance, the case of a pair of terms, t_i and t_j, where the class characterized by t_i is a subclass of the class characterized by t_j. If a document satisfies t_i, it surely also satisfies t_j. On the other hand, if it satisfies t_j, we are not sure that it also satisfies t_i; it may be expected to satisfy t_i in only, say, half of the cases. Thus, rather than representing the similarity between t_i and t_j symmetrically by, say, $s_{ij} = s_{ji} = 0.8$, we may increase the precision of the knowledge represented by, say, $s_{ij} = 1.0$, and $s_{ji} = 0.5$. This requires that we allow the matrix S (and thus the matrix E) to be asymmetric.[5] This changes the graph diagramming S to a directed graph; however, the formulas outlined above do also satisfy such cases. Adding precision to the similarity knowledge by use of asymmetric similarities is a possibility that the more "advanced" local domain expert may wish to explore.

6 Implementational issues

In this section, we discuss aspects of importance for implementations of the computational method described in section 4.1, including possible modification as suggested in section 5. In section 6.1, we analyze the complexity of the computations and suggest improvements. In section 6.2, we outline an efficient method for interfacing the matching method to traditional information retrieval systems.

6.1 Computational complexity and improvements

Complexity

Computation of the matrix E by formula 1 requires $n-2$ multiplications of two $n \times n$ matrices, and each element in a product matrix requires n basic operations for evaluation of the max operator. Thus, the matrix E is computed by formula 1 in $O(n^4)$ time, and $O(n^2)$ space.

[5] The asymmetric similarities may be seen as possibilistic necessities. Thus, s_{ij} represents $N(t_j \mid t_i)$, the degree of necessity that a document is in the class of t_j, given it is in the class of t_i.

Given the matrix E, which is computed only when the initial similarities are updated, the degree to which a document matches a query is computed by the formulas 2 and 3 in $O(n^2)$ time and $O(n^2)$ space. The computation of an answer, as defined by formula 4, requires this degree to be computed for each document in the collection queried. Thus, given the matrix E, and a query to a document collection Ω, the answer is computed in $O(n^2 ||\Omega||)$ time and $O(n^2)$ space, where $||\Omega||$ is the cardinality of the set Ω.

Improvements

Since $S^k = E$ for $k \geq n-1$, we have in particular $E = S^{2^m}$ for $m = \lceil \log_2(n-1) \rceil$. Thus, the matrix E can be computed by m matrix multiplications:

$$E = (\ldots(S^2)^2 \ldots)^2 \quad \text{(squaring m times)}$$

This method computes the matrix E in $O(n^3 \log_2 n)$ time. [6]

The storage required by the matrix E can be reduced as follows. Let α_L be a value selected as a lower bound for thresholds used in the local group, *i.e.* for the value of α applied in formula 4. If $e_{ij} < \alpha_L$, we are sure that no path from t_i to t_j can contribute to the degree of match for documents included in answers. Thus, we will still get the same answers by setting $e_{ij} = 0$ when $e_{ij} < \alpha_L$. The resulting matrix is typically sparse and may therefore be represented efficiently by avoiding storage of the zeros, for instance by the set $\{(i, j, e_{ij}) \mid e_{ij} \geq \alpha_L\}$ organized as a hash file using (i, j) as the key. Further, since E is reflexive, we can avoid storage of the diagonal elements e_{ii} $(i = 1, \ldots, n)$. Finally, if S, and therefore also E, is symmetric, we can avoid to store one of a pair e_{ij} and e_{ji}; thus, we can choose to store only elements below the diagonal, *i.e.* $\{e_{ij} \mid i > j\}$.

6.2 Interfacing to traditional host systems

In using the matching method with traditional external informations retrieval systems or host systems, the time complexity $O(n^2 ||\Omega||)$ for answering a query is too high for practical use. The collection Ω may contain millions of documents[7], and the number of documents retrieved is typically a major cost factor. In the following, we outline an efficient method for using the retrieval system offered by the host to obtain the answer as defined by $A_S(Q, \Omega, \alpha)$ in formula 4.

The idea is first to broaden the query (by use of the effective term similarities) to retrieve a set Ω' which contains the answer. The set Ω' may be seen as preliminary answer which has the recall of the final answer but typically a much lower precision. Then, at the local site, we pose the original query to Ω' to obtain $A_S(Q, \Omega', \alpha)$ as the final answer $A_S(Q, \Omega, \alpha)$. We notice that the recall of the preliminary answer is maintained in the final answer, whereas the precision is likely to be much improved. The procedure is outlined in the following.

The user query Q, here considered as the set of query terms, is converted as follows into a Boolean query Q', which only uses terms from the primary vocabulary A. For the conjunctive query interpretation (formula 3) we "expand" each query term q into the primary vocabulary by replacing q by the disjunction of those primary terms whose similarity to q is at least the threshold α. Thus, the original user query, $Q = \{q_1, \ldots, q_m\}$, is converted into the following Boolean query

[6] Some more efficient algorithms and their characteristics are described in [8].

[7] By "documents" we mean here the document surrogates which are actually stored at the host.

to be used as the broadened query:

$$Q' = Q_1 \wedge \cdots \wedge Q_m \tag{7}$$

such that, for $i = 1, ..., m$,

$$Q_i = (c_1 \vee \cdots \vee c_{n_i})$$

where

$$\{c_1, \ldots, c_{n_i}\} = A_i = \{t \in A \mid e(t, q_i) \geq \alpha\}$$

The set A_i may be obtained as follows from the matrix E. Let k be the index of q_i in the local vocabulary T, *i.e.* $t_k = q_i$. Then A_i is obtained from the k'th column, row 1 through h (since $A = \{t_1, \ldots, t_h\}$), by

$$A_i = \{t_j \in T \mid j \leq h \wedge e_{jk} \geq \alpha\}$$

The constructed Boolean query Q' is given the syntactic form required by specific query language supported by the information retrieval system. In this form, it is submitted to retrieve the set Ω' to the local site, where $A_S(Q, \Omega', \alpha)$ is computed.

We notice, that in situations where both queries and documents are unweighted, the set Ω' retrieved by Q' is in fact the final answer $A_S(Q, \Omega, \alpha)$. In such cases, formula 7 simply defines the translation of the user query, using the local vocabulary, to the equivalent Boolean query, using only the primary vocabulary maintained at the host.

7 Conclusion

We have shown how an official vocabulary for information retrieval may be tailored locally to an end-user environment. The knowledge about the similarity between the user terms and the official (primary) terms is maintained in a local knowledge base by a local domain expert. We have outlined a method for matching documents to user queries in computation of answers to such queries. We have shown some central characteristics of the matching method, and discussed computational improvements for an efficient implementation of the method.

References

[1] M.J. Bates. An exploratory paradigm for online information retrieval. In: B.C. Brookes (ed), *Intelligent Information Systems for the Information Society.* North-Holland: Elsevier Science Publishers B.V., 1986, pp. 91–99.

[2] J.C. Bezdek, G. Biswas, and L. Huang. Transitive closures of fuzzy thesauri for information-retrieval systems. *International Journal of Man-Machine Studies* 25:343–356 (1986).

[3] P.P. Bonissone. Summarizing and propagating uncertain information with triangular norms. *International Journal of Approximate Reasoning* 1:71–101 (1987).

[4] G.W. Furnas, T.K. Landauer, L.M. Gomez, and S.T. Dumais. The vocabulary problem in human-system communication. *Communications of the ACM* 30:964–971 (1987).

[5] T.R. Girill. Online access aids for documentation: a bibliographic outline. *SIGIR Forum* 18(2-4):24-27 (1985).

[6] B. Kosko. Fuzzy knowledge combination. *International Journal of Intelligent Systems* 1:293–320 (1986).

[7] H.L. Larsen. Knowledge representation in IRIS, an Information Retrieval Intermediary System. *Proceedings of the 7'th International Workshop on Expert Systems & their Application (AVIGNON'87)*, Avignon, France, 1987, Vol. 1, pp. 529–548.

[8] H.L. Larsen, and R.R. Yager. Efficient computation of transitive closures. *Fuzzy Sets and Systems* (To appear).

[9] H. Mili, and R. Rada. Merging thesauri: principles and evaluation. *IEEE Transactions on Pattern Analysis and Machine Intelligence* 10(2):204-220 (1988).

[10] G.S. Pedersen, and H.L. Larsen. The design of an information retrieval assistant system. *Proceedings of the 4th Annual ESPRIT Conference (ESPRIT'87)*, Brussels, 1987. North- Holland, 1987, pp. 688-699.

[11] Y.-C. Wang, J. Vandendorpe, and M. Evens. Relational thesauri in information retrieval. *Journal of the American Society of Information Science* 36(1):15-27 (1985).

[12] R.R. Yager. On a general class of fuzzy connection. *Fuzzy Sets and Systems* 3:235-242 (1980).

3

GROUP DECISION MAKING UNDER FUZZINESS

MEANS AND SOCIAL WELFARE FUNCTION IN FUZZY BINARY RELATION SPACES

Sergei Ovchinnikov
San Francisco State University
San Francisco, CA 94132, U.S.A.

ABSTRACT

An approach to the social choice problem based on multiple valued logic is suggested. It is assumed that individual preferences are classical binary relations, whereas the social preference is a fuzzy binary relation. The role of Lukasiewicz logic in modelling fuzzy transitive relations is discussed. We argue that this logic is a natural choice for constructing fuzzy social preferences. The paper is concerned with the Arrow's theory of social welfare functions. Standard conditions of Unrestricted Domain, Pareto, Independence of Irrelevant Alternatives, and Nondictatorship are employed. We show that certain means in fuzzy binary relation spaces are social welfare functions satisfying conditions of the classical Arrow's Possibility Theorem. It is also shown that collective choice rules given by means are completely characterized by the conditions of Unrestricted Domain, Anonymity, Neutrality, and Positive Responsiveness.

Keywords: Decision Making, Multiple-Valued Logic, Fuzzy Relations

1. INTRODUCTION

In this paper we are concerned with Arrow's approach to social choice problem based on his theory of social welfare functions. Main result in this area is the Arrow's Possibility Theorem (see Arrow [1], or, for example, Sen's monograph [13]).

Let $A = \{a, b, c, \ldots\}$ be a finite set of alternative social states, and $N = \{1, 2, 3, \ldots, n\}$ be a set of n persons. We assume that A has at least three elements. The preference relation of the *i*th individual is an ordering R_i, i.e., R_i is a reflexive, complete, and transitive binary relation on A. Let R refer to the social preference relation. A *collective choice rule* is a function F such that for any n-tuple $< R_1, \ldots, R_n >$ of individual preferences a social preference relation R is determined,

$$R = F(R_1, \ldots, R_n).$$

We are concerned here only with *decisive* collective choice rules, i.e., R is assumed to be a reflexive and complete binary relation. If, in addition, the social preference R is an ordering, F is called a *social welfare function* (SWF).

For any preference relation R (individual or social), a *strict* preference relation P and an

J. Kacprzyk and M. Fedrizzi (eds.), Multiperson Decision Making Using Fuzzy Sets and Possibility Theory, 143–154.

indifference relation I are defined by

$$aPb \leftrightarrow aRb \;\&\; \sim bRa$$

and

$$aIb \leftrightarrow aRb \;\&\; aRb,$$

for all a and b in A.

Particular collective choice rules are usually described in the social choice theory by means of their properties. Some very well known properties are listed below.

Condition U (*Unrestricted Domain*): F is defined for all logically possible combinations of individual preferences.

Condition P (*Pareto Principle*): $[(\forall i)\ aP_ib] \rightarrow aPb$, for all a and b in A.

Condition I (*Independence of Irrelevant Alternatives*): Suppose S is a subset of A. For any binary relation R on A we denote the restriction of R on S by R_S. Let R and R′ be social preferences determined by F corresponding respetively to n-tuples $< R_1, ..., R_n >$ and $< R'_1, ..., R'_n >$. If, for all i, $(R_i)_S = (R'_i)_S$, then $R_S = R'_S$.

Condition D (*Nondictatorship*): There is no individual i such that for every element in the domain of F

$$aP_ib \rightarrow aPb$$

for all a and b in A.

The classical Arrow's Possibility Theorem [1] states that there is no SWF satisfying conditions U, P, I, and D.

In this paper, we employ the same conditions on SWFs, but assume that social preferences are models of a binary predicate in the multiple valued logic of Lukasiewicz. The individual preferences are still assumed to be usual binary relations.

From the formal logic point of view, individual and social preferences are interpretations of a binary predicate $\mathbb{R}$ satisfying conditions

$(\forall x)\ \mathbb{R}(x,x)$ *reflexivity*

$(\forall x)(\forall y)\ \mathbb{R}(x,y) \vee \mathbb{R}(y,x)$ *completeness*

$(\forall x)(\forall y)(\forall z)\ (\mathbb{R}(x,y) \wedge \mathbb{R}(y,z)) \rightarrow \mathbb{R}(x,z)$ *transitivity*

In the classical binary logic, any interpretation (model) of $\mathbb{R}$ is an ordering. In a multiple valued logic, an interpretation of $\mathbb{R}$ is a fuzzy binary relation on A, i.e., a function R(a, b) defined on A×A with values in [0, 1]. Properties of R depend on interpretations of propositional connectives $\wedge$, $\vee$, and $\rightarrow$, and the universal quantifier $\forall$.

The paper is organized as follows:

A brief overview of operations on the unit interval [0, 1] is presented in section 2. We define triangular norms (t−norms), triangular conorms (t−conorms), and negation functions, and formulate their basic properties. These functions are used as interpretations for propositional connectives $\wedge$, $\vee$, and $\rightarrow$. We also introduce generalized means in section 2.

In section 3, a first-order language for preference relations is described. We also introduce interpretations for propositional connectives based on t−norms and negation functions. Properties of fuzzy preference relations are established. It is shown that the classical system of Lukasiewicz interpretations for logical connectives is a natural choice for modelling fuzzy social preference relations.

Our approach to the Arrow's theory of social welfare functions is presented in section 4.

The only new assumption is that the society uses Lukasiewicz logic for modelling preferences. We prove that certain means in fuzzy binary relation spaces are SWFs satisfying conditions U, P, I, and D.

Finally, in section 4, we provide a complete characterization of fuzzy collective choice rules given by means in terms of conditions of Unrestricted Domain, Anonymity, Neutrality, and Positive Responsiveness.

2. OPERATIONS ON THE UNIT INTERVAL

This section contains a brief overview of t−norms, t−conorms, negation functions, and means. A good overview of t−norms and conorms is found in [12]. Negation functions and their properties are introduced in [15]. Proofs are found in [12] , [15], and [10].

A *t−norm* T is defined as a function T: $[0,1]\times[0,1] \rightarrow [0,1]$ satisfying the following properties:
(i) $T(\alpha,1) = \alpha$,
(ii) $T(\alpha,\beta) \leq T(\gamma,\delta)$, if $\alpha \leq \gamma$ and $\beta \leq \delta$,
(iii) $T(\alpha,\beta) = T(\beta,\alpha)$, and
(iv) $T(\alpha,T(\beta,\gamma)) = T(T(\alpha,\beta),\gamma)$,
for all α, β, γ, and δ in $[0,1]$.

An *Archimedian* t−norm is a t−norm satisfying
(v) $T(\alpha,\alpha) < \alpha$, for all α in $(0,1)$.

Let $\mathbb{R}^+ = [0,+\infty]$. A t−norm is continuous and Archimedian if and only if

$$T(\alpha,\beta) = f(g(\alpha) + g(\beta))$$

where

a) g is a continuous and strictly decreasing function from $[0,1]$ to $\mathbb{R}^+$, such that $g(1) = 0$,

b) f is a continuous from $\mathbb{R}^+$ onto $[0,1]$, such that $f(\alpha) = g^{-1}(\alpha)$ on $[0,g(0)]$, and $f(\alpha) = 0$ for all $\alpha \geq g(0)$, or, equivalently,

$$T(\alpha,\beta) = g^{-1}(\min(g(\alpha)+g(\beta)), g(0))$$

We say that a t−norm T has *zero divisors* if it satisfies
(vi) $T(\alpha,\beta) = 0$ for some positive α and β.

A t−norm T is *strict* if it is continuous and strictly increasing in each place. A strict t−norm has no zero divisors; it is always Archimedian. A 'canonical' example of a strict t−norm is the product

$$\Pi(\alpha,\beta) = \alpha\cdot\beta.$$

If T is a continuous Archimedian t−norm, then it has zero divisors if and only if $g(0) < +\infty$.
A 'canonical' example of a t−norm with zero divisors is given by the Lukasiewicz t−norm

$$W(\alpha,\beta) = \max\{\alpha + \beta - 1, 0\}.$$

We call a strictly increasing function ϕ from the unit interval onto itself an *automorphism* of the unit interval. Any automorphism of the unit interval is a continuous function satisfying boundary conditions $\phi(0) = 0$ and $\phi(1) = 1$.

Theorem 2.1. A t−norm T is a continuous Archimedian t−norm with zero divisors if and only if there exists an automorphism ϕ of the unit interval $[0,1]$, such that

$$T(\alpha,\beta) = W^{\phi}(\alpha,\beta) = \phi^{-1}(W(\phi(\alpha),\phi(\beta))).$$

The automorphism ϕ from Theorem 2.1 is a *generator* of T.

A *negation* N is defined as a strictly decreasing function N: [0,1] → [0,1] satisfying

$$N(N(\alpha)) = \alpha, \text{ for all } \alpha \in [0,1].$$

Thus defined negation is a continuous function satisfying boundary conditions $N(0) = 1$ and $N(1) = 0$.

A standard example of a negation is given by $N(\alpha) = 1 - \alpha$.

<u>Theorem</u> <u>2.2.</u> N is a negation function if and only if there exists an automorphism ϕ of the unit interval, such that

$$N(\alpha) = N^{\phi}(\alpha) = \phi^{-1}(1 - \phi(\alpha)).$$

The automorphism ϕ from Theorem 2.2 is a *generator* of N.

Suppose a t−norm T and a negation function N are given. Then the function S given by

$$S(\alpha,\beta) = N(T(N(\alpha), N(\beta)))$$

is the *t−conorm* of T. For example, if N is the standard negation $N(\alpha) = 1 - \alpha$, the t−conorm of W is given by

$$W^*(\alpha,\beta) = \min\{\alpha + \beta, 1\}.$$

A triple < T, S, N > is called a De Morgan triple [6]. We will use functions T, S, and N as interpretations for propositional connectives $\wedge$, $\vee$, and $\sim$, respectively.

For a given automorphism ϕ of the unit interval the associative mean of n numbers α_1, ..., α_n in [0, 1] is defined by ([5])

$$\mathbb{M}_{\phi}(\alpha_1, ..., \alpha_n) = \phi^{-1}\Big(\frac{1}{n}\sum_{k=1}^{n}\phi(\alpha_k)\Big).$$

For example, when ϕ is the identity automorphism, we have the standard arithmetic mean; for $\phi(\alpha) = \alpha^r$ we have an *elementary* mean

$$\mathbb{M}_r(\alpha_1, ..., \alpha_n) = \Big(\frac{1}{n}\sum_{k=1}^{n}\alpha_k^r\Big)^{1/r}.$$

3. FUZZY PREFERENCE RELATIONS

Let L be the usual first order language with identity based on the following symbols:

a) variables − x, y, z, u, w, ... ;
b) n-ary function symbols and n-ary predicate symbols;
c) propositional connectives $\sim$, $\wedge$;
d) quantifiers $\forall$ and $\exists$.

The only *logical* predicate symbol is =. All other function and predicate symbols are *nonlogical* symbols. Formulas in L are constructed in the usual way. We write $\mathbb{F}(x,y,z,...)$ to indicate that the free variables of the formula $\mathbb{F}$ are among the variables x, y, z, ... Propositional connectives $\vee$ and $\to$ are defined by means of $\sim$ and $\wedge$ as follows:

$\mathbb{A} \vee \mathbb{B}$ is an abbrevation of $\sim((\sim\mathbb{A}) \wedge (\sim\mathbb{B}))$;

$\mathbb{A} \to \mathbb{B}$ is an abbrevation of $(\sim\mathbb{A}) \vee \mathbb{B}$.

We also introduce a nonlogical predicate symbol $\mathbb{R}$ and nonlogical axioms:

i) $(\forall x)\ \mathbb{R}(x,x)$ *reflexivity*
ii) $(\forall x)(\forall y)\ (x \neq y) \to \mathbb{R}(x,y) \vee \mathbb{R}(y,x)$ *completeness*

iii) $(\forall x)(\forall y)(\forall z)\ (x\mathbb{R}y \wedge y\mathbb{R}z) \rightarrow x\mathbb{R}z$ *transitivity*

Suppose that a De Morgan triple $<T, S, N>$ is given. We use the following interpretations for propositional connectives:

$$\alpha \wedge \beta = T(\alpha, \beta), \text{ and}$$

$$\sim\alpha = N(\alpha)$$

for all $\alpha, \beta \in [0,1]$.
Then, in accordance with our definitions,

$$\alpha \vee \beta = N(T(N(\alpha), N(\beta))) = S(\alpha, \beta)$$

and

$$\alpha \rightarrow \beta = S(N(\alpha), \beta) = N(T(\alpha, N(\beta))),$$

for all $\alpha, \beta \in [0,1]$.

Since t−norms and conorms are associative functions, we will not use parenthesis in expressions like $\alpha \wedge \beta \wedge \gamma$ or $\alpha \vee \beta \vee \gamma$.

The function $\forall$ is defined for any non-empty set $X \subseteq [0,1]$ by $\forall X = \mathit{inf}X$.

A *interpretation* for L is a triple $<A, Id, R>$, where A is a non-empty set, Id is a characteristic function of the identity relation on A, and R is a function $A \times A \rightarrow [0,1]$.

For each formula $\mathbb{F}(x, y, z, ...)$ and interpretation of x, y, z, ... as elements a, b, c, ... in A, we define a value function $F(a,b,c,...)$ with codomain $[0,1]$ by induction on the formulas using interpretations for propositional connectives introduced above. For instance, $R(a,b)$ is simply the function in the triple $<A, Id, R>$. If, say, $\mathbb{F}(x,y)$ is a formula such that $F(a,b)$ is defined for all a and b in A, then define

$$(\sim F)(a,b) = \sim(F(a,b)),$$

$$(F \wedge G)(a,b) = F(a,b) \wedge G(a,b),$$

$$(\forall y\ F)(a) = \forall\{F(a,b) \mid b \in A\} = \inf\{F(a,b) \mid b \in A\}, \text{ etc.}$$

We say that $F(a, b, c, ...)$ is *true* if $F(a, b, c, ...) = 1$ (designated value is 1, cf. [11]); in particular, our axioms and properties are assumed to be true formulas.

We call $R(a,b)$ introduced above a *fuzzy preference relation* on the set A. We assume that this relation satisfies the following conditions which are interpretations of the nonlogical axioms i), ii), and iii):

$$\forall\{R(a,a) \mid a \in A\} = 1,$$

$$\forall\{R(a,b) \vee R(b,a) \mid a \neq b,\ a,b \in A\} = 1, \text{ and}$$

$$\forall\{(R(a,b) \wedge R(b,c)) \rightarrow R(a,c) \mid a,b,c \in A\} = 1,$$

or, equivalently,

$R(a,a) = 1,$ *reflexivity*

$R(a,b) \vee R(b,a) = 1$ *completeness*

$[(R(a,b) \wedge R(b,c)) \rightarrow R(a,c)] = 1$ *transitivity*

for all a, b, and c in A ($a \neq b$ in the second condition).

Since

$$R(a,b) \wedge R(b,c)) \rightarrow R(a,c) = \sim(R(a,b) \wedge R(b,c)) \vee R(a,c) = \sim(R(a,b) \wedge R(b,c) \wedge \sim R(a,c))$$

the transitivity condition is equivalent to

$$[R(a,b) \wedge R(b,c) \wedge \sim R(a,c)] = 0 \qquad \textit{transitivity}$$

We define a fuzzy *indifference relation* I on A by

$$I(a,b) = R(a,b) \wedge R(b,a),$$

for all a and b in A, and a fuzzy *strict preference relation* P on A by

$$P(a,b) = R(a,b) \wedge (\sim R(b,a)),$$

for all a and b in A (compare with the definitions for binary relations in Introduction).

The following theorem shows that, in the case of strict t−norms, the only fuzzy weak preference relations are classical preference relations.

Theorem 3.1. Suppose T in <T, S, N> is a strict t−norm. Let A be a set and R a reflexive transitive fuzzy binary relation on A. Then $R(a,b) \in \{0,1\}$.

Proof. By transitivity of R, $R(a,b) \wedge R(b,b) \wedge \sim R(a,b) = 0$ (substituting c = b) for all a and b in A. Since R is a reflexive fuzzy binary relation, $R(b,b) = 1$. Therefore, $R(a,b) \wedge \sim R(a,b) = 0$, or $T(R(a,b), \sim R(a,b)) = 0$. Since T has no zero divisors, we conclude that either $R(a,b) = 0$, or $R(a,b) = 1$. □

Naturally, in the framework of fuzzy social preferences, one would want to have a "sufficiently large" set of such relations; at least we should demand that, for any $\alpha \in [0,1]$ and a, b $\in$ A, there is a fuzzy relation R such that $R(a,b) = \alpha$. It is obvious from the proof of Theorem 3.1, that the following condition

$$(*) \qquad \alpha \wedge \sim\alpha = 0, \quad \text{for all } \alpha \in [0,1]$$

is a necessary condition for the existence of reflexive transitive fuzzy binary relations on A. Of course, this condition cannot be satisfied in the case of strict t−norms.

The last condition can be written in the following equivalent forms

$$\sim(\alpha \wedge \sim\alpha) = 1 \qquad \textit{Law of Contradiction}$$

and

$$\alpha \vee \sim\alpha = 1 \qquad \textit{Law of the Excluded Middle}$$

for all $\alpha \in [0,1]$.

In the rest of the paper we shall assume that T and N are such functions that condition (∗) is satisfied. Then T is obviously a t−norm with zero divisors. We also assume that T is a continuous Archimedian t−norm. Then, by Theorem 2.1, there is an automorphism ϕ such that

$$T(\alpha,\beta) = W^{\phi}(\alpha,\beta) = \phi^{-1}(\max\{\phi(\alpha) + \phi(\beta) - 1, 0\}).$$

Suppose that N is generated by the same automorphism ϕ. Then elements of the De Morgan triple < T, S, N > have the following representations

$$T(\alpha,\beta) = \phi^{-1}(\max\{\phi(\alpha) + \phi(\beta) - 1, 0\}),$$

$$S(\alpha,\beta) = \phi^{-1}(\min\{\phi(\alpha) + \phi(\beta), 1\}),$$

$$N(\alpha) = \phi^{-1}(1 - \phi(\alpha)).$$

Lemma 3.1. $\alpha \wedge (\sim\beta) = 0$ iff $\alpha \leq \beta$.

Proof. By definition,

$$\alpha \wedge (\sim\beta) = T(\alpha, N(\beta)) = \phi^{-1}(\max\{\phi(\alpha) + \phi(N(\beta)) - 1, 0\}) =$$

$$\phi^{-1}(\max\{\phi(\alpha) + 1 - \phi(\beta) - 1, 0\}) = \phi^{-1}(\max\{\phi(\alpha) - \phi(\beta), 0\}).$$

The assertion of the proposition follows immediately.

□

In particular, $\alpha \wedge (\sim\alpha) = 0$, for all $\alpha \in [0,1]$. Therefore, the condition $(*)$ is satisfied along with the Law of Contradiction and the Law of the Excluded Middle.

We will need a different representation for the transitivity property of fuzzy binary relations.

Theorem 3.2. The transitivity property is equivalent to

$$R(a,b) \wedge R(b,c) \leq R(a,c)$$

for all a, b, and c in A.

Proof. $1 = [(R(a,b) \wedge R(b,c)) \rightarrow R(a,c)] = \sim(R(a,b) \wedge R(b,c) \wedge (\sim R(a,c)))$ is equivalent to $R(a,b) \wedge R(b,c) \wedge (\sim R(a,c)) = 0$. By Proposition 3.1, it is equivalent to $R(a,b) \wedge R(b,c) \leq R(a,c)$.

□

Using representations for elements of <T, S, N>, we write properties of a fuzzy preference relation R as follows:

$R(a,a) = 1$ *reflexivity*

$\phi(R(a,b)) + \phi(R(b,a)) \geq 1$ *completeness*

$\phi(R(a,b)) + \phi(R(b,c)) - 1 \leq \phi(R(a,c))$ *transitivity*

for all a, b, and c in A.

Let us consider a special case when ϕ is the identity automorphism of $[0,1]$, i.e., $\phi(\alpha) = \alpha$, for all $\alpha \in [0,1]$. Then we have the following interpretations for propositional connectives:

$\alpha \wedge \beta = \max\{\alpha + \beta - 1, 0\}$,

$\sim\alpha = 1 - \alpha$,

$\alpha \vee \beta = \min\{\alpha + \beta, 1\}$, and

$\alpha \rightarrow \beta = \min\{1 - \alpha + \beta, 1\}$.

These are Lukasiewicz's interpretations for propositional connectives (see the discussion in [4] in connection with definitions in [7] and [11]). Then a fuzzy preference relation R satisfes properties

$R(a,a) = 1$ *reflexivity*

$R(a,b) + R(b,a) \geq 1$ *completeness*

$R(a,b) + R(b,c) - 1 \leq R(a,c)$ *transitivity*

for all a, b, and c in A.

Suppose R is an interpretation of $\mathbb{R}$ based on <T, S, N>. Then $\phi(R)$ is the interpretation of $\mathbb{R}$ based on the standard Lukasiewicz logic. Therefore, Lukasiewicz interpretations seem to be a natural choice for the first step towards developing a general theory of fuzzy preference relations. In what follows, we assume that fuzzy preference relations are interpretations of $\mathbb{R}$ in the framework of Lukasiewicz logic.

4. FUZZY SOCIAL PREFERENCES IN THE ARROW'S FRAMEWORK

In this section we show that there exist fuzzy social welfare functions satisfying conditions U, P, I, and D. First, we reformulate these conditions, taking in account that social preferences are fuzzy binary relations.

Condition U (*Unrestricted Domain*): F is defined for all logically possible combinations of individual preferences.

Condition P (*Pareto Principle*): $[(\forall i)\ P_i(a,b) = 1] \rightarrow [P(a,b) = 1]$, for all a and b in A.

Condition I (*Independence of Irrelevant Alternatives*): Suppose S is a subset of A. For any fuzzy binary relation R on A we denote the restriction of R on S by R_S. Let R and R′ be social preferences determined by F corresponding respetively to n-tuples $< R_1, ..., R_n >$ and $< R'_1, ..., R'_n >$. If, for all i, $(R_i)_S = (R'_i)_S$, then $R_S = R'_S$.

Condition D (*Nondictatorship*): There is no individual i such that for every element in the domain of F

$$[P_i(a,b) = 1] \rightarrow [P(a,b) = 1]$$

for all a and b in A.

Note, that these conditions are exactly the standard ones. We simply use characteristic functions of individual preferences instead of preferences themselves.

Let ϕ be an automorphism of the unit interval. For any given n-tuple $< R_1, ..., R_n >$ of individual preferences we define a fuzzy social binary relation R by

$$R(a,b) = M_\phi(R_1(a,b), ..., R_n(a,b)) = \phi^{-1}\Big(\frac{1}{n}\sum_{k=1}^{n}\phi(R_k(a,b))\Big).$$

For technical reasons, we prefer to work with the inverse automorphism $\psi = \phi^{-1}$ instead of ϕ itself. Then

$$R(a,b) = \psi\Big(\frac{1}{n}\sum_{k=1}^{n}\psi^{-1}(R_k(a,b))\Big),$$

or, since, for all k, $R_k(a,b) \in \{0,1\}$,

$$R(a,b) = \psi\Big(\frac{1}{n}\sum_{k=1}^{n}R_k(a,b)\Big). \tag{4.1}$$

We define a *ψ-transform* R^ψ of a fuzzy binary relation R by $R^\psi(a,b) = \psi^{-1}(R(a,b))$. First, we prove the following

Lemma 4.1. The ψ-transform R^ψ of a fuzzy binary relation R defined by (4.1) is a fuzzy preference relation.

Proof. We must proof that R^ψ is a reflexive, complete, and transitive fuzzy binary relation. Indeed,

i) $R^\psi(a,a) = \frac{1}{n}\sum_{k=1}^{n}R_k(a,a)) = 1$, since, for all k, $R_k(a,a) = 1$, for all a in A.

ii) $R^\psi(a,b) + R^\psi(b,a) = \frac{1}{n}\sum_{k=1}^{n}R_k(a,b) + \frac{1}{n}\sum_{k=1}^{n}R_k(b,a) = \frac{1}{n}\sum_{k=1}^{n}[R_k(a,b) + R_k(b,a)] \geq 1$,

since $\max\{R_k(a,b), R_k(b,a)\} = 1$, for all k, and a and b in A.

iii) $R^\psi(a,b) + R^\psi(b,c) - 1 = \frac{1}{n}\sum_{k=1}^{n}R_k(a,b) + \frac{1}{n}\sum_{k=1}^{n}R_k(b,c) - 1 =$

$= \frac{1}{n}\sum_{k=1}^{n}[R_k(a,b) + R_k(b,c) - 1] \leq \frac{1}{n}\sum_{k=1}^{n}R_k(a,c) = R^\psi(a,c)$,

since, for all k, R_k is a transitive binary relation.

□

In the particular case of the identity automorphism, $\psi = \text{id}$, Lemma 4.1 asserts that (4.1) defines a social welfare function. Our next theorems show that there are nontrivial automorphisms ψ such that social binary relations defined by (4.1) are fuzzy preference relations.

Theorem 4.1. Suppose ψ satisfies the following conditions

1) For all α, $\beta \in [0,1]$,

$$\alpha + \beta \geq 1 \quad \text{implies} \quad \psi(\alpha) + \psi(\beta) \geq 1 \tag{4.2}$$

2) For all α, β, and γ in $[0,1]$,

$$\alpha \geq \gamma,\ \beta \geq \gamma,\ \alpha+\beta-1 \leq \gamma \quad \text{imply} \quad \psi(\alpha)+\psi(\beta)-1 \leq \psi(\gamma) \tag{4.3}$$

Then R defined by (4.1) is a fuzzy preference relation.

Proof. We must proof that R is a reflexive, complete, and transitive fuzzy binary relation. By the definition of ψ-transform, $R(a,b) = \psi(R^{\psi}(a,b))$.
i) Obviously, R is a reflexive fuzzy binary relation.
ii) Let $\alpha = R^{\psi}(a,b)$ and $\beta = R^{\psi}(b,a)$. Then $\alpha + \beta \geq 1$, since ψ-transform of R is a complete fuzzy binary relation (by Lemma 4.1). Completeness of R follows from (4.2).
iii) Let $\alpha = R^{\psi}(a,b)$, $\beta = R^{\psi}(b,c)$, and $\gamma = R^{\psi}(a,c)$. To prove transitivity, we must show that

$$\psi(\alpha)+\psi(\beta)-1 \leq \psi(\gamma) \tag{4.4}$$

If $\alpha < \gamma$ or $\beta < \gamma$, then (4.4) is trivially satisfied. Otherwise, $\alpha \geq \gamma$ and $\beta \geq \gamma$. We also have $\alpha + \beta - 1 \leq \gamma$ since ψ-transform is a transitive fuzzy relation (by Lemma 4.1). Now (4.4) follows immediately from (4.3).

□

Theorem 4.2. Suppose that R defined by (4.1) is a fuzzy preference relation for any $n \geq 2$. Then ψ satisfies conditions (4.2) and (4.3).

Proof. Since ψ is a continuous function, it suffices to prove (4.2) and (4.3) for rational values of α, β, and γ. Let a, b, and c be any three different elements in A, and R_1, R_2, and R_3 be three individual preferences such that

$$a P_1 b P_1 c, \quad b P_2 c P_2 a, \quad \text{and} \quad c P_3 a P_3 b.$$

For a given $n \geq 2$ and non-negative integers p, q, and r, such that $p+q+r = n$, we define $\lambda = p/n$, $\mu = q/n$, and $\nu = r/n$. Then $\lambda+\mu+\nu = 1$. Let us consider an n-tuple $<R'_1, \ldots, R'_n>$ such that $R'_i = R_1$, for $i \leq p$, $R'_i = R_2$, for $p+1 \leq i \leq p+q$, and $R'_i = R_3$, for $p+q+1 \leq i \leq n$. Then, from (4.1), we have

$$R(a,b) = \psi(\lambda+\nu),\ R(b,a) = \psi(\mu),\ R(b,c) = \psi(\lambda+\mu),\ \text{and}\ R(a,c) = \psi(\lambda). \tag{4.5}$$

Obviously, for any three non-negative rational numbers λ, μ, and ν, such that $\lambda+\mu+\nu = 1$, there is an n-tuple of individual preferences, such that the social preference R satisfies (4.5).

Let α and β be two rational numbers in $[0,1]$, such that $\alpha + \beta = 1$. Let $\lambda = \alpha$, $\mu = \beta$, $\nu = 0$. Then, from (4.5),

$$\psi(\alpha) + \psi(\beta) = R(a,b) + R(b,a) \geq 1,$$

since R is a complete fuzzy relation. (4.2) follows immediately from the last inequality, since ψ is an increasing function.

Let α, β, and γ be rational numbers in $[0,1]$, such that $\alpha \geq \gamma$, $\beta \geq \gamma$, and $\alpha+\beta-1 \leq \gamma$. We define $\lambda = \gamma$, $\mu = \alpha - \gamma$, $\nu = 1 - (\lambda + \mu) = 1 - \alpha$. Then, by (4.5) and transitivity of R,

$$\psi(\lambda+\nu) + \psi(\lambda+\mu) - 1 \leq \psi(\lambda),$$

or

$$\psi(\gamma + 1 - \alpha) + \psi(\alpha) - 1 \leq \psi(\gamma).$$

But $\psi(\gamma + 1 - \alpha) \geq \psi(\beta)$, since $\alpha+\beta-1 \leq \gamma$. Therefore, condition (4.3) is proven.

□

Conditions (4.2) and (4.3), as the following lemma asserts, can be written in a rather simpler form.

Lemma 4.2. Conditions (4.2) and (4.3) together are equivalent to the following system of an equation and inequlity

$$\psi(\alpha) + \psi(1 - \alpha) = 1, \tag{4.6}$$

$$\psi(\alpha + \beta) \leq \psi(\alpha) + \psi(\beta), \tag{4.7}$$

for all α, $\beta \in [0,1]$.

We assume, obviously, that $\alpha + \beta \leq 1$ in the last inequality.

Proof. i) Suppose (4.2) and (4.3) are satisfied. Let $\gamma = 0$ in (4.3). Then $\psi(\alpha) + \psi(\beta) \leq 1$ for $\alpha + \beta \leq 1$. Together with (4.2) it implies that $\psi(\alpha) + \psi(\beta) = 1$ for $\alpha + \beta = 1$. Therefore, (4.6) is satisfied.

Suppose α and β are given and $\alpha + \beta \leq 1$. Then, obviously, $\alpha + \beta \geq \alpha$, $1 - \beta \geq \alpha$, and $(\alpha + \beta) + (1 - \beta) - 1 \leq \alpha$. Then, by (4.3),

$$\psi(\alpha + \beta) + \psi(1 - \beta) - 1 \leq \psi(\alpha),$$

or

$$\psi(\alpha + \beta) \leq \psi(\alpha) + 1 - \psi(1 - \beta) = \psi(\alpha) + \psi(\beta)$$

by (4.6).

ii) Suppose now that (4.6) and (4.7) are satisfied. Then (4.2) follows from (4.6), since ψ is an increasing function.

Let α, β, and γ be numbers in $[0,1]$, such that $\alpha \geq \gamma$, $\beta \geq \gamma$, and $\alpha + \beta - 1 \leq \gamma$. Then $\alpha \leq \gamma + (1 - \beta) \leq 1$, and we have

$$\psi(\alpha) \leq \psi(\gamma + (1 - \beta)) \leq \psi(\gamma) + \psi(1 - \beta) = \psi(\gamma) + 1 - \psi(\beta),$$

by (4.7) and (4.2). Therefore, (4.3) is satisfied.

□

We summarize our results as follows:

Theorem 4.3. The fuzzy relation R defined by

$$R(a,b) = \psi\left(\frac{1}{n} \sum_{k=1}^{n} R_k(a,b)\right)$$

is a fuzzy preference relation for any n if and only if ψ satisfies conditions

$$\psi(\alpha) + \psi(1 - \alpha) = 1,$$

and

$$\psi(\alpha + \beta) \leq \psi(\alpha) + \psi(\beta),$$

for all α, $\beta \in [0,1]$.

Theorem 4.3 can be reformulated as follows:

Theorem 4.4. Formula (4.1) defines a fuzzy social welfare function if and only if ψ satisfies conditions (4.6) and (4.7).

It is easy to verify that the SWF given by (4.1) satisfies conditions U, P, I, and D. Therefore, in the framework developed in this paper, there are SWFs satisfying conditions of the Arrow's Possibility Theorem.

Obviously, the arithmetic mean is the only elementary mean generating SWFs. A nontrivial example of ψ satisfying (4.6) and (4.7) is given by

$$\psi(\alpha) = \alpha + \tau \cdot \sin(2\pi\alpha),$$

where τ is a positive number less than $1/2\pi$.

5. AN AXIOMATIZATION OF MEANS

A collective choice rule F defined by

$$R(a,b) = F(R_1, ..., R_n) = \psi(\tfrac{1}{n}\sum_{k=1}^{n} R_k(a,b))$$

has many important properties. Function F is symmetric (*anonymity*), R(a, b) is expressed in terms of $R_k(a,b)$ and the formula does not depend on particular choice of elements a and b (*neutrality*), and, finally, R(a, b) is a strictly increasing function of $R_k(a,b)$ (*positive responsiveness*). Conditions of anonymity, neutrality, and positive resonsiveness play an important role in social choice theory. Particularily, these conditions completly describe the *simple majority rule* (see May's original paper [6], and, for example, discussion in [13]). Of course, the majority rule cannot be expressed by any mean; therefore, in order to charaterize means, we must employ conditions which are stronger than standard May's conditions.

Let $R = F(R_1, ..., R_n)$ be a collective choice rule, where R_k are individual preferences and R is a reflexive and complete fuzzy binary relation. We introduce the following conditions:

Condition A (*Anonymity*): Function F is a symmetric function of individual preferences.

Condition N (*Neutrality*): Let $<R_1, ..., R_n>$ and $<R'_1, ..., R'_n>$ be two n-tuples of individual preferences such that, for all a, b, c, and d in A,

if, for all k, $R_k(a,b) = R'_k(c,d)$, then $R(a,b) = R'(c,d)$.

Condition S (*Positive Responsiveness*): If, for some a and b in A and k, individual k changes his strict preference for b over a any way in favor of a, than the value of R strictly increases.

We also assume that the condition of Unrestricted Domain holds. It can be shown that these conditions (with exception of Anonymity, which is exactly the same) are indeed stronger than classical ones; we do not prove it here.

Theorem 5.1. A fuzzy collective choice rule F satisfies conditions U, A, N, and S if and only if there exists automprphism ψ such that

$$R(a,b) = F(R_1, ..., R_n) = \psi(\tfrac{1}{n}\sum_{k=1}^{n} R_k(a,b)). \tag{5.1}$$

Proof. i) Let F be given by (5.1). Then conditions U, A, and N are obviously satisfied. Suppose that, for some k, $P_k(b,a) = 1$. Then $R_k(a,b) = 0$. Any change of R_k in favor of a yields $R_k(a,b) = 1$. Therefore, condition S is satisfied, since ψ is a strictly increasing function.

ii) Suppose now that F is a fuzzy collective choice rule satisfying conditions U, A, N, and S. For each ordered pair (a, b), F defines a function $\mathbb{F}_{ab}$ with codomain [0, 1] such that

$$R(a,b) = \mathbb{F}_{ab}(R_1, ..., R_n).$$

Let $<R_1, ..., R_n>$ and $<R'_1, ..., R'_n>$ be any two n-tuples such that, for all k, $R_k(a,b) = R'_k(a,b)$. Then, by condition N, $R(a,b) = R'(a,b)$. It means that there is a function F_{ab} such that

$$R(a,b) = F_{ab}(R_1(a,b), ..., R_n(a,b)).$$

Let (a, b) and (c, d) be two ordered pairs ($a \neq b$ and $c \neq d$). Let π be a permutation of A such that $\pi(a) = c$ and $\pi(b) = d$. For any binary relation Q we define $Q^{\pi}(x,y) = Q(\pi^{-1}(x), \pi^{-1}(y))$. Consider now two n-tuples $<R_1, ..., R_n>$ and $<R'_1, ..., R'_n>$ where $R'_k = R_k^{\pi}$. Then, for all k,

$$R'_k(c,d) = R_k^{\pi}(c,d) = R_k(a,b)$$

and, by condition N, $R(a,b) = R'(c,d)$. We have

$$F_{ab}(R_1(a,b), ..., R_n(a,b)) = R(a,b) = R'(c,d) =$$

$$F_{cd}(R'_1(c,d), ..., R'_n(c,d)) = F_{cd}(R_1(a,b), ..., R_n(a,b)).$$

Sinse condition U is assumed, we conclude that $F_{ab} = F$, for all a and b, where F is a real valued function of boolean variables. Therefore, for all a and b in A,

$$R(a,b) = F(R_1(a,b), ..., R_n(a,b)).$$

By conditions A and S, F is a strictly increasing function of N where N is the number of individuals k with $R_k(a,b) = 1$. But $N = \sum_k R_k(a,b)$. Therefore, there exists an automorphism ψ such that R is given by (5.1).

□

6. CONCLUDING REMARKS

1) It follows from our results that Arrow's axioms are consistent if the society uses Lukasiewicz logic for modelling its preferences. This result was also established in [14]. Arithemetic means of individual preferences are considered, for example, in [2], [3], and [9].

2) The arithmetic mean is essentially a pair-comparison matrix. Therefore, classical pair-comparison techniques can be applied for decision-making based on SWF's introduced in this paper.

3) From the point of view developed in the paper the main source of the "paradox" in the classical theory is the gap between axiom requirements and a model chosen for the solution. For instance, the society is treated just as another individual in the classical theory. Our approach could be possibly used for modelling "pluralistic" social choice functions.

REFERENCES

[1] K.J. Arrow, Social Choice and Individual Values (Willey, New York, 1963).

[2] J.C. Bezdek, B. Spillman, and R. Spillman, Fuzzy relation spaces for group decision theory: an application, Fuzzy Sets and Systems, 2, 1979, 5-14.

[3] J.M. Blin and A.B. Whinston, Fuzzy sets and social choice, J. of Cybernetics, 3(4), 1974, 28-36.

[4] R. Giles, Lukasiewicz logic and fuzzy set theory, Int. J. Man-Machine Studies, 8, 1976, 313-327.

[5] G.H. Hardy, J.E. Littlewood, and G. Polya, Inequalities (Cambridge, The Univ. Press,1934).

[6] E.P. Klement, Operations on fuzzy sets and fuzzy numbers related to triangular norms, in: Proc. of the 11th ISMVL, University of Oklahoma, 1981, 218-225.

[7] J. Lukasiewicz and A. Tarski, Investigations in the sentential calculus, in: A. Tarski, Logic, Semantics and Metamathematics (Oxford, Clarendon Press, 1956).

[8] K.O. May, A set of necessary and sufficient conditions for simple majority decision, Econometrica, 20, 1952, 680-684.

[9] H. Nurmi, Approaches to collective decision making with fuzzy preference relations, Fuzzy Sets and Systems, 6, 1981, 249-259.

[10] S. Ovchinnikov and M. Roubens, On strict preference relations, Fuzzy Sets and Systems, to appear.

[11] N. Recher, Many-Valued Logic (McGraw Hill, New York, 1969).

[12] B. Schweizer and A. Sclar, Probabilistic Metric Spaces (North-Holland, Amsterdam, 1983).

[13] A.K. Sen, Collective Choice and Social Welfare (Holden Day, San Francisco, 1970).

[14] H.J. Skala, Arrow's impossibility theorem: some new aspects, in: H.W. Gottinger and W. Leinfellner (eds.), Decision Theory and Social Ethics (Reidel, Dordrecht, 1978).

[15] E. Trillas, Sobre funciones de negacion en la teoria de conjuntos difusos, Stochastica, vol. III, no. 1, 1979, 47-84.

AGGREGATION OF FUZZY PREFERENCES

C. R. Barrett and Prasanta K. Pattanaik

Department of Economics, University of Birmingham,
Birmingham B15 2TT, United Kingdom

Abstract: The paper discusses the problem of aggregating fuzzy individual preferences into fuzzy group preferences. The problem is posed in the 'ordinal' fuzzy set theoretic framework. The structure of aggregation rules satisfying unanimity and independence is explored, given the assumption of maxmin transitivity for group as well as individual preferences. Characterizations are given for certain types of aggregation rules based on consensus pooling and for generalised rank-based rules.

Key words: fuzzy aggregation rule; consensus pooling; and generalised rank-based rules.

1. INTRODUCTION

In this paper, we explore certain problems relating to the aggregation of fuzzy individual preferences into fuzzy group preferences. While most of the earlier papers[1] dealing with this problem have used the more familiar 'cardinal' fuzzy set theory, we pose the problem within the 'ordinal' framework of a version of Goguen's (1967) L-fuzzy set theory. Within this framework, we analyse the structure of different classes of aggregation rules satisfying the properties of unanimity and independence (see Section 2 for definitions of these properties).

We start by analysing the structure of aggregation rules satisfying unanimity and independence, when individual and group preferences are assumed to satisfy maxmin transitivity. We show that, given the assumption of maxmin transitivity, under such aggregation rules some individuals will enjoy a type of power which may be considered undesirable in some ways. Thus, our first result is reminiscent of the well known impossibility results of Arrow (1963) and Gibbard (1969) in exact social choice theory. However, unlike some earlier works dealing with aggregation of fuzzy individual preferences (see, for example, Barrett, Pattanaik and Salles (1986)), we do

---

1. See, among others, Blin and Whinston (1973), Blin (1974), Nurmi (1981), Dimitrov (1983), Leclerc (1984), Tanino (1984), Barrett, Pattanaik and Salles (1986) and Subramanian (1988).

J. Kacprzyk and M. Fedrizzi (eds.), Multiperson Decision Making Using Fuzzy Sets and Possibility Theory, 155–162.

not concentrate on impossibility results. Instead, we seek to characterize some important subclasses of the class of aggregation rules satisfying unanimity and independence. One such subclass is the class of aggregation rules based on consensus pooling. These are characterized, given the assumption of maxmin transitivity for group as well as individual preferences. Because of the inherent indecisiveness of consensus pooling and because, though transitivity conditions are widely applied, the intuitive basis for their application in this context is not altogether clear[2], it is worth considering alternatives to consensus pooling. Therefore, we go on to consider a wider class of aggregation rules satisfying unanimity and independence – the class of what we call the generalised rank-based rules. We provide a characterization of these aggregation rules.

Section 2 of the paper gives the notation and definitions. We prove our main results in Sections 3 and 4. We conclude in Section 5.

2 NOTATION AND DEFINITIONS

Let X be a set of alternatives ($|X| \geq 3$). Let L ($|L| \geq 2$) be a finite set, and let $\succcurlyeq$ be an exact antisymmetric ordering over L, $\succ$ being the asymmetric factor of $\succcurlyeq$. Specifically, let $L = \{d_1, \ldots, d_m\}$, where $d_m \succ d_{m-1} \succ \cdots \succ d_1$. Let L be the index set for fuzzy binary relations over X. Thus, a <u>fuzzy binary relation</u> over X is defined to be a function $P: X^2 \to L$.

L is to be interpreted as the set of possible 'degrees of belonging': for all $d, d' \in L$, if $d \succ d'$, then, intuitively, d represents a higher degree of belonging than d'. d_1 and d_m are to be interpreted, respectively, as 'definitely not belonging' and 'definitely belonging'. In choosing to use a completely ordered set L as the set of 'degrees of belonging', we have adopted a version of Goguen's (1967) ordinal L–fuzzy set theoretic framework rather than the more usual cardinal framework of fuzzy set theory.[3]

<u>Notation 2.1</u> Ω denotes the set of fuzzy binary relations over X satisfying:

(2.1) (<u>irreflexivity</u>) for all $P \in \Omega$ and all $x \in X$, $P(x, x) = 0$;

and (2.2) (<u>antisymmetry</u>) for all $P \in \Omega$ and for all distinct $x, y \in X$, $P(x, y) = d_m$ implies $P(y, x) = d_1$.

$\bar{\Omega}$ will denote the set of all $P \in \Omega$ which satisfy:

(2.3) (<u>max-min transitivity</u>) for all $x, y, z \in X$, $P(x, z) \geq \min[P(x, y), P(y, z)]$.

Ω is to be interpreted as the set of fuzzy <u>strict</u> preference relations over X. (2.1) is then the condition that an alternative is clearly not preferred to itself, and (2.2) the condition that, when one alternative is clearly preferred to another, the latter alternative is clearly not preferred to the former. Even within the ordinal framework of Goguen's L–fuzzy set theoretic framework used here, a variety of

---

2. See Dasgupta and Deb (1988).

3. For an alternative ordinally fuzzy framework, see Basu, Deb and Pattanaik (1988).

transitivity conditions are available. We have introduced the frequently used condition of max–min transitivity which will sometimes be imposed on fuzzy strict preference in our analysis.

Let $N = \{1, \ldots, n\}$ index a finite set of individuals.

Definition 2.2 A fuzzy aggregation rule (FAR) is a function $F: \hat{\Omega}^n \to \Omega$, where $\Phi \neq \hat{\Omega} \subseteq \Omega$

For ease of notation, let a given FAR map $(P_1, \ldots, P_n)$ to P, $(P'_1, \ldots, P'_n)$ to P', and so on. We will interpret P_i as the fuzzy preference relation of the i–th individual ($i \in N$), P as the corresponding preferences of the society, etc.

A non–empty subset of N will be referred to as a coalition. We will be concerned with the 'power' of coalitions.

Definition 2.3 Let the FAR be given. Consider any distinct $x, y \in X$ and any $d, d' \in L$.

(2.3.1) A coalition C is upper [lower] d–decisive with respect to (x, y) iff, for all $(P_1, \ldots, P_n) \in \hat{\Omega}^n$, $P_i(x, y) \geqslant d$ $[d \geqslant P_i(x, y)]$ for all $i \in C$ implies $P(x, y) \geqslant d$ $[d \geqslant P(x, y)]$.

(2.3.2) A coalition C is (d, d')–decisive with respect to (x, y) iff, for all $(P_1, \ldots, P_n) \in \hat{\Omega}^n$, $P_i(x, y) = d$ for all $i \in C$ and $P_i(x, y) = d'$ for all $i \in N - C$ implies $P(x, y) = d$.

We say that a coalition C is upper [lower] d–decisive iff C is upper [lower] d–decisive with respect to all $(a, b) \in X^2$ $(a \neq b)$. Again, we say that C is d–decisive iff C is both upper and lower d–decisive, and C is (d, d')–decisive iff C is (d, d')–decisive with respect to all $(a, b) \in X^2$ $(a \neq b)$.

We now introduce some properties of an FAR.

Definition 2.4 An FAR satisfies:

(2.4.1) unanimity iff N is decisive;

(2.4.2) anonymity iff for all distinct $x, y \in X$, all $d, d' \in L$, and all coalitions C and C', if $|C| = |C'|$, and C is upper [lower] d–decisive with respect to (x, y) [alternatively, (d, d')–decisive with respect to (x, y)], then C' is upper [lower] d–decisive with respect to (x, y) [alternatively, (d, d')–decisive with respect to (x, y)];

(2.4.3) monotonicity iff, for all distinct $x, y \in X$ and all $d, d' \in L$ such that $d' \succ d$, if a coalition C is lower d–decisive with respect to (x, y), then C is lower d'–decisive with respect to (x, y);

(2.4.4) tightness iff, for all distinct $x, y \in X$ and for all $d_j \in L - \{d_m\}$, if a coalition C is not lower d_j–decisive with respect to (x, y), then N – C is upper d_{j+1}–decisive with respect to (x, y);

and (2.4.5) independence iff, for all $x, y \in X$ and for all $(P'_1, \ldots, P'_n)$, $(P_1, \ldots, P_n) \in \hat{\Omega}^n$, if $[P_i(x, y) = P'_i(x, y)$ and $P_i(y, x) = P'_i(y, x)]$ for all $i \in N$, then $P(x, y) = P'(x, y)$.

Remark 2.5 Unanimity is the condition that society's preference for one alternative over another is in a minimal sense representative of individual preferences. Anonymity is a 'democratic' condition – the 'power' of a coalition is determined just by the number of its members. Monotonicity requires that, should a coalition have the power to ensure that society's preference for one alternative over another is no greater than a certain degree, it would have the power to ensure the weaker result that this preference is no greater than a higher degree. According to tightness, if a coalition cannot ensure an 'issue' of the form $d \geqslant P(x, y)$, then its complement can ensure the complementary issue $P(x, y) \succ d$. Independence is the familiar condition that society's preferences between two alternatives depend only on individual preferences concerning the same two alternatives.

3. THE STRUCTURE OF FARs SATISFYING UNANIMITY AND INDEPENDENCE

Throughout this section, we assume that:

(3.1) $\bar{\Omega}^n$ is the domain of the FAR and $\bar{\Omega}$ is the range of the FAR.

Proposition 3.1 Let assumption (3.1) be satisfied, and let the FAR satisfy unanimity and independence. Then for all $x, y, x', y' \in X$ and all $(P_1, \ldots, P_n), (P'_1, \ldots, P'_n) \in \bar{\Omega}^n$, if $[P_i(x, y) = P'_i(x', y')$ and $P_i(y, x) = P'_i(y', x')]$ for all $i \in N$, then $P(x, y) = P'(x', y')$.

Proof: Let $z \in X - \{x, y\}$. Consider any $(P_1, \ldots, P_n) \in \bar{\Omega}^n$. Let $(P'_1, \ldots, P'_n) \in \bar{\Omega}^n$ be such that, for all $i \in N$, $[P'_i(x, y) = P_i(x, y)$ & $P'_i(y, x) = P_i(y, x)]$, $P'_i(y, z) = d_m$ and $[P'_i(x, z) = P_i(x, y)$ & $P'_i(z, x) = P_i(y, x)]$. (It can be easily checked that this specification of $(P'_1, \ldots, P'_n)$ is consistent with (3.1).) By independence, we have $P'(x, y) = P(x, y)$. By unanimity, we have $P'(y, z) = d_m$. Hence, by max–min transitivity of P', we have $P'(x, z) \geqslant P(x, y)$. Therefore, by independence, it follows that

(3.2) for all $(P_1, \ldots, P_n), (P'_1, \ldots, P'_n) \in \bar{\Omega}^n$, if for all $i \in N$, $[P'_i(x, z) = P_i(x, y)$ & $P'_i(z, x) = P_i(y, x)]$, then $P'(x, z) \geqslant P(x, y)$.

Obtaining the reverse inequality in an exactly similar way, we have:

(3.3) for all $(P_1, \ldots, P_n), (P'_1, \ldots, P'_n) \in \bar{\Omega}^n$, if for all $i \in N$, $[P'_i(x, z) = P_i(x, y)$ & $P'_i(z, x) = P_i(y, x)]$, then $P'(x, z) = P(x, y)$.

Similarly, it can be shown that

(3.4) for all $(P_1, \ldots, P_n), (P'_1, \ldots, P'_n) \in \bar{\Omega}^n$, if for all $i \in N$, $[P'_i(z, y) = P_i(x, y)$ & $P'_i(y, z) = P_i(y, x)]$, then $P'(z, y) = P(x, y)$.

The proof of Proposition 3.1 can now be completed following the familiar Arrow (1963) technique of switching alternatives. ■

Theorem 3.2 Let assumption (3.1) be fulfilled, and let the FAR satisfy unanimity and independence. Then there exists $k \in N$ such that for all distinct $a, b \in X$, and all $(P_1, \ldots, P_n) \in \bar{\Omega}^n$, $P_k(a, b) \geqslant P(a, b)$.

Proof: We show that, for all $d \in L$, if $d = P_k(a, b)$, then $d \geqslant P(a, b)$. If $d = d_m$, then the conclusion follows immediately. Let $d \in L - \{d_m\}$, and consider distinct $x, y, z \in X$. By unanimity, there exists a coalition which is (d, d_m)-decisive with respect to (x, z). Let C be a smallest of these coalitions and let $k \in C$. Let $(P_1, \ldots, P_n) \in \bar{\Omega}^n$ be as follows:

	$P_i(x, y)$	$P_i(y, x)$	$P_i(y, z)$	$P_i(z, y)$	$P_i(x, z)$	$P_i(z, x)$
$i \in C-\{k\}$:	--	--	d	d_1	d	d_1
$i \in N-C$:	--	--	d_m	d_1	d_m	d_1
k:	d	--	d_m	d_1	d	d_1

In the above table -- denotes that one can fill in the blank space by any indicator taken from $\{d_1, ..., d_m\}$. Note that this specification of $(P_1, ..., P_n) \in \bar{\Omega}^n$ is consistent with max–min transitivity of P_i $(i \in N)$. Further note that, by the specification of C, we have $d = P(x, z)$. Since C is a smallest of coalitions which are (d, d_m)–decisive with respect to (x, z), by Proposition 3.1 C is a smallest of coalitions which are (d, d_m)–decisive with respect to (y, z). Noting unanimity, it follows that $P(y, z) \succ d$. Given $P(y, z) \succ d$ and $d = P(x, z)$, max–min transitivity of P requires that $d \geqslant P(x, y)$. This is true irrespective of the way in which we fill in the blank spaces for $P_i(x, y)$ $(i \in N - \{k\})$ in the table specifying $(P_1, ..., P_n) \in \bar{\Omega}^n$. Hence, by Proposition 3.1, for all $a,b \in X$ and all $(P_1, ..., P_n) \in \bar{\Omega}^n$, $P_k(a, b) \geqslant P(a, b)$. ∎

Remark 3.3 From Theorem 3.2 it follows that, if assumption (3.1) is fulfilled, and the FAR satisfies unanimity and independence, then there exists a coalition C such that, for all $k \in C$, all distinct $x,y \in X$, and all $(P_1, ..., P_n) \in \bar{\Omega}^n$, if $P_k(x, y) = 0$, then $P(x, y) = 0$. C loosely resembles the 'oligarchy' of Gibbard's (1969) result in exact social choice theory. Theorem 3.2 is also reminiscent of Theorem 3.5 of Barrett, Pattanaik and Salles (1986) who used the 'cardinal' version of fuzzy set theory and imposed somewhat different restrictions on the domain and the range of the aggregation rule.

An important class of FARs, satisfying unanimity and independence, is based on what is known in the literature as consensus pooling. We provide a characterization of an important member of this class.

Definition 3.4 An FAR is based on consensus pooling iff, for all distinct $x,y \in X$ and all $(P_1, ..., P_n) \in \bar{\Omega}^n$, $P(x, y) = \min[P_i(x, y), i \in N]$.

Remark 3.5 In the special case where P_i $(i \in N)$ and P are assumed to be exact, it is clear that an FAR based on consensus pooling coincides with the 'Pareto extension rule', well known in the exact theory of social choice.[4]

Theorem 3.6 An FAR $F: \bar{\Omega}^n \rightarrow \bar{\Omega}$ is based on consensus pooling iff it satisfies unanimity, independence and anonymity.

Proof: The proof of necessity is obvious. We prove sufficiency.

Let $F: \bar{\Omega}^n \rightarrow \bar{\Omega}$ satisfy unanimity, independence and anonymity. Then, by Theorem 3.2, it follows that there exists $k \in N$ such that, for all distinct $x,y \in X$ and for all $(P_1, ..., P_n) \in \bar{\Omega}^n$, $P_k(x, y) \geqslant P(x, y)$. By anonymity, it follows that, for all distinct $x,y \in X$, for all $(P_1, ..., P_n) \in \bar{\Omega}^n$ and for all $k \in N$, $P_k(x, y) \geqslant P(x, y)$. Hence, for all distinct $x,y \in X$ and all $(P_1, ..., P_n) \in \bar{\Omega}^n$, $\min[P_i(x, y), i \in N] \geqslant P(x, y)$. Since, by unanimity, we cannot have $\min[P_i(x, y), i \in N] \geqslant P(x, y)$, it follows that for all distinct $x,y \in X$ and all $(P_1, ..., P_n) \in \bar{\Omega}^n$, we must have $\min[P_i(x, y), i \in N] = P(x, y)$. ∎

With consensus pooling, society may be indecisive. Suppose the individual preferences are such that, for every $(x, y) \in X^2$, $P_i(x, y) = 0$ for some $i \in N$. Clearly, under these circumstances, consensus pooling will be of little value for

4. See Sen (1970).

decision making. Given that the likelihood of this situation may increase with the size of N, it is worth examining alternatives to consensus pooling. This is what we do in the next section.

4 GENERALISED RANK-BASED RULES

FARs based on consensus pooling, which we introduced in the preceding section, constitute a special case of a fairly wide class of fuzzy aggregation rules which we characterize in this section. First we need more notation.

Notation 4.1 Given distinct $x,y \in X$ and $(P_1, \ldots, P_n) \in \bar{\Omega}^n$, let t_j denote $|\{i \in N: d_j \geqslant P_i(x, y)\}|$, for all $j \in \{1, \ldots, m\}$.

Remark 4.2 Given distinct $x,y \in X$ and $(P_1, \ldots, P_n) \in \hat{\Omega}^n$, it is clear that $t_1 \leqq t_2 \leqq \ldots \leqq t_m = n$.

Definition 4.3 An FAR is a generalised rank-based rule iff for all distinct $x,y \in X$, there exist integers $w_1, \ldots, w_m$ such that $[n \geqq w_1 \geqq w_2 \geqq \ldots \geqq w_m = 0]$ and, for all $(P_1, \ldots, P_n) \in \hat{\Omega}^n$, $[P(x, y) = d_k$, where k is the smallest integer in $\{1, \ldots, m\}$ such that $t_k \geqq w_k]$. (See Notation 4.1.)

Remark 4.4 Given Remark 4.2, it is clear that, for all non-negative integers $w_1, \ldots, w_m$ such that $[n \geqq w_1 \geqq w_2 \geqq \ldots \geqq w_m = 0]$, k, as specified in Definition 4.3, will exist, and hence Definition 4.3 describes a well defined FAR. It is clear that consensus pooling constitutes a special case where $w_j = 1$ for all $j \in \{1, \ldots, m\}$ and all distinct $x,y \in X$. Another special case is the 'median rule' where, assuming n to be odd, $w_j = (n + 1)/2$ for all $j \in \{1, \ldots, m\}$ and all distinct $x,y \in X$. Consensus pooling and the median rule are examples of what may be called simple rank-based rules which constitute a subclass of the class of generalised rank-based rules. Note that a generalised rank-based rule may treat one pair of alternatives differently from another. However, if the generalised rank-based rule satisfies 'neutrality', i.e. its structure does not vary as between different pairs of alternatives, then we have a simple rank-based rule when, for all coalitions, lower d-decisiveness for some $d \in L$ always implies lower decisiveness.

Theorem 4.5 An FAR is a generalised rank-based rule if and only if it satisfies unanimity, anonymity, monotonicity and tightness.

Proof: The necessity part of the proof is straightforward. We give only the proof of sufficiency.

Let the FAR satisfy unanimity, anonymity, monotonicity and tightness. Consider distinct $x,y \in X$. For all $j \in \{1, \ldots, m-1\}$, let C_j be a smallest coalition which is lower d_j-decisive with respect to (x, y) (by unanimity and the finiteness of the number of possible coalitions, such a coalition exists), and let $w_j \equiv |C_j|$. By anonymity, it is clear that, for all $j \in \{1, \ldots, m-1\}$, every coalition of size w_j or larger, but no coalition of size smaller than w_j, is lower d_j-decisive with respect to (x, y). Let $w_m = 0$. By monotonicity, $n \geqq w_1 \geqq w_2 \geqq \ldots \geqq w_{m-1} \geqq w_m = 0$.

Given $(P_1, \ldots, P_n) \in \hat{\Omega}^n$, let k be the smallest integer in $\{1, \ldots, m\}$ such that $t_k \geqq w_k$ (by Remark 4.4 such k exists). Since for all $j \in \{1, \ldots, m-1\}$, coalitions of size w_j, or larger, are lower d_j-decisive with respect to (x, y), and since it is obvious that every coalition is lower d_m–decisive with respect to (x, y), given $t_k \geqq w_k$, it is clear that $d_k \geqslant P(x, y)$. If k = 1, then, noting $d_k \geqslant P(x, y)$, we have $P(x, y) = d_k$. Suppose $k \epsilon \{2, ..., m\}$. Since $t_{k-1} < w_{k-1}$, $\{i \epsilon N: d_{k-1} \geqslant P_i(x, y)\}$ is not lower d_{k-1}–decisive with respect to (x, y). Therefore, by tightness, $N - \{i \epsilon N: d_{k-1} \geqslant P_i(x, y)\} = \{i \epsilon N: P_i(x, y) \geqslant d_k\}$ is upper d_k–decisive with respect to (x, y). Hence, $P(x, y) \geqslant d_k$. However, as noted earlier, $d_k \geqslant P(x, y)$. Therefore, $P(x, y) = d_k$. This completes the proof. ∎

5 CONCLUSION

Within the framework of Goguen's (1967) L–fuzzy set theory, we have explored several aspects of the problem of aggregating fuzzy individual preferences into fuzzy group preferences. However, fuzzy group preferences, by themselves, may not constitute the desired final product of such an aggregation exercise. One may like to generate, from fuzzy group preferences, exact group choice which, after all, is frequently the focus of our interest. This is a problem which needs separate investigation.

REFERENCES

Arrow, K. J. (1963), Social Choice and Individual Values; second edition. New York: John Wiley and Sons.

Barrett, C. R. and P. K. Pattanaik (1989), 'Fuzzy Sets, Preference and Choice: Some Conceptual Issues", Bulletin of Economic Research, 41, 229–253

Barrett, C. R., P. K. Pattanaik and M. Salles (1986), "On the Structure of Fuzzy Social Welfare Functions", Fuzzy Sets and Systems, 19, 1–10.

Barrett, C. R., P. K. Pattanaik and M. Salles (1987), "On Choosing Rationally when Preferences are Fuzzy", mimeograph; forthcoming in Fuzzy Sets and Systems.

Basu, K. (1984), "Fuzzy Revealed Preference Theory", Jounral of Economic Theory, Vol. 32, 212–227.

Basu, K., R. Deb and P. K. Pattanaik (1988), "Soft Sets: An Ordinal Reformulation of Vagueness with Some Applications to the Theory of Choice", mimeograph.

Blin, J. M. and A. B. Whinston (1973), "Fuzzy Sets and Social Choice", Journal of Cybernetics", 4.

Blin, J. M. (1974), "Fuzzy Relation in Group Decision Theory", Journal of Cybernetics", 4.

Dasgupta, M. and R. Deb (1988), "Fuzzy Choice Functrions", mimeograph.

Dimitrov, V. (1983), "Group Choice under Fuzzy Information", Fuzzy Sets and Systems, 9, 25–39.

Gibbard, A. (1969), "Intransitive Social Indifference and the Arrow Dilemma", mimeograph.

Goguen, J. A. (1967), "L-fuzzy Sets", Journal of Mathematical Analysis and Applications, 18.

Leclerc, B. (1984), "Efficient and Binary Consensus Function on Transitively Valued Relations", Mathematical Social Sciences, 8, 46-61.

Nurmi, H. (1981), "Approaches to Collective Decision Making with Fuzzy Preference Relations", Fuzzy Sets and Systems, 6, 249-259.

Ovchinnikov, S. V. and V. M. Ozernoy (1988), "Using Fuzzy Binary Relations for Identifying Non-inferior Decision Alternatives", mimeograph.

Sen, A. K. (1970), Collective Choice and Social Welfare. Amsterdam: North-Holland Publishing Company.

Subramanian, S. (1987), "The Liberal Paradox with Fuzzy Preferences", Social Choice and Weflare, 4, 213-223.

Tanino, T. (1984), "Fuzzy Preference Orderings in Group Decision Making", Fuzzy Sets and Systems. 12, 117-131.

SINGLE-PEAKEDNESS IN WEIGHTED AGGREGATION OF FUZZY OPINIONS IN A FUZZY GROUP

J. Montero (*)
Faculty of Mathematics
Complutense University
Madrid 28040 (Spain)

Abstract: In this paper it is considered a formal approach to the problem of aggregating individual opinions in a fuzzy group, when alternatives can be represented in a real hyper-space and each individual defines his/her fuzzy set of non rejectable alternatives. On one hand, weighted aggregation rule for consensus opinion is axiomatically justified. On the other hand, it is shown a sufficient condition for the stability of such consensus solution.
Keywords: aggregation rules, group decision making, fuzzy opinions.

1. INTRODUCTION

Every society is faced with the problem of opinion aggregation each time its individuals define different judgments or different preference attitudes. Amalgamating them into a consensus represents in this way a key point for the development of any group of persons. However, as shown by Arrow (1951, 1964), there is no general methodology for aggregating crisp individual preferences through a "social welfare function" satisfying some natural rationality conditions. Two ways for avoiding such a result have been proposed in the past: relaxing the concept of solution by considering, for example, "social choice functions" (Sen, 1970), or constraining the preferences domain of individual by assuming, for example, that such crisp preferences verify "single-peakedness" property (Black, 1958). This paper is based on the last approach, but allowing a continuous strength of preferences.

The three seminal books cited above suppose that each

(*) Research supported by Dirección General de Investigación Científica y Técnica (National Grant number PB88-0137)

J. Kacprzyk and M. Fedrizzi (eds.), Multiperson Decision Making Using Fuzzy Sets and Possibility Theory, 163–171.

individual i defines a crisp binary preference relation R_i on the set of feasible alternatives X (xP_iy will denote that xR_iy holds but not yR_ix). Black's single-peakedness means that there exists a strict ordering S on X such that all individual preference orderings can be represented by a preference curve with only one peak (with one or two elements), in such a way that on each side of such a peak it slopes downwards: formally, for each individual i and all distinct alternatives $x,y,z\in X$ the relation xP_iz holds whenever xR_iy for some alternative y between x and z (xSySz or zSySx). Then it was shown that the method of majority decision leads to a consensory transitive ordering on the set of alternatives. Inada (1964) has pointed out that single-peakedness with respect to the entire set of alternatives is not necessary: single-peakedness with respect to every triple of alternatives is enough.

2. FORMAL DESCRIPTION OF THE MODEL

Let $\beta:G\to[0,1]$ be a fixed fuzzy group of individuals (experts), G having two elements at least; $\beta(i)>0$, $\forall i\in G$, is the degree of competence of individual i (see Cholewa, 1985, for a discussion of the concept of competence), and let us assume that the set of (at least two) feasible alternatives X can be represented as a convex subset of the real hyper-space $\mathbb{R}^k$. This does not seem a serious restriction, since most practical problems are multidimensional in nature according a finite number of single characteristics, each one usually represented in the real line, but it requires a continuous degree in such a set of feasible alternatives. Individuals will be supposed to give their opinions through a fuzzy preference, that is, a fuzzy set of non-rejectable alternatives $\mu_i:X\to[0,1]$, where $\mu_i(x)$ represents the degree of membership of alternative x in the set of solutions, as given by individual i. Following Montero (1985), we must be able to define an aggregation rule, that is, a correspondence

$$* : (\mathcal{F}(X)x\mathcal{P}(G)) \ x \ (\mathcal{F}(X)x\mathcal{P}(G)) \to (\mathcal{F}(X)x\mathcal{P}(G))$$

which assigns to each pair of opinions $\mu_A,\mu_B\in\mathcal{F}(X)$ from two disjoint and non-empty groups $A,B\in\mathcal{P}(G)$ an opinion $\mu_{A\cup B}\in\mathcal{F}(X)$ of the union group $A\cup B\in\mathcal{P}(G)$, satisfying the associativity and commutativity:

$$((\mu_A,A)*(\mu_B,B))*(\mu_C,C) = (\mu_A,A)*((\mu_B,B)*(\mu_C,C))$$

$$(\mu_A,A)*(\mu_B,B) = (\mu_B,B)*(\mu_A,A)$$

for arbitrary disjoint and non-empty groups $A,B,C\in\mathcal{P}(G)$.

In this way, opinion μ_A of a group A will be represented by the pair (μ_A,A) and

$$(\mu_A, A) * (\mu_B, B) = (\mu_{A\cup B}, A\cup B) \qquad \forall A,B\neq\emptyset,\ A\cap B=\emptyset$$

It can be assumed that the aggregated opinion $\mu_{A\cup B}$ depends on the members of both groups A and B through their sizes, *card*(A) and *card*(B), and their associated competence β_A and β_B, obtained by aggregating the individual competences. That is, by applying an associative and commutative correspondence

$$@ : ([0,1] x \mathcal{P}(G)) \ x \ ([0,1] x \mathcal{P}(G)) \rightarrow ([0,1] x \mathcal{P}(G))$$

$$(\beta_A, A) @ (\beta_{A'}, B) = (\beta_{A\cup B'}, A\cup B) \qquad \forall A,B\neq\emptyset,\ A\cap B=\emptyset$$

which assigns an aggregated competence $\beta_{A\cup B}\in[0,1]$ of the union group $A\cup B\in\mathcal{P}(G)$ of two disjoint and non-empty groups $A,B\in\mathcal{P}(G)$, and it can be assumed also that $\beta_{A\cup B}$ depends on A and B only through their competence (β_A and β_B, respectively) and their size ((*card*(A) and *card*(B), respectively).

Obviously, input data of our aggregation problem must be all the individual opinions $\{\mu_{\{i\}}\equiv\mu_i\}_{i\in G}$ and their associated individual competences $\{\beta_{\{i\}}=\beta(i)\}_{i\in G}$, and both aggregation rules must be each one ethical and connected in a rational way (as shown in Montero, 1988a, not every opinion aggregation rule * is compatible with any fixed competence aggregation rule @). For example, the following conditions (i), (ii), (iii) and (i'), (ii'), (iii') can be easily accepted for @ and *, respectively:

(i) $\beta_{A\cup B}=\beta_{A'\cup B'}$ holds for any non-empty groups being $A\cap B=A'\cap B'=\emptyset$ and such that

$$card(A)=card(A'),\ \beta_A=\beta_{A'}$$
$$card(B)=card(B'),\ \beta_B=\beta_{B'}$$

(ii) $\beta_{A\cup B}=\beta_B$ holds for any non-empty groups ($A\cap B=\emptyset$) such that $\beta_A=\beta_B$.

(iii) If $\beta_A\geq\beta'_A$ and $\beta_B\geq\beta'_B$ for two non-empty and disjoint groups A and B, then $\beta_{A\cup B}\geq\beta'_{A\cup B}$.

(i') $\mu_{A\cup B}\equiv\mu_{A'\cup B'}$ holds for any non-empty groups being $A\cap B=A'\cap B'=\emptyset$ and such that $\mu_A\equiv\mu_{A'}$ and $\mu_B\equiv\mu_{B'}$ and

$$card(A)=card(A'),\ \beta_A=\beta_{A'}$$
$$card(B)=card(B'),\ \beta_B=\beta_{B'}$$

(ii') $\mu_{A\cup B}(x)=\mu_B(x)$ holds for any non-empty groups ($A\cap B=\emptyset$) such that $\mu_A(x)=\mu_B(x)$.

(iii') If $\mu_A(x)\geq\mu'_A(x)$ and $\mu_B(x)\geq\mu'_B(x)$ hold for two non-empty and disjoint groups A and B, then $\mu_{A\cup B}(x)\geq\mu'_{A\cup B}(x)$.

Under these conditions for both aggregation rules it make sense to define a "decisiveness": a group A with competence

β_A is said to be δ-decisive in competence over a group B ($A \cap B = \emptyset$) with competence β_B ($\beta_A \neq \beta_B$) if $\beta_{A \cup B} = \delta . \beta_A + (1-\delta) . \beta_B$; a group A with opinion μ_A is said to be δ-decisive in opinion over a group B ($A \cap B = \emptyset$) with opinion μ_B and relative to a fixed alternative $x \in X$ ($\mu_A(x) \neq \mu_B(x)$) if

$$\mu_{A \cup B}(x) = \delta . \mu_A(x) + (1-\delta) . \mu_B(x)$$

3. ETHICAL PROPERTIES OF THE WEIGHTED MEAN RULE

The following result has been proved in Montero (1988a), and it can be considered as an axiomatic justification for the weighted mean rules.

Theorem 1. Let us consider opinion aggregation rules * verifying conditions (i'), (ii') and (iii'). Then, the minimum ratio opinion decisiveness $\delta/\beta(i)$ of individuals leads to the maximum for the weighted opinion aggregation rule such that

$$\mu_{A \cup B}(x) = (card(A) . \beta_A . \mu_A(x) + card(B) . \beta_B . \mu_B(x)) / (card(A \cup B) . \beta_{A \cup B})$$

for any non-empty and disjoint groups $A, B \in G$, β being the aggregated weighted competence given by

$$\beta_{A \cup B} = (card(A) . \beta_A + card(B) . \beta_B) / card(A \cup B)$$

for any non-empty and disjoint groups $A, B \in G$. Moreover, this competence aggregation rule @ maximizes the minimum ratio competence decisiveness $\delta/\beta(i)$ of individuals, and it verifies conditions (i), (ii) and (iii).

It must be pointed out that the concept of an aggregation rule considered here generalizes an analogous aggregation operation of Fung and Fu (1975), avoiding a restrictive result obtained by them (and thus we can see that the main cause of such restrictive result is just that their aggregation rule does not depend on the size of the aggregated groups). Other axioms will lead to different aggregation rules, including those "mixed" rules of Fung and Fu (see Montero, 1988b, for an analysis in the context of multicriteria aggregation, and Dubois and Prade, 1985, for a relation with various fuzzy set aggregation connectives).

In any case, the weighted aggregation rule of Theorem 1 verifies ethical conditions translated into this context from classical impossibility theorems: the universal domain has been assumed by definition, the anonymity will be assured due to condition (i'), the unanimity will be assured by condition (ii'), the non-negative response and independence of irrelevant alternatives will hold due to condition (iii') and even the neutrality holds. Some formal definitions of all these ethical conditions are written below. Alternative definitions can be found in the literature (see, e.g., Dubois and Koning, 1989).

Anonymity: Let us consider $A, A', B, B' \in \mathcal{P}(G)$ being non-empty groups such that $A\cap B=A'\cap B'=\emptyset$, with $\beta(A)=\beta(A')$, $\beta(B)=\beta(B')$ and $card(A)=card(A')$, $card(B)=card(B')$. If $\mu_A\equiv\mu_{A'}$ and $\mu_B\equiv\mu_{B'}$ hold, then

$$(\mu_A,A)*(\mu_B,B) = (\mu_{A\cup B},A\cup B)$$
$$(\mu_{A'},A')*(\mu_{B'},B') = (\mu_{A'\cup B'},A'\cup B')$$

must verify that $\mu_{A\cup B}\equiv\mu_{A'\cup B'}$.

Independence of Irrelevant Alternatives: Let $A,B\in\mathcal{P}(G)$ be disjoint and non-empty groups and let us suppose that

$$\mu_A(x)=\mu'_A(x)\ \ \forall x\in Y\subset X,\ \mu_B(x)=\mu'_B(x)\ \ \forall x\in Y\subset X$$

Then

$$(\mu_A,A)*(\mu_B,B) = (\mu_{A\cup B},A\cup B)$$
$$(\mu'_A,A)*(\mu'_B,B) = (\mu'_{A\cup B},A\cup B)$$

must verify that

$$\mu_{A\cup B}(x)=\mu'_{A\cup B}(x) \qquad \forall x\in Y\subset X$$

Neutrality: Let $A,B\in\mathcal{P}(G)$ be disjoint and non-empty groups, and $P:X\to X$ be an arbitrary one-to-one mapping. If for each opinion $\mu\in\mathcal{F}(X)$ we define μ^P such that

$$\mu^P(x)=\mu(P(x)) \quad \forall x\in X$$

then

$$(\mu^P_A,A)*(\mu^P_B,B) = (\mu^P_{A\cup B},A\cup B)$$

Non-Negative Response: If $A,B\in\mathcal{P}(G)$ are disjoint and non-empty groups and

$$\mu_A(x)\geq\mu'_A(x)\ \ \forall x\in X,\ \mu_B(x)\geq\mu'_B(x)\ \ \forall x\in X$$

with at least one strict inequality, then

$$(\mu_A,A)*(\mu_B,B)=(\mu_{A\cup B},A\cup B),\ \ (\mu'_A,A)*(\mu'_B,B)=(\mu'_{A\cup B},A\cup B)$$

ought to verify

$$\mu_{A\cup B}(x)\geq\mu'_{A\cup B}(x) \qquad \forall x\in X$$

Unanimity: Let $A,B\in\mathcal{P}(G)$ be disjoint and non-empty groups with identical fuzzy preferences $\mu_A\equiv\mu_B\equiv\mu\in\mathcal{F}(X)$. Then

$$(\mu_A,A)*(\mu_B,B) = (\mu,A\cup B)$$

Analogous ethical conditions (depending only on the size of the aggregated groups, as shown in Montero, 1985) could be imposed to competence aggregation operations @, and it is easy to see that they are satisfied by weighted competence aggregation of Theorem 1.

Finally, we must point out a clear criticism to this approach, since it is assumed that the assignments are commensurable, and furthermore, if $\mu_i(x)>\mu_j(x)$ it is understood that individual i prefers alternative x more than individual j does. The problem of interpersonally comparing preference intensities is, as pointed out by Pattanaik (1971), part of the broader and controversial problem of the possibility of knowing other minds; but though in the present state of our knowledge there does not exist any precise method of measuring and interpersonally comparing the intensities of individual preferences, it can hardly be denied that ethically it is desirable to take them into account. An interesting attempt in this sense is that of Vansnick (1986).

4.- CONCAVITY OF INDIVIDUAL PROFILES

In our context, single-peakedness means that alternatives are ordered in the real hyper-space $\mathbb{R}^k$ of k characteristics in such a way that, as we go from left to the right in each characteristic, every individual preferences increases up to a peak , and then decreases after we pass it. As shown in Montero (1989), such a property will hold when the considered intensity is concave from an analytic point of view:

Definition 1. Let $\mu:X\to[0,1]$ be a fuzzy set of feasible alternatives defined over a convex subset of the real hyper-space $X\subset\mathbb{R}^k$. Then μ is said to be "concave" if

$$\mu(\lambda.x+(1-\lambda).y) \geq \lambda.\mu(x)+(1-\lambda).\mu(y) \quad \forall x,y\in X,\ \forall\lambda\in[0,1]$$

It is clear that a single-peaked representation on the real line is in its spirit close to Black's single-peakedness and even to Inada's single-caveness (Inada, 1964), a condition less restrictive than single-peakedness which allows for indifference plateaus, and also assuring a consensory transitive solution under the crisp majority rule. Though it is clear that single-peaked intensity and single-caved intensity are not always concave, the introduction of concavity is justified due to some interesting properties which will appear when individual intensities are concave compatible in our context.

Definition 2. A profile of individual intensities $\{\mu_i:X\to[0,1]\}_{i\in G}$ defined over a convex set $X\subset\mathbb{R}^k$ of alternatives verifies the property of "concavity" if they are all concave.

In this way, if concavity is assumed, an important but desirable restriction on the family of admissible preference patterns has been introduced, since each individual preference must be inside the set $\mathcal{C}(X)\subset\mathcal{F}(X)$ of concave opinions. Therefore, if it considered that the restrictions on individual preferences must be just the restrictions on social preferences, our aggregation rule * must be defined as an associative and commutative correspondence

$$* : (\mathcal{C}(X)\mathrm{x}\mathcal{P}(G)) \mathrm{ x } (\mathcal{C}(X)\mathrm{x}\mathcal{P}(G)) \to (\mathcal{C}(X)\mathrm{x}\mathcal{P}(G))$$

verifying the above ethical conditions.

It must be pointed out that single-peakedness of individual intensities does not assure a social single-peaked opinion under the weighted mean rule. Such a closure theorem can be assured under concavity (its mathematical proof is trivial).

Theorem 2. Let us assume that the concavity property holds, and let us consider the weighted aggregation rule as given in Theorem 1. Then, aggregated opinions are concave for any disjoint and non-empty groups $A,B\in\mathcal{P}(G)$.

Concavity of a social preference pattern can be understood as a condition which assures an equilibrium and stability in group decision making: a best alternative is around one point, and a manipulation or measurement errors (if they are not too big) can modify in fact the solution, but new solution will not be in any case too far from the initial solution.

The concave individual intensities mean in fact that people are inclined to reach consensus. The single-peakedness of individual intensities seems to be a natural assumption when dealing with single characteristic problems, but if they are non-concave, they represent very clear individual preferences, with people rejecting other alternatives different than their own best alternatives:binary intensities (those verifying $\mu(x)\in\{0,1\}$ $\forall x\in X$) will never be concave unless they are constant (that is, $\mu(x)=\mu(y)$ $\forall x,y\in X$). For example, if there is an individual $i\in G$ such that $\mu(x_i)=1$ but $\mu(y)=0$, $\forall y\neq x_i$, for a fixed alternative $x_i\in X$, our common sense tells us that a satisfactory consensory solution will be very difficult due to such a crisp (absolutely clear) opinion, and in fact there is no representation making such a preference concave. In some way we can conclude that Theorem 2 reveals how "spreaded" intensities (in the sense of concavity) can easily be aggregated according to the weighted mean rule, and points out the expected difficulties in aggregating crisp (or too clear) opinions.

In any case, it also must be pointed out that concavity requires a sensitivity of the decision makers, since around best alternatives the intensity must be strictly decreasing (there is no identical intensities at each side of the peak).

A dual result can be obtained when the values of our intensity preferences μ are understood as degrees of rejecting each alternative by the individuals (like a fuzzy veto instead of a fuzzy preference). It shows the difficulties in aggregating vetos of an arbitrary shape, and its mathematical proof is also trivial.

Theorem 3. Let us consider the weighted aggregation rule * as given in Theorem 1, and let us suppose that each individual opinion verifies convexity, that is,

$$\mu(\lambda.x+(1-\lambda).y)\leq\lambda.\mu(x)+(1-\lambda).\mu(y) \quad \forall x,y\in X,\ \forall\lambda\in[0,1]$$

Then, an aggregated opinion for any pair of disjoint and non-empty groups in G also verifies convexity.

Under the concavity condition, the weighted aggregation will be between moderate alternatives (that is, non-extreme alternatives). Such a property seems in principle desirable for the stability of any social system.

5.- FINAL COMMENTS

If our k descriptive characteristics have been fixed, the

representation on the real hyper-space of the set of alternatives is obviously not necessarily unique. Hence, if such real representations in a convex subset X of $\mathbb{R}^k$ exist, we must look for one of them where all the individual opinions are concave. And it has been pointed out that the existence of such concave-compatible representation is not assured for arbitrary profiles of individual fuzzy preferences (trivial examples can be given by considering single-peaked intensity preferences with constant intensities in extreme alternatives).

Moreover, in practice only a finite number of alternatives will be evaluated at the first step. Some reasonable real representations could then be considered in order to sequentially search for better alternatives: for example, if alternatives are evaluated according to two characteristics, each one represented by a real number between 0 and 10, the individual intensities could be initially estimated only for the alternatives {(2.j,2.k)/j,k=0,1,...,5}, that is, for 36 alternatives covering as a net the whole space of alternatives [0,10]x[0,10]; if concavity holds for this real representation of individual preferences, then a best alternative must be looked for around the peaks obtained by the weighted mean aggregation; for example, if only one maximum is reached in alternative (4,6), it will be enough to consider in the next step only the subspace of alternatives [3,5]x[5,7], where another net of points could be established in order to repeat the procedure. If concavity does not hold, perhaps a concave representation could be tried by considering appropriate order preserving one-to-one mappings from the initial evaluation space [0,10] of each characteristic into different real intervals, or perhaps by considering a deeper transformation (alternatives could be described in terms of different but more appropriate characteristics).

REFERENCES

K.J. Arrow (1951, 1964), *Social Choice and Individual Values*, Wiley, New York.

D. Black (1958), *The Theory of Committees and Elections*, Cambridge University Press, Cambridge.

W. Cholewa (1985), *Aggregation of Fuzzy Opinions: an Axiomatic Approach*, Fuzzy Sets Syst. 17, 249-258.

D. Dubois and J.L. Koning (1989), *Social Choice Axioms for Fuzzy Sets Aggregation*, Fuzzy Sets Syst. (to appear).

D. Dubois and H. Prade (1985), *A review of Fuzzy sets Aggregation Connectives*, Inf. Sci. 36, 85-121.

L.W. Fung and K.S. Fu (1975), *An Axiomatic Approach to Rational Decision Making in a Fuzzy Environment*; in: L.A. Zadeh, K.S. Fu, K. Tanaka and M. Shimura (Eds.), *Fuzzy Sets and their Applications to Cognitive and Decision Processes*, Academic Press, New York, pp. 227-256.

K. Inada (1964), *A Note on Simple Majority Rule*, Econometrica

32, 525-531.
J. Montero (1985), *A note on Fung-Fu's Theorem,* Fuzzy Sets Syst. 17, 259-269.
J. Montero (1988a), *Aggregation of Fuzzy Opinions in a Non-Homogeneous Group*, Fuzzy Sets Syst. 25, 15-20.
J. Montero (1988b), *An Axiomatic Approach to Fuzzy Multicriteria Analysis;* in: M.M. Gupta and T. Yamakawa (Eds.), *Fuzzy Logic in Knowledge-Based Systems, Decision and Control*, North-Holland, Amsterdam, pp. 259-269.
J. Montero (1989), *Weighted Aggregation and Single Peaked Intensities*, Workshop on Aggregation and Best Choices on Imprecise Opinions, Brussels.
P.K. Pattanaik (1971), *Voting and Collective Choice*, Cambridge University Press, Cambridge.
A.K. Sen (1970), *Collective Choice and Social Welfare*, Holden-Day, San Francisco.
J.C. Vansnick (1987), *Intensity of Preference;* in: Y. Sawaragi, K. Inoue and H. Nakayama (Eds.), *Toward Interactive and Intelligent Decision Support Systems* (Volume 2), Springer-Verlag, Berlin, pp. 220-229.

ON GROUP DECISION MAKING UNDER FUZZY PREFERENCES

Tetsuzo Tanino
Department of Information Technology
Okayama University
Tsushima, Okayama 700, Japan

Abstract: In this paper some results on group decision making under fuzzy preferences are reviewed. Since preferences of men are often "fuzzy" and the direct interpersonal comparison of preferences is not easy, the above topic is inevitable and quite important. First we discuss how to represent fuzzy preferences. They might be represented by fuzzy choice sets, fuzzy binary relations or fuzzy utility functions. Secondly group decision making situations are classified according to the above levels in representing both individual and group preferences. Some methods are explained in each category.

Keywords: group decision making, fuzzy preferences

1. INTRODUCTION

Since human judgments including preferences are often vague, fuzzy theory plays an important role in decision making. Several authors have provided interesting results on group decision making or social choice theory with the help of fuzzy theory (Blin and Whinston, 1973; Blin, 1974; Bezdek et al., 1978, 1979; Kuz'min and Ovchinnikov, 1980a, b; Nurmi, 1981; Dimitrov, 1983; Kim, 1983; Tanino, 1984, 1988; Seo and Sakawa, 1985; Cholewa, 1985; Kacprzyk, 1985a, b, 1986; Barrett et al., 1986; Zahariev, 1987; Fedrizzi and Kacprzyk, 1988; Montero de Juan, 1988; Switalski, 1988; Roubens, 1989; and some others). They have

J. Kacprzyk and M. Fedrizzi (eds.), Multiperson Decision Making Using Fuzzy Sets and Possibility Theory, 172–185.

proved that fuzzy theory provides a more flexible framework for discussing group decision making.

The purpose of this paper is to classify several results on group decision making under fuzzy preferences according to the levels of representation of fuzzy preferences. As is well known, nonfuzzy preferences can be represented as the set of preferred alternatives (choice set), preference relations (orderings), or (cardinal) utility functions. In this paper, therefore, we analogously consider the following three types of representations of fuzzy preferences:

1) fuzzy choice set,
2) fuzzy preference relation,
3) fuzzy utility function.

The details of those representations will be discussed in Section 2, where some relationships between nonfuzzy and fuzzy preferences are also dealt with.

Situations of group decision making can be classified into several categories according to the above levels of representation of preferences. We concentrate on the cases in which individual preferences are fuzzy and several methods in the above cases are reviewed in Section 3.

2. REPRESENTATION OF FUZZY PREFERENCES

In this section we discuss some ways of representing fuzzy preferences. Let X be a set of alternatives over which preferences are defined.

2.1. Representation of nonfuzzy preferences

It is known that a nonfuzzy (crisp) preference attitude of a decision maker over the set X may be represented in the following three levels (Tanino et al., 1981) (See Table 1):

nonfuzzy			fuzzy		
choice set	relation	utility fun.	choice set	relation	utility fun.
subset	preference relation	real-valued function	fuzzy subset	fuzzy preference relation	fuzzy real-valued function

Table 1 Representation of preferences

1) Choice set : A set of preferred alternatives C is shown by the decision maker. This set is often a singleton as in usual voting.

2) Preference relation : A preference relation of the decision maker (typically a partial order, a weak order or a linear order) is given. The theory of social choice in welfare economics has been based on this type of representation.

3) Utility function : A real-valued function reflecting the decision maker's preference u is identified. Since an ordinal utility function is essentially equivalent to a weak order, we mean by utility functions cardinal ones.

We should note that the first level includes only a small amount of information on the preference attitude of the decision maker, while the second level includes more and the third level the most. In fact, it is immediate to derive a weak order from a utility function and also to derive a set of preferred (or nondominated) alternatives from a weak order (or a partial order).

2.2. Representation of fuzzy preferences

Similarly to the nonfuzzy case, fuzzy preferences may be represented in the three levels shown in Table 1. Each level includes the nonfuzzy case as a special case.

2.2.1. Fuzzy choice set

This is the case when the degree of preference of each alternative is provided by the decision maker. Hence the total preference attitude is described by a fuzzy subset of X, i.e., by a membership function μ on X, whose value $\mu(x)$ denotes the degree to which x is chosen as a preferred (desirable) alternative.

2.2.2. Fuzzy preference relation

A fuzzy binary relation on X is used to represent the preference attitude of the decision maker. A fuzzy binary relation is a fuzzy set on the product set $X \times X$, of which membership function is denoted by r. In order that a fuzzy binary relation reflects a preference, it should satisfy some conditions as an ordinary weak order satisfies reflexivity and transitivity. The following conditions are often considered:

reciprocity: $r(x, y) + r(y, x) = 1 \quad \forall x, y$

max-min transitivity: $r(x, z) \geq \min(r(x, y), r(y, z)) \quad \forall x, y, z$

max-max transitivity: $r(x, z) \geq \max(r(x, y), r(y, z))$

$$\forall x, y, z$$

restricted max-min transitivity: $r(x, y) \geqq 0.5,\ r(y, z) \geqq 0.5$
$$\Rightarrow \quad r(x, z) \geqq \min(r(x, y), r(y, z)) \qquad \forall x, y, z$$

restricted max-max transitivity: $r(x, y) \geqq 0.5,\ r(y, z) \geqq 0.5$
$$\Rightarrow \quad r(x, z) \geqq \max(r(x, y), r(y, z)) \qquad \forall x, y, z$$

additive transitivity: $r(x, y) + r(y, z) - 0.5 = r(x, z)$
$$\forall x, y, z$$

multiplicative transitivity:
$$\frac{r(y, x)}{r(x, y)} \cdot \frac{r(z, y)}{r(y, z)} = \frac{r(z, x)}{r(x, z)} \qquad \forall x, y, z$$

We should also note that a fuzzy binary relation can be described by a square matrix whose (i, j) element is $r(x_i, x_j)$ when the set of alternatives X is finite (i.e. $X = \{x_1, x_2, \ldots, x_m\}$). In this case, its average fuzziness and average certainty were introduced by Bezdek et al. (1978).

2.2.3. Fuzzy utility function

As an extension of ordinary utility functions, we may consider fuzzy utility functions. A fuzzy utility function is defined as a fuzzy mapping which associates a fuzzy subset of the utility values space (usually the space of real numbers $\mathbb{R}$) with each alternative. Hence it is characterized by a function ν from $X \times \mathbb{R}$ to [0, 1], where $\nu(x; t)$ denotes the degree to which the utility value of the alternative x is equal to t. Since identification of the function ν is fairly difficult, we might assume a simple form of ν (e.g. a triangular form as in Seo and Sakawa 1985).

2.2.4. From higher levels to lower levels

As in the nonfuzzy case, the higher the level of representation is, the more information on the preference it provides. For example, when a fuzzy preference relation described by a function $r: X \times X \rightarrow [0, 1]$ is given, a membership function μ of a fuzzy subset of X can be defined by

$$\mu(x) = \min_{\substack{y \in X \\ y \neq x}} r(x, y)$$

or, if X is finite (let m be the number of alternatives), by

$$\mu(x) = \frac{1}{m-1} \sum_{\substack{y \in X \\ y \neq x}} r(x, y)$$

Discussions in detail along this line can be seen in, for example, Kacprzyk (1986), Switalski (1988), Roubens (1989), and so on.

When a fuzzy utility function ν is given and is normalized so that

$$\{t \mid \nu(x;t) > 0\} \subset [0,1],$$

we may define a membership function μ by

$$\mu(x) = \int_0^1 t\,\nu(x;t)\,dt \Big/ \int_0^1 \nu(x;t)\,dt$$

Here the denominator is assumed not to be 0. This corresponds to the operation taking the mean value of a probability density. When ν is nonfuzzy and the set $\{t \mid \nu(x;t) > 0\}$ is a singleton $\{t_0\}$, then $\mu(x) = t_0$. Another definition of μ is given by

$$\mu(x) = \arg\max_t \nu(x;t)$$

(If the right-hand side is not a singleton, take the median). We may also define a fuzzy preference relation, for example, by

$$r(x,y) = (\mu(x) - \mu(y) + 1)/2$$

This fuzzy relation satisfies reciprocity.

2.3. Relationships between nonfuzzy and fuzzy preferences

In this subsection we discuss relationships between nonfuzzy and fuzzy preferences. Since fuzziness in preferences may be considered reflecting degrees of (certainty of) preferences, it is not so strange that fuzzy preferences have close relationships with nonfuzzy utility functions or preference relations. Those relationships will be helpful in deriving some aggregation rules under fuzzy individual preferences.

2.3.1. Nonfuzzy utility functions and fuzzy choice sets

Both a nonfuzzy utility function u and a membership function μ of a fuzzy choice set are real-valued functions defined on the set X, though the values of μ are restricted in the interval [0,1]. Therefore, by normalizing u appropriately, we may identify u and μ.

2.3.2. Nonfuzzy utility functions and fuzzy preference relations

When a nonfuzzy utility function u is given as a difference scale on X and is normalized so that $\max_X u(x) - \min_X u(x) \leq 1$, we may define a fuzzy preference relation by

$$r(x,y) = (u(x) - u(y) + 1)/2$$

This fuzzy relation satisfies reciprocity and additive transitivity and hence restricted max-min transitivity (Tanino, 1984).

Another fuzzy preference relation is defined as follows when a nonfuzzy utility function is a ration scale and takes only positive values.

$$r(x,y) = u(x)/(u(x)+u(y))$$

This fuzzy relation satisfies reciprocity and multiplicative transitivity (Tanino, 1984).

2.3.3. Nonfuzzy preference relations and fuzzy preference relations

Given a fuzzy preference relation r, a class of nonfuzzy preference relations $\gtrsim_\alpha$, where $\alpha \in [0.5, 1]$, is defined by

$$x \gtrsim_\alpha y \quad \Leftrightarrow \quad r(x,y) \geq \alpha$$

If r satisfies reciprocity, $\gtrsim_\alpha$ is connected for $\alpha = 0.5$ and irreflexive for $\alpha > 0.5$. If, moreover, r satisfies restricted max-min transitivity, then $\gtrsim_\alpha$ is transitive for any $\alpha \in [0.5, 1]$. In other words, if the given fuzzy relation r is reciprocal and restrictedly max-min transitive, then $\gtrsim_\alpha$ is a weak order for $\alpha = 0.5$ and a strict partial order for $\alpha > 0.5$.

When the number of alternatives is finite, introducing an appropriate distance into the space of fuzzy preference relations may also enable us to derive nonfuzzy preference relation. Typical such distances are the l_2-metric by Bezdek et al. (1978, 1979) and the l_1-metric by Kuz'min and Ovchinnikov (1980a, b).

2.3.4. Nonfuzzy choice sets and fuzzy preference relations

Deriving a nonfuzzy choice set (i.e. the most preferred alternative) from a given fuzzy preference relation is very important in group decision making, since we may first find a group fuzzy preference relation from individual preferences and then a social outcome (Nurmi, 1981; Kacprzyk, 1985a, b, 1986; Zahariev, 1987).

The familiar concepts are the α-consensus winner and the minimax consensus winner by Nurmi (1981). Let α be a number such that $0.5 < \alpha \leq 1$. An alternative x is said to be an α-consensus winner if

$$r(x,y) \geq \alpha \qquad \text{for any } y \neq x$$

The α-consensus winner is unique if it exists and if r is reciprocal. Let

$$s(y) = \max_x r(x,y)$$

When

$$s(x) = \min_y s(y) = \min_y \max_x r(x,y),$$

the alternative x is said to be a minimax consensus winner. It always exists but is not necessarily unique.

3. GROUP DECISION MAKING UNDER FUZZY INDIVIDUAL PREFERENCES

In this section we discuss several methods for group decision making under fuzzy individual preferences. Throughout this section we use the following notations:

$X = \{x, y, z, \ldots\}$: set of alternatives

$N = \{1, 2, \ldots, n\}$: group consisting of n individuals

C : nonfuzzy choice set

$\succsim$: nonfuzzy preference relation

u : nonfuzzy utility function

μ : membership function of a fuzzy choice set

r : membership function of a fuzzy preference relation

ν : membership function of a fuzzy utility function

C, $\succsim$, u, μ, r, and ν with the superscript i imply the preferences of the individual i, and those without the superscript denote the preferences of the whole group.

3.1. Classification of group decision making situations

According to the classification of the levels of both individual and group preferences, we may consider several kinds of approaches aggregating individual preferences into a group one. The summary is shown in Table 2. In Table 2, × implies that the level of representation of individual preferences is not enough to obtain a group preference in the corresponding level. ← implies that the group preference in the concerned level can be obtained from that in higher levels. For example, in the nonfuzzy case, we can construct a group utility function from individual utility functions and it leads to a weak order or a choice set of the group. ↑ implies that the results can be obtained by considering nonfuzzy individual preferences as special cases of fuzzy preferences. Each number in Table 2 denote the section in which the correspnding category is dealt with.

Ordinary group decision theory has been dealing with the case in which both individual and group preferences are nonfuzzy. A large amount of excellent results have been obtained (Sen, 1970; Fishburn, 1973; Keeney and Kirkwood, 1975; and so on). Moreover, nonfuzzy individual preferences may be regarded as a special case of fuzzy

individual \ group		nonfuzzy choice set	nonfuzzy relation	nonfuzzy utility fun.	fuzzy choice set	fuzzy relation	fuzzy utility fun.
nonfuzzy	choice set	←	ranking by votes	×	↑	↑	×
	relation	social choice function	social welfare function	×	↑	↑ Blin method	×
	utility fun.	←	←	group utility function	similar to 3.2.1	E C R method	×
fuzzy	choice set	←	←	3.2.1	3.2.2	3.2.2	×
	relation	3.3.1	3.3.2	3.3.2	3.3.3	3.3.4	×
	utility fun.	←	←	←	←	←	3.4

Table 2 Classification of group decision making methods

individual preferences. In this paper, therefore, we concentrate on the case in which individual preferences are fuzzy.

3.2. Group decision making based on individual fuzzy choice sets

First we consider the case when individual preferences are described by choice sets. We suppose that n fuzzy subsets of X specified by membership functions μ^i (i=1, 2, ..., n) are given.

3.2.1. Nonfuzzy group preferences

As pointed out in subsection 2.3.1, the functions μ^i can be regarded as utility functions on X. Therefore we may formally derive a nonfuzzy group utility function u in several ways. Typical examples are the following:

$$u(x) = \sum_{i=1}^{n} w_i \mu^i(x), \qquad w_i \geq 0, \quad (\sum_{i=1}^{n} w_i = 1),$$

$$u(x) = \min_i \mu^i(x)$$

The group preference relation and the group choice set are immediately obtained from the above utility function.

3.2.2. Group fuzzy choice sets and fuzzy preference relations

The nonfuzzy group utility function defined above may be regarded as the membership function μ of the group choice set. Some fuzzy preference relations may be derived from the nonfuzzy group utility functions as in subsection 2.3.2.

3.3. Group decision making based on individual fuzzy preference relations

In this subsection individual preferences are supposed to be represented by membership functions r^i $(i=1,2,\ldots,n)$ on $X \times X$. Most of the existing results on group decision making (social choice) under fuzzy preferences are dealing with this case.

3.3.1. Group nonfuzzy choice sets

In this case a group choice set (social outcome) is derived directly from the individual fuzzy preference relations. Fuzzy α-core, α-minimax set and so on are known well in this category (Nurmi 1981).

Suppose that a number $q \leqq n$ which implies majority in the group is given. Then the set defined by

$$C_\alpha = \{x \in X \mid \text{there exists no } y \in X \text{ such that } r^i(y,x) > \alpha \text{ for at least } q \text{ individuals}\}$$

for $\alpha \in [0.5, 1)$ is called the fuzzy α-core (Nurmi, 1981). The fuzzy α-core is a set of alternatives which are not sufficiently (at least to degree α) defeated by the required majority q.

For $\alpha \in (0, 0.5)$, let

$$w_\alpha(x,y) = \#\{i \mid r^i(x,y) < \alpha\}$$

$$v_\alpha(x) = \max_{y \in X} w_\alpha(x,y)$$

Then the α-minimax set is defined by

$$M_\alpha = \operatorname*{argmin}_X v_\alpha(x)$$

(Nurmi, 1981). The α-minimax set contains alternatives that are less preferred at lease to degree α by no more than max $v_\alpha(x)$ individuals.

Kacprzyk (1985a,b, 1986) defined the Q-minimax set, α/Q-minimax set and s/Q-minimax set as extensions of the minimax set and the α-minimax set. He introduced linguistic quantifiers which are defined as fuzzy sets in [0,1]. Let Q be a linguistic quantifier such as 'most' whose membership function is q_Q. Let

$$h^i(x,y;\alpha)=\begin{cases}1 & \text{if } r^i(x,y)<\alpha\leqq 0.5\\ 0 & \text{otherwise}\end{cases}$$

i.e. $h^i(x,y;\alpha)=1$ means that individual i preferes y over x. And analogously to the above, let

$$h^i(x;\alpha)=\frac{1}{m-1}\sum_{y\neq x} h^i(x,y;\alpha),$$

$$h(x;\alpha)=\frac{1}{n}\sum_{i=1}^{n} h^i(x;\alpha)$$

$$t_Q(x;\alpha)=q_Q(h(x;\alpha)).$$

Then, the set

$$M_\alpha = \operatorname*{argmin}_x \; t_Q(x;\alpha)$$

is called the α/Q-minimax set. It is a set of alternatives that are to the least extent sufficiently (at least to degree $1-\alpha$) defeated by Q individuals. As for the s/Q-minimax set, see Kacprzyk (1985a,b, 1986).

When the group fuzzy preference relation is obtained, some preferred alternatives such as the α-consensus winner and the minimax consensus winner explained in subsection 2.3.4 can be also derived.

3.3.2. Group nonfuzzy preference relations and utility functions

Since group nonfuzzy preference relations and utility functions can be derived from group fuzzy choice sets or fuzzy preference relations, we should refer to those cases.

3.3.3. Group fuzzy choice sets

Assume that the number of alternatives is finite (= m). Then the most natural group fuzzy choice set is derived by the relation

$$u(x)=\frac{1}{n}\sum_{i=1}^{n}\frac{1}{m-1}\sum_{y\neq x} r^i(x,y)$$

Some other solution concepts were provided by Kacprzyk (1985a,b, 1986) as extensions of Nurmi's results. Let

$$h^i(x,y;\alpha)=\begin{cases}1 & \text{if } r^i(x,y)<\alpha\leqq 0.5\\ 0 & \text{otherwise}\end{cases}$$

as before. Then

$$k^i(y;\alpha)=\frac{1}{m-1}\sum_{x\neq y} h^i(x,y;\alpha)$$

is to what extent individual i is not against y; from 0 for certainly against, to 1 for certainly not against, through all intermediate values. Next

$$k(y;\alpha)=\frac{1}{n}\sum_{i=1}^{n} k^{i}(y;\alpha)$$

is to what extent all the individuals are not against y, and the fuzzy set deefined by

$$\mu(y)=q_{Q}(k(y;\alpha))$$

is called the fuzzy α/Q-core. $\mu(y)$ is to what extent Q individuals are not against y. The fuzzy α/Q-core is an extension of the α-core. The fuzzy Q-core is defined as a special case of the fuzzy α/Q-core. Kacprzyk also proposed the concept of fuzzy s/Q-core.

Some other group fuzzy choice sets can be defined through group fuzzy preference relations. Let r be a group fuzzy preference relation and define

$$g(x,y;\alpha)=\begin{cases}1 & \text{if } r(x,y)>\alpha\geqq 0.5\\ 0 & \text{otherwise}\end{cases}$$

and

$$g(x;\alpha)=\frac{1}{m-1}\sum_{y\neq x} g(x,y;\alpha)$$

is the extent to which x is sufficiently (at least to degree α) preferred. Then

$$\mu(x)=q_{Q}(g(x;\alpha))$$

is the extent to which x is sufficiently preferred over Q other alternatives. This fuzzy set is called the fuzzy α/Q-consensus winner. The fuzzy Q-consnsus winner is a special case and the s/Q-consensus winner was also proposed. The details can be seen in Kacprzyk (1985a,b,1986).

3.3.4. Group fuzzy preference relations

A function aggregating (nonfuzzy) individual preference relations into a (nonfuzzy) group preference relation is called a social welfare function. Arrow's general impossibility theorem denied the existence of social welfare functions satifying several rational conditions. Barrett et al. (1986) considered fuzzy social welfare functions which aggregate individual fuzzy preference relations into group fuzzy preference relations, and obtained an extension of Arrow's theorem. Their conclusion is as follows: All fuzzy aggregation rules which have non-narrow domains and which satisfy the fuzzy counterparts of the familiar properties of independence of irrelevant alternatives and Pareto criterion, are characterized by a distribution of veto power which would be generally considered undesirable.

The most simple group fuzzy preference relation derived from individual fuzzy preference relations is defined by

$$r(x,y)=\frac{1}{n}\sum_{i=1}^{n} r^{i}(x,y)$$

However, the obtained fuzzy preference relation does not generally satisfy max-min transitivity even when so do individual preference relations.

The following fuzzy group preference relation was proposed by Tanino (1984):

$$r(x,y)=\begin{cases}\dfrac{1}{2}\dfrac{\sum_{i=1}^{n}(r^{i}(x,y)-0.5)}{\sum_{i=1}^{n}|r^{i}(x,y)-0.5|} & \text{if } x\neq y\\ 0.5 & \text{if } x=y\end{cases}$$

This relation satisfies restricted max-min transitivity if every r^{i} is additively transitive and it reflects the diversity of individual preferences in the group. Tanino also proposed another fuzzy preference relation which satisfies restricted max-min transitivity provided that every r^{i} is multiplicatively transitive (Tanino 1984).

3.4. Group decision making based on individual fuzzy utility functions

Finally in this section we suppose that fuzzy preferences of the individuals are represented by the functions ν^{i} $(i=1,2,\ldots,n)$. In this case we may obtain a group fuzzy utility function by aggregating ν^{i}'s into ν. Aggregation rules are similar to those for utility functions, i.e., for example,

$$\nu(x;t)=\sum_{i=1}^{n} w_{i}\nu_{i}(x;t),\quad w_{i}\geq 0,\quad \sum_{i=1}^{n} w_{i}=1,$$

or

$$\nu(x;t)=\min_{i}\ \nu_{i}(x;t)$$

for each $x\in X$. Other types of group preferences can be derived from the above utility functions.

4. CONCLUSIONS

In this paper some methods in group decision making (social choice theory) under fuzzy individual preferences. They have been categorized on the basis of the levels of representating fuzzy preferences. Though results on axiomatic approaches to social welfare functions and social choice functions are omitted, several practical methods (aggregation rules) for obtaining group preferences have been reviewed.

REFERENCES

Barrett C.R., P.K. Pattanaik and M. Salles (1986) On the structure of fuzzy social welfare functions, Fuzzy Sets and Syst. 19, 1-10.

Bezdek J.C., B. Spillman and R. Spillman (1978) A fuzzy relation space for group decision theory, Fuzzy Sets and Syst. 1, 255-268.

Bezdek J.C., B. Spillman and R. Spillman (1979) Fuzzy relation spaces for group decision theory: An application, Fuzzy Sets and Syst. 2, 5-14.

Blin J.M. (1974) Fuzzy relations in group decision theory, J. Cybernet. 4, 17-22.

Blin J.M. and A.P. Whinston (1973) Fuzzy sets and social choice, J. Cybernet. 3, 28-33.

Cholewa W. (1985) Aggregation of fuzzy opinions - an axiomatic approach, Fuzzy Sets and Syst. 17, 249-258.

Dimitrov V.N. (1983) Group choice under fuzzy information, Fuzzy Sets and Syst. 9, 25-40.

Dubois D. and H. Prade (1980) Fuzzy Sets and Systems : Theory and Applications, Academic Press, New York.

Fedrizzi M. and J. Kacprzyk (1988) On measuring consensus in the setting of fuzzy preference relations, in J. Kacprzyk and M. Roubens (Eds.) : Non-Conventional Preference Relations in Decision Making, Springer-Verlag, Berlin, pp. 129-141.

Fishburn P.C. (1973) The Theory of Social Choice, Princeton Univ. Press, Princeton, NJ.

Kacprzyk J. (1985a) Group decision-making with a fuzzy majority via linguistic quantifiers, Part I: a consensory-like pooling, Cybernet. and Syst. 16, 119-129.

Kacprzyk J. (1985b) Group decision-making with a fuzzy majority via linguistic quantifiers, Part II: a competitive-like pooling, Cybernet. and Syst. 16, 131-144.

Kacprzyk J. (1986) Group decision making with a fuzzy linguistic majority, Fuzzy Sets and Syst. 18, 105-118.

Keeney R.L. and C. Kirkwood (1975) Group decision making using cardinal social welfare functions, Management Sci. 22, 430-437.

Kim J.B. (1983) Fuzzy rational choice functions, Fuzzy Sets and Syst. 10, 37-43.

Kuz'min V.B. and S.V. Ovchinnikov (1980a) Group decisions I. In arbitrary spaces of fuzzy binary relations, Fuzzy Sets and Syst. 4, 53-62.

Kuz'min V.B. and S.V. Ovchinnikov (1980b) Design of group decisions II. In spaces of partial order fuzzy relations, Fuzzy Sets and Syst. 4, 153-165.

Montero de Juan F.J. (1988) Aggregation of fuzzy opinions in a nonhomogeneous group, Fuzzy Sets and Syst. 25, 15-20.

Nurmi H. (1981) Approaches to collective decision making with fuzzy preference relations, Fuzzy Sets and Syst. 12, 117-131.

Roubens M. (1989) Some properties of choice functions based on valued binary relations, Europ. J. Oper. Res. 40, 309-321.

Sen A.K. (1970) Collective Choice and Social Welfare, Holden-Day, San Francisco.

Seo F. and M. Sakawa (1985) Fuzzy multiattribute utility analysis for collective choice, IEEE Trans. Syst. Man Cybernet. SMC-15, 45-53.

Switalski Z. (1988) Choice functions associated with fuzzy preference relations, in J. Kacprzyk and M. Roubens (Eds.): Non-Conventional Preference Relations in Decision Making, Springer-Verlag, Berlin, pp. 106-118.

Tanino T. (1984) Fuzzy preference orderings in group decision making, Fuzzy Sets and Syst. 12, 117-131.

Tanino T. (1988) Fuzzy preference relations in group decision making, in J. Kacprzyk and M. Roubens (Eds.): Non-Conventional Preference Relations in Decision Making, Springer-Verlag, Berlin, pp. 54-71.

Tanino T., H. Nakayama and Y. Sawaragi (1981) On methodology for group decision support, in J.N. Morse (Ed.): Organizations: Multiple Agents with Multiple Criteria, Springer-Verlag, Berlin, pp.409-423.

Zahariev S. (1987) An approach to group choice with fuzzy preference relations, Fuzzy Sets and Syst. 22, 203-213.

GROUP DECISION MAKING WITH FUZZY AND NON-FUZZY EVALUATIONS

Stoichko Zahariev
Academy of Social Sciences and Management-Branch Varna
9000 Varna, P.O. Box 65, Bulgaria

Abstract. This paper presents an approach to group choice which is nearest in the sense of Hamming distance to the individual estimations. Each individual estimation is a multiple-criteria judgement obtained on the basis of fuzzy and non-fuzzy evaluations of the criteria and the alternatives. As such are considered fuzzy preference relations, expressing the estimations of the alternatives with respect to each criterion and non-fuzzy weights of the criteria or non-fuzzy numbers, expressing the degrees to which the alternatives satisfy each criterion and fuzzy relation and fuzzy relations, expressing the signification of the criteria. The choice of the "best" alternative is made on the basis of Orlovsky's approach, but maximizing another choice function.

Keywords: Group choice, Multiple-criteria estimation, Hamming distance Choice function, Best alternative.

1. INTRODUCTION

By judgements of alternatives two kinds of procedures are mainly used: pairwise comparison and ranking. The pairwise comparison leads very often to valued preferences, which provide a preference relation on the set of alternatives. In many cases an appropriate tool to capture the inherent imprecision of human value judgements may be fuzzy preference relations. The direct ranking is preferred by experts when the number of alternatives to be compared is relatively small and the criteria are wel determined. Most often the significance of the ranked alternatives is expressed additionally with weights which are crisp (non-fuzzy) numbers.

This paper deals with an approach to group decision making on the basis of experts' judgements which can be: fuzzy preference relations, expressing the estimation of the alternatives with respect to several criteria and non-fuzzy weights of the criteria, expressing their significance or non-fuzzy numbers, expressing the degrees to which the alternatives satisfy the criteria and fuzzy relations, expressing the significance of the criteria.

J. Kacprzyk and M. Fedrizzi (eds.), Multiperson Decision Making Using Fuzzy Sets and Possibility Theory, 186–197.

Group choice is performed on the basis of a natural assumption that it has to be as near as possible to every single expert's estimation. For this purpose all experts' estimations are reduced to fuzzy preference relations and the individual multiple-criteria choices as well as the group choice are determined with an appropriate nearness measure for fuzzy preference relations. This approach is an extension of Kemeny and Snell's (1962) idea for finding the group choice, based on non-fuzzy order relations.

To determine the "best" decision alternative or a selection of a small number of nondominated alternatives on the basis of the group choice, which is a fuzzy preference relation, we use Orlovsky's (1978) approach, but maximizing another function of strict nondominated alternatives than that proposed by him. As it will be shown, this is necessary because Orlovsky's choice function leads to certain contradictions.

2. DETERMINATION OF THE INDIVIDUAL ESTIMATIONS OF THE ALTERNATIVES

Let $A=\{a_1,a_2,\dots,a_p\}$ be the set of alternatives, $E=\{e_1,e_2,\dots,e_q\}$ the set of experts and $C=\{c_1,c_2,\dots,c_r\}$ the set of criteria. We assume that the experts assess each alternative with respect to each criterion. Therefore the significance of the criteria is judged by the experts too. In this paper we consider the following two pairs of individual estimations: fuzzy preference relations, expressing the judgements of the alternatives, with respect to each criterion and non-fuzzy weights, expressing the significance of the criteria or non-fuzzy numbers, expressing the degrees to which the alternatives satisfy the criteria and fuzzy relations, expressing the significance of the criteria.

Let R^k_m be a fuzzy preference relation, expressing the judgement of the alternatives made by expert e_k, $k=1,2,\dots,q$, with respect to the criterion c_m, $m=1,2,\dots,r$. Its membership function is

$$\mu_{R^k_m} : \to [0,1] \tag{1}$$

such that $\mu_{R^k_m}(a_i,a_j)\in[0,1]$, $i,j=1,2,\dots,p$, denotes the degree of preference of a_i over a_j as perceived by expert e_k. By definition $\mu_{R^k_m}(a_i,a_i)=1$, which means that each alternative is not less preferred than itself.

Let w^k_m be a weight of criterion c_m, determined by expert e_k. We assume that the criteria weights are non-fuzzy numbers from the unit interval, such that $w^k_m\in[0,1]$ denotes the significance of criterion c_m as perceived by expert e_k.

As multiple-criteria individual estimation of expert e in this case we will consider a preference relation R_k^M which is nearest to the fuzzy preference relations r_m^k with the membership function

$$\mu_{r_m^k}(a_i,a_j)=w_m^k.\mu_{R_m^k}(a_i,a_j) \tag{2}$$

By definition, $\mu_{r_m^k}(a_i,a_i)=1$.

Let us now consider how the individual multiple-criteria estimations will be determined if the judgement procedures are exchanged.

For this purpose we will assume that the alternatives' estimations $\mu_m^k(a_i)$ with respect to each criterion are non-fuzzy numbers from the unit interval, such that $\mu_m^k(a_i)\in[0,1]$ denotes the degree to which alternative a_i satisfies criterion c_m as perceived by expert e_k.

By $\mathcal{R}^k$ we will denote a fuzzy preference relation expressing the judgements of the criteria by expert e_k. Its membership function is

$$\mu_{\mathcal{R}^k} : \rightarrow [0,1] \tag{3}$$

such that $\mu_{\mathcal{R}^k}(c_m,c_n)\in[0,1]$ denotes the degree of preference (significance) of c_m over c_n as perceived by expert e_k. By definition, $\mu_{\mathcal{R}^k}(c_m,c_m)=1$.

This preference relation induces on the set of the alternatives preference relations $p_{m,n}^k$ to be determined below.

We will consider the degrees to which the alternatives satisfy a given criterion as membership degrees of the alternatives in a fuzzy set which expresses this criterion. So, criterion c_m can be presented by a fuzzy set c_m^k as follows:

$$c_m^k=\{\mu_m^k(a_1)/a_1+\mu_m^k(a_2)/a_2+...+\mu_m^k(a_p)/a_p\} \tag{4}$$

But each ordered pair of fuzzy sets (c_m^k,c_n^k), where c_m^k and c_n^k are characterized by their membership functions μ_m^k and μ_n^k, induces on the set of the alternatives a fuzzy relation $\imath_{m,n}^k$ with the membership function

$$\mu_{\imath_{m,n}^k}(a_i,a_j)=\mu_m^k(a_i)\wedge\mu_n^k(a_j) \tag{5}$$

By definition, $\mu_{\mathcal{R}^k_{m,n}}(a_i,a_i)=1$.

On the other hand an ordered pair (c^k_m,c^k_n) belongs to a fuzzy preference relation with the same degrees as the ordered pair (c_m,c_n) belongs to the fuzzy preference relation $\mathcal{R}^k$. Therefore we assume that each ordered pair (c_m,c_n) induces on the set of the alternatives a fuzzy preference relation with the membership function

$$\mu_{p^k_{m,n}}(a_i,a_j)=\mu_{\mathcal{R}^k}(c_m,c_n).\mu_{\mathcal{R}^k_{m,n}}(a_i,a_j) \tag{6}$$

By definition $\mu_{p^k_{m,n}}(a_i,a_i)=1$.

As a multiple-criteria individual estimation of expert e_k in this case we will consider a fuzzy preference relation P^M_k which is nearest to the fuzzy preference relations $p^k_{m,n}$. An appropriate nearness measure of fuzzy preference relations will be proposed in the next section.

3. CHOICE OF A NEARNESS MEASURE OF FUZZY PREFERENCE RELATIONS

When choosing a nearness measure of fuzzy preference relations, we follow an axiomatic approach used by Kemeny and Snell(1962) for finding a nearness measure of non-fuzzy weak ordering. According to this approach we first have to define axioms that express the requirements for the nearness measure of fuzzy preference relations. Next we have to prove that the nearness measure is uniquely defined by the chosen system of axioms. Finally, we have to develop a formula for calculating the values of that nearness measure which satisfies the system of axioms and consequently appears to be the only possible one among the given set of requirements.

Before doing so, however, we will define a fuzzy preference relation between two preference relations. This makes it possible to introduce an analogue of a line and on this basis to formulate the axiom that expresses one of the requirements for the nearness measure. We will say that a fuzzy preference relation R_3 lies between fuzzy preference relations R_1 and R_2 if the following condition is satisfied

$$R_1 \cap R_2 \subseteq R_3 \subseteq R_1 \cup R_2 \tag{7}$$

Let R_1, R_2 and R_3 be three fuzzy preference relations that express estimations of three experts. If for each ordered pair of alternatives (a_i,a_j) it turns out that the degree to which the second expert prefers alternative a_i to alternative a_j lies within the interval determined by the degrees at which the first and the third expert prefer a_i to a_j, i.e.

$$\min\{\mu_{R_1}(a_i,a_j),\mu_{R_3}(a_i,a_j)\}\le\mu_{R_2}(a_i,a_j)\le\max\{\mu_{R_1}(a_i,a_j),\mu_{R_3}(a_i,a_j)\} \quad (8)$$

then we naturally assume that the nearness (difference) in the estimations of the first and the third expert is as large as the sum of the nearnesses (differences) in the estimations of the first and the second, and the second and the third experts.

Let us denote the nearness measure between two fuzzy preference relations by d. Then the above mentioned requirement for d can be formulated as follows.

Axiom 1. If R_2 lies between R_1 and R_3, then $d(R_1,R_3)=d(R_1,R_2)+ d(R_2,R_3)$.

The next requirement for the nearness measure, that we will state as an axiom, expresses a natural condition that if estimations of two experts differ only in the degree of preference for the alternatives for one ordered pair, then the nearness measure of estimations should be equal to the module of the difference in the degrees of preference for the alternatives for this pair.

Axiom 2. If R_1 and R_2 differ only in the value of the membership degree of one ordered pair (a_g,a_h), i.e. $\mu_{R_1}(a_g,a_h)\neq\mu_{R_2}(a_g,a_h)$ and $\mu_{R_1}(a_i,a_j)=\mu_{R_2}(a_i,a_j)$ for $i,j=1,2,\dots,g-1,h+1,\dots,p$, then $d(R_1,R_2)=|\mu_{R_1}(a_g,a_h)-\mu_{R_2}(a_g,a_h)|$.

But if we consider each fuzzy preference relation as a point in the n or (or n-1)-dimensional Euclidean space, then it can easily be seen that these two axioms determine the 1-norm in R^{nxn}. Hence, it directly follows that d is the only distance that satisfies the above axioms and

$$d(R_1,R_2)=\Sigma|\mu_{R_1}(a_i,a_j)-\mu_{R_2}(a_i,a_j)| \quad (9)$$

which is the Hamming distance.

It has already been pointed out that the group preference relation has to be as near as possible to the fuzzy preference relations that express the experts' estimations. This leaves the question of how to find a fuzzy relation that is nearest to the given fuzzy preference relations.

As the nearest fuzzy relation R^N to the fuzzy preference relations $R_1,R_2,\dots,R_q$ we will consider

$$R^N=\min_R \sum_{k=1}^{q} d(R,R_k) \quad (10)$$

Let $R_1, R_2, \ldots, R_q$ be fuzzy preference relations that express the experts' estimations and let us assume that

$$\mu_{R_1}(a_i,a_j) \le \mu_{R_2}(a_i,a_j) \le \ldots \le \mu_{R_q}(a_i,a_j)$$

We will determine the fuzzy relation R^M as follows

$$\mu_{R^M}(a_i,a_j)=\mu_{R_k}(a_i,a_j) \quad \text{where} \quad k=\frac{1}{2}(q+1) \quad \text{if} \quad q \quad \text{is} \quad \text{odd}$$

$$\mu_{R^M}(a_i,a_j)=\frac{1}{2}(\mu_{R_k}(a_i,a_j)+\mu_{R_{k+1}}(a_i,a_j)) \quad \text{where } k=\frac{1}{2}q \quad \text{if} \quad q \text{ is even} \tag{11}$$

The fuzzy relation that is nearest in the sense of the Hamming distance to the given fuzzy preference relations is determined by the following theorem.

Theorem 1. R^M is fuzzy relation that is nearest to the fuzzy preference relations $R_1, R_2, \ldots, R_q$ in the above sense.

The proof of the theorem is given in Zahariev(1987).

4. DETERMINATION OF THE BEST ALTERNATIVE

In the previous section we have described how to determine multiple-criteria individual estimations as well as group estimation using the nearness ıeasure proposed by us. But in practice the necessity often arises to point out a best decision alternative or a selection of a small number of nondominated alternatives on the basis of the group choice. For this purpose we will follow the approach used by Orlovsky(1978), but maximizing another choice function of strict nondomination than that proposed by him.

A choice function $C(A)$ is a mapping from A into [0,1] and $C(A)(a_i)$ is the degree to which an alternative $a_i \in A$ is the "best" element in the set A. When no confusion is possible, $C(A)(a_i)$ will be denoted $C(a_i)$.

For a given preference relation R, Orlovsky introduce a strict preference relation R^S

$$\mu_{R^S}(a_i,a_j)=\max\{\mu_R(a_i,a_j)-\mu_R(a_j,a_i),0\} \tag{12}$$

in order to define the "nondomination" choice function

$$ND^R(a_i)=\min_j[1-\mu_{R^S}(a_j,a_i)]=1-\max_j \mu_{R^S}(a_j,a_i) \tag{13}$$

The value $ND^R(a_i)$ is introduced by him as a degree to which alternative a_i is dominated by no other alternative in A. In the fuzzy set context $ND^R(a_i)$, i=1,2,...p, determines the fuzzy set of nondominated alternatives. As reasonable choices are considered those elements of the set of nondominated alternatives which give a value of $ND^R(a_i)$ close enough to the value of

$$\max_i ND^R(a_i)=1-\min_i \max_j \mu_{R^S}(a_j,a_i) \tag{14}$$

As pointed out by Orlovsky himself $ND^R(a_i)$ is in fact introduced to determine the degree to which alternative a_i is strictly dominated by no other alternative in A. Therefore, the choice function $ND^R(a_i)$ should be called "strict nondomination". Under this name it is mentioned by some other authors too (Switalski, 1988; Roubens, 1989).

Let us consider now a preference relation R which expresses the group choice and is reflexive and transitive.

R	a_1	a_2	a_3	a_4
a_1	1	0.27	0.27	0.25
a_2	0.32	1	0.40	0.57
a_3	0.37	0.20	1	0.22
a_4	0.40	0.17	0.52	1

A fuzzy relation R is said to be
reflexive if $\mu_R(a_i,a_i)=1$ for every a_i in A,
irreflexive if $\mu_R(a_i,a_i)=0$ for every a_i in A,
symmetric if $\mu_R(a_i,a_j)=\mu(a_j,a_i)$ for all a_i and a_j in A,
antisymmetric if $\min\{\mu_R(a_i,a_j),\mu_R(a_j,a_i)\}=0$ for all a_i and a_j in A and
transitive if $\min\{\mu_R(a_i,a_j),\mu_R(a_j,a_k)\}=\mu_R(a_i,a_k)$ for all a_i,a_j and a_k in A.

According to Orlovsky, the corresponding strict preference relation R^S is

R^S	a_1	a_2	a_3	a_4
a_1	0	0	0	0
a_2	0.05	0	0.2	0.4
a_3	0.1	0	0	0
a_4	0.15	0	0.3	0

and hence the fuzzy set of nondominated elements is

$$X^{ND}=0.85/a_1+1/a_2+0.7/a_3+0.6/a_4$$

Following Orlovsky, as the "best" alternative we should choose a_2. In this case a problem arises if on the basis of the fuzzy set of nondominated alternatives two or more alternatives have to be chosen. Here, for example, a_2 and a_1 must be specified as such. But if relation R^S is considered, one can see that to the sub-set of "best" alternatives belong alternative a_2, which strictly dominates all other alternatives, as well as alternative a_1 which is strictly dominated by all other alternatives of A. On the other hand, relation R^S is antisymmetric, irreflexive and transitive (Orlovsky, 1978). Thus, it is a fuzzy strict order and determines a non-fuzzy order a_2, a_4, a_3, a_1 (Kaufmann, 1975). On the basis of this, however, alternatives a_2 and a_4 have to be chosen, which is in contradiction to Orlovsky's choice too. As shown in Zahariev (1987) there are some problems in choosing the "best" alternative, too, if the preference relation R is reflexive only.

It can easily be shown, that the contradictions by interpreting the results obtained using Orlovsky's "nondomination" choice function arise as it is actually not "strict nondomination" but "nonstrict domination". For this purpose we will make some definitions.

The "domination" choice function will be defined as follows

$$D^R(a_i)=\min_j \mu_R(a_i,a_j) \tag{15}$$

where R is a fuzzy preference relation and $D^R(a_i)$ is the degree of domination of a_i.

The "nondomination" choice function is defined by us as follows

$$ND^R(a_i)=\min_j \mu_{R}(a_i,a_j) \tag{16}$$

where $ND^R(a_i)$ is the degree to which a_i is nondominated and $\mu_{R^N}(a_i,a_j)=1-\mu_R(a_i,a_j)$. This definition of "nondomination" choice function is the same as by Switalski(1988) and Roubens (1989).

The "strict domination" choice function will be defined as

$$SD^R(a_i)=\min_j \mu_{R^S}(a_i,a_j) \tag{17}$$

where R^S is the strict preference relation corresponding to R and $SD^R(a_i)$ is the degree of strict domination of a_i.

There are different definitions of R^S. In addition to Orlovsky's definition, in the literature the following definition can often be met (Roubens and Vincke, 1985)

$$\mu_{R^S}(a_i,a_j)=\min[\mu_R(a_i,a_j),1-\mu_R(a_j,a_i)] \tag{18}$$

On the basis of the definitions of the "nondomination" and "strict domination" choice functions we can now define the "nonstrict domination" choice function as follows

$$NSD^R(a_i)=\min_j \mu_{R^{NS}}(a_i,a_j) \tag{19}$$

where $NSD^R(a_i)$ is the degree to which a_i is nonstrictly dominated, and $\mu_{R^{NS}}(a_i,a_j)=1-\mu_{R^S}(a_j,a_i)$. It can be immediately seen that this is Orlovsky's "nondomination" choice function.

The "strict nondomination" choice function will be defined as

$$SND^R(a_i)=\min_j \mu_{R^{SN}}(a_i,a_j) \tag{20}$$

where $SND^R(a_i)$ is the degree to which a_i is strictly nondominated and $\mu_{R^{SN}}(a_i,a_j)=\mu_{R^S}(a_i,a_j)$.

Virtually, $\mu_{R^{SN}}(a_i,a_j) = \max\{\mu_{R^N}(a_i,a_j)-\mu_{R^N}(a_j,a_i),0\} = \max\{1-\mu_R(a_j,a_i)-1+\mu_R(a_i,a_j),0\} = \max\{\mu_R(a_i,a_j)-\mu_R(a_j,a_i),0\} = \mu_R(a_i,a_j)$ if Orlovsky's definition of strict preference is used or $\mu_{R^{SN}}(a_i,a_j) = \min\{\mu_{R^N}(a_i,a_j), 1-\mu_{R^N}(a_j,a_i)\} = \min\{1-\mu_R(a_j,a_i), 1-1+\mu_R(a_i,a_j)\} = \min\{\mu_R(a_i,a_j),1-\mu_R(a_j,a_i)\}=\mu_{R^S}(a_i,a_j)$ if Roubens and Vincke's definition of strict preference is used.

When determining the "best" alternative we will use the "strict nondomination" choice function defined with R^S as proposed by Roubens and Vincke. Orlovsky's definition is "stronger" than that of Roubens and Vincke, and by its application a part of the information included in R is lost. This can make the choice of the sub-set of "best" alternatives difficult. To illustrate this, let us consider again the same example.

Using Orlovsky's definition of R^S we obtain

R^S	a_1	a_2	a_3	a_4
a_1	0	0	0	0
a_2	0.05	0	0.2	0.4
a_3	0.1	0	0	0
a_4	0.15	0	0.3	0

and hence $SND^R(a_1)=SND^R(a_3)=SND^R(a_4)=0$ and $SND^R(a_2)=0.05$. It can be immediately seen that to the subset of "best" alternatives belongs the alternative a_2 only. This result is in accordance with the fact that here R^S determines a non-fuzzy order.

Using the Roubens and Vincke's definition of R^S we obtain

R^S	a_1	a_2	a_3	a_4
a_1	0	0.27	0.27	0.25
a_2	0.32	0	0.40	0.57
a_3	0.37	0.20	0	0.22
a_4	0.40	0.17	0.52	0

and hence $SND^R(a_1)=0.25$, $SND^R(a_2)=0.32$, $SND^R(a_3)=0.20$ and $SND^R(a_4)=0.17$. Here the "best" alternative is again a_2 but now we have the possibility to choose more than one alternative. If we have to choose, for example, two alternatives a_2 and a_1 should be pointed out as such.

Now we can specify an algorithm for finding the group choice and on its basis the "best" alternative:

1. Determine the weighed preference relations r^k_m or $p^k_{m,n}$ on the set of alternatives A.

2. Arrange the values of the membership function for each pair of the fuzzy preference relations r^k_m or $p^k_{m,n}$ in the ascending order.

3. Determine the individual multiple-criteria estimations as fuzzy preferences R^M_k or P^M_k which are nearest to the fuzzy preference relations r^k_m or $p^k_{m,n}$, respectively.

4. Arrange the values of the membership function for each pair of the fuzzy preference relations R^M_k or P^M_k in an ascending order.

5. Determine the group choice as a fuzzy preference relation R^M or P^M which is nearest to the fuzzy preference relations R^M_k or P^M_k, respectively.

6. Determine, using the "strict nondomination" choice function $R^{MS} = \min\{\mu_R(a_i,a_j), 1-\mu_R(a_j,a_i)\}$ or $P^{MS} = \min\{\mu_{P^M}(a_i,a_j), 1-\mu_{P^M}(a_j,a_i)\}$ the "best" alternative or the sub-set of the "best" alternatives.

Let us consider an example which illustrates this algorithm. Assume that p=3. q=3 and r=2. Further on we will assume that the individual estimations of the alternatives with respect to each criterion are presented by the relations:

R_1^1	a_1	a_2	a_3
a_1	1	0.6	0.8
a_2	0.3	1	0.7
a_3	0.5	0.3	1

R_1^2	a_1	a_2	a_3
a_1	1	0.7	0.5
a_2	0.9	1	0.7
a_3	0.7	0.9	1

R_1^3	a_1	a_2	a_3
a_1	1	0.4	0.5
a_2	0.7	1	0.3
a_3	0.7	0.8	1

R_2^1	a_1	a_2	a_3
a_1	1	0.7	0.9
a_2	0.4	1	0.8
a_3	0.6	0.4	1

R_2^2	a_1	a_2	a_3
a_1	1	0.6	0.8
a_2	0.8	1	0.5
a_3	0.4	0.3	1

R_2^3	a_1	a_2	a_3
a_1	1	0.6	0.8
a_2	0.7	1	0.3
a_3	0.7	0.8	1

and the individual estimations of the criteria are $w_1^1=0.8$; $w_1^2=0.6$; $w_1^3=0.6$; $w_2^1=0.4$; $w_2^2=0.4$ and $w_2^3=0.8$.

The weighed fuzzy preference relations are

r_1^1	a_1	a_2	a_3
a_1	1	0.48	0.64
a_2	0.24	1	0.56
a_3	0.40	0.24	1

r_1^2	a_1	a_2	a_3
a_1	1	0.42	0.30
a_2	0.54	1	0.42
a_3	0.42	0.54	1

r_1^3	a_1	a_2	a_3
a_1	1	0.24	0.30
a_2	0.12	1	0.18
a_3	0.42	0.48	1

r_2^1	a_1	a_2	a_3
a_1	1	0.28	0.36
a_2	0.16	1	0.32
a_3	0.24	0.16	1

r_2^2	a_1	a_2	a_3
a_1	1	0.24	0.32
a_2	0.32	1	0.20
a_3	0.16	0.12	1

r_2^3	a_1	a_2	a_3
a_1	1	0.48	0.64
a_2	0.56	1	0.24
a_3	0.56	0.64	1

and hence the individual multiple-criteria estimations

R_1^M	a_1	a_2	a_3
a_1	1	0.42	0.30
a_2	0.24	1	0.42
a_3	0.42	0.48	1

R_2^M	a_1	a_2	a_3
a_1	1	0.28	0.36
a_2	0.32	1	0.24
a_3	0.24	0.16	1

The group choice R^M and the fuzzy strict preference relation R^{MS} are

R^M	a_1	a_2	a_3
a_1	1	0.35	0.33
a_2	0.28	1	0.34
a_3	0.33	0.32	1

R^{MS}	a_1	a_2	a_3
a_1	0	0.35	0.33
a_2	0.28	0	0.34
a_3	0.33	0.32	0

Thus $SND^{R^M}(a_1)=0.33, SND^{R^M}(a_2)=0.28, SND^{R^M}(a_3)=0.32$ and the "best" alternative is a_1.

5. CONCLUSIONS

An approach to determine the weighed preference relations expressing the alternative judgements with regard to each criterion was proposed. On the basis of some natural requirements, an aggregation rule for finding the individual multiple-criteria estimations as well as the group estimation was considered. The choice of the "best" alternative or the selection of a sub-set of "best" alternatives was made on the basis of the "strict nondomination" choice function. This function was also shown to be more appropriate than the "nondomination" choice function proposed by Orlovsky.

REFERENCES

Kaufmann A. (1975) Introduction to the Theory of Fuzzy Subsets, Volume I. Academic Press, New-York San Francisco London

Kemeny J.G. and I.L. Snell (1965) Preference Rankings, Mathematical Models in the Social Sciences. Blaisdell, New-York

Orlovsky S.A. (1978) Decision-Making with a Fuzzy Preference Relation. Fuzzy Sets and Systems 1, 155-167

Roubens M. (1989) Some Properties of Choice Functions Based on Valued Binary Relations. EJOR 40, 309-321

Roubens M. and P. Vincke (1985) Preference Modelling. Springer Verlag, Berlin

Switalski Z. (1988) Choice Function Associated with Fuzzy Preference Relations, in J. Kacprzyk and M. Roubens (eds.): Non-conventional Preference Relations in Decision Making. Springer Verlag, Berlin

Zahariev S. (1987) An Approach to Group Choice with Fuzzy Preference Relations. Fuzzy Sets and Systems 22, 203-213

ON CONSTRUCTION OF THE FUZZY MULTIATTRIBUTE RISK FUNCTION FOR GROUP DECISION MAKING

Fumiko Seo

Kyoto Institute of Economic Research
Kyoto University
Yoshida-honmachi, Sakyoku, Kyoto 606
JAPAN

Abstract. A fuzzy extension of multiattribute risk analysis for the multiple-agents decision problem is presented. Multiattribute risk analysis is concerned with decision making under hazardous environments. This paper presents a method for fuzzy evaluation of the multiattribute risk function, which embodies an aggregation device based on membership functions of the fuzzy set theory. The construction of the fuzzy multiattribute risk function is intended to be consistent with the multiattribute risk attitude in the multiattribute utility function and also with the expected utility hypothesis in a modified form. The possibilistic approach is used for treating a conceptual imprecision that accrues from a multiplicity of evaluation and multiattribute risk attitudes.

Keywords. fuzzy multiattribute risk function, multiattribute utility analysis, group decision making, expectation hypothesis, possibility theory, fuzzy multiattribute aversion function.

1. INTRODUCTION

Multiattribute risk analysis is concerned with decision making under hazardous environments which often occur in modern high technology society. In this paper, we discuss a method for fuzzy evaluation of the multiattribute risk function which accrues from multiperson decision making.

In the evaluation of a multiattribute risk function, fuzziness is often caused by a multiplicity of decision makers. Variation in the evaluation can not be measured with full precision and uniqueness because of variety in human sensitiveness to the hazardous events which usually is assessed with linguistic ambiguity. This paper extends some

J. Kacprzyk and M. Fedrizzi (eds.), Multiperson Decision Making Using Fuzzy Sets and Possibility Theory, 198–218.

results for construction of the multiattribute risk function to the fuzzy evaluation due to the existence of the multiple decision makers.

In Section 2, construction of the multiattribute risk function (MRF) for the individual decision maker is presented, where the consistency of the MRF with the multiattribute utility function (MUF) is discussed. In Section 3, a device for constructing the fuzzy multiattribute risk function (FMRF) which embodies collective choice is discussed in the correspondence to the construction of the fuzzy multiattribute utility function (FMUF). In Section 4, possibilistic estimation for FMRF is discussed, where the concept of the fuzzy multiattribute aversion function (FMAF) is derived in a form different from the FMRF. In Section 5, the effectiveness of the expectation hypothesis for risk evaluation is examined in terms of the FMAF. In Section 6, concluding remarks are presented.

2. REPRESENTATION OF THE MULTIATTRIBUTE RISK FUNCTION

The multiattribute risk function is constructed for assessing a risky event with multiple attributes which represent their probable consequences as uncertain quantities combined with their probabilities (Table 1).

Table 1 Representation of a risky event

Event A		State $E_1 \cdots E_j \cdots E_n$	Probability E_j
Single attribute outcome	X	$x_1 \cdots x_j \cdots x_n$	p_j
Multiple attribute outcome	X_1	$x_{11} \cdots x_{1j} \cdots x_{1n}$	p_{1j}
	$\vdots$	$\vdots \quad \vdots \quad \vdots$	$\vdots$
	X_i	$x_{i1} \cdots x_{ij} \cdots x_{in}$	p_{ij}
	$\vdots$	$\vdots \quad \vdots \quad \vdots$	$\vdots$
	X_m	$x_{m1} \cdots x_{mj} \cdots x_{mn}$	p_{mj}
Probability		$p_1 \cdots p_j \cdots p_n$	p_j

Note.

(1) Risk profile: $p_j \triangleq (p_{1j}, p_{2j}, \cdots p_{ij}, \cdots p_{mj}),\ j = 1,\ldots,n.$

(2) Uncertain consequences for an attribute x_i:

$$x_i \triangleq (x_{i1}, x_{i2}, \ldots, x_{ij}, \ldots, x_{in}),\ i = 1,\ldots,m.$$

(3) For a state j, $p_i \triangleq p_{ij}$.

(4) x_{ij} shows a consequence of an event A when a state j occur.

Consider an event A (as an outcome of an action) with an n-chance fork whose state j, j = 1,...,n, is assessed in terms of the multiple attributes. Denote a set of m-attributes for the event A as $X \triangleq (X_1,\ldots,X_m)$ and a set of their quantities as $x \triangleq (x_1,\ldots,x_m)$. The probability function $p_i(x_i)$ is assessed subjectively for each attribute x_i, i = 1,...,m, in terms of the n-chance fork. Uncertain consequences for an attribute x_i, i = 1,...,m, form a set of random variables:

$$x_i \triangleq (x_{i1},\ldots, x_{in}), \tag{2.1}$$

A set of the consequences in a state j,

$$\Omega \triangleq \{x_j\} \triangleq \{(x_{1j},\ldots,x_{mj})\}, \qquad j = 1,\ldots,n, \tag{2.2}$$

defines a reference set for the event A. A set of the probability p_{ij}, i = 1,...,m,

$$p_j \triangleq (p_{1j},\ldots,p_{mj}), \tag{2.3}$$

defines a multiattribute risk profile of this event when a state j occurs.

A multiattribute risk function (MRF) of an event A is defined on the multiattribute risk profile p_j and represented as

$$R(p_j) \triangleq R(p_{1j},\ldots,p_{mj}). \tag{2.4}$$

Properties of MRF are as follows.

(i) MRF represents a multidimensional adverse utility function that represents a negative human preference, or an aversion. A higher order of magnitude of the MRF-value indicated a higher degree of the aversion, or displeasure, for the uncertain consequences of an event A in terms of their probabilities.

(ii) Assessment for the MRF is subjectively performed on each element p_{ij} in the risk profile p_j when a state j occurs, which takes a value between 0 and 1, and $p_{ij} > 0$ and $\sum_{j=1}^{n} p_{ij} = 1$. $R(p_j) = 0$ when $p_j = 0$, and $R(p_j) = 1$ when $p_j = 1$. Notice that p_{ij} is an epistemic probability and is judgmentally assessed for the x_{ij}-value which an attribute x_i will take when the state j occurs.

(iii) Assessment for the MRF includes the value trade-offs between the attributes, which are assessed in terms of their probabilities.

Now we consider the representation theorem of MRF under the following assumptions.

<u>Assumption 1</u> (Event independence). The occurrence of a set of attribute values x_j, j = 1,...,n, in (2.2) as uncertain consequences of an event A is independent of each other in the reference set Ω.

Assumption 2 (Aversion independence). Aversion to the occurrence of an attribute x_i is independent of that to the other attributes $\bar{x}_i$ in a state j.

Assumption 3 (Trade-off independence). A trade-off between a pair of attributes is assessed independently of the trade-offs for other pairs of the attributes in a state j.

Theorem 1 (representation of MRF). Under the Assumptions 1-3, the multiattribute risk function for a state j of an event A is represented in a form,

$$R(p_{1j},\ldots,p_{mj}) = \sum_{i=1}^{m} k_i r_i(p_{ij}) + K \sum_{\substack{i=1\\ s>i}}^{m} k_i k_s r_i(p_{ij}) r_s(p_{sj})$$

$$+ K^2 \sum_{\substack{i=1\\ s>i\\ \ell>s}}^{m} k_i k_s k_\ell r_i(p_{ij}) r_s(p_{sj}) r_\ell(p_{\ell j})$$

$$+ \ldots + K^{m-1} k_1 k_2 \ldots k_m r_1(p_{1j}) r_2(p_{2j}) \ldots r_m(p_{mj}), \tag{2.5}$$

where $0 < k_i < 1$. K is a solution of $1 + K = \prod_{i=1}^{m} (1 + Kk_i)$, where $K \neq 0$ and $K > -1$. $0 \leqslant r_i \leqslant 1$ and $0 \leqslant R \leqslant 1$.

Theorem 2 (Multiplicative form of MRF). The MRF (2.5) is represented in the multiplicative form,

$$R(p_{1j},\ldots,p_{mj}) = \frac{1}{K} [\prod_{i=1}^{m} (Kk_i r_i(p_{ij}) + 1) - 1], \tag{2.6}$$

when $\sum_{i=1}^{m} k_i \neq 1$.

Proof of Theorems 1 and 2 is straightforward on the same line as that for the multiattribute utility function (MUF) derived by Keeney and Raiffa (1976).

The $r_i(p_{ij})$ is a singleattribute risk function (SRF) for a state j and is assessed on a value of the probability function $p_i(x_i)$, and $\partial r_i/\partial p_{ij} > 0$. $r_i(p_{ij}) = 0$ when $p_{ij} = 0$, and $r_i(p_{ij}) = 1$ when $p_{ij} = 1$.

It has been known that the multiplicative form of the multiattribute utility function with the negative coefficient parameters implies the multiattribute risk averse attitude (Richard 1975). In terms of the singleattribute utility function (SUF), the risk averse attitude is consistent with the preference function having the classical Bernoulli type defined on the entire domain, gains and losses (Figure 1).

In terms of the SRF, the risk seeking attitude in the positive domain, where the negative preference is assessed, corresponds to the

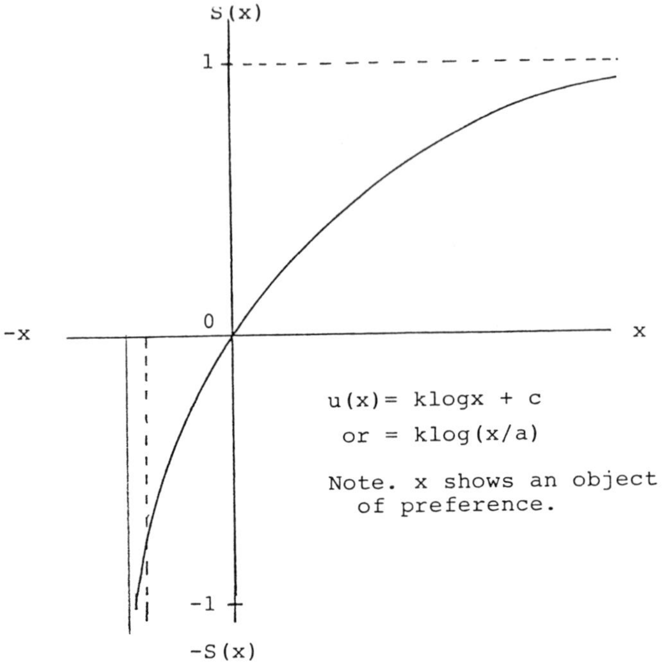

Figure 1. Representation of the preference function S(x) in a classical Bernoulli type.

risk averse attitude in the negative domain of the preference function (Figure 2). The existence of a critical level and the asymptotic approach to it in the risk evaluation for losses correspond to the existence of a saturation level and the asymptotic approach to it in the utility assessment for gains. In the assessment of the preference function in an entire domain of gains and losses, the negative utility assessment for losses will rapidly approach -1.0 beyond a critical level for an attribute, although the utility assessment for gains is bounded to 1.0. The mapping of this preference function onto the SRF $r_i(p_{ij})$ derives the risk proneness property. In other words, the value of SRF generally has a steeply increasing property corresponding to an increase of the probability for a risky attribute. This reflects the critical effect in the risk evaluation due to the existence of irreversibility in hazardous events. Thus the multiplicative form of MRF (2.6) with the positive coefficient parameters, which embodies the risk prone attitude, is consistent with the multiplicative type of MUF with the negative coefficient parameters, which embodies the risk averse attitude, when the monotonicity of the singleattribute functions SRF and SUF is presumed. Similarly, the multiplicative type of MRF with the negative coefficient parameters, which shows the risk averse attitude, is also consistent with the multiplicative type of MUF with the positive

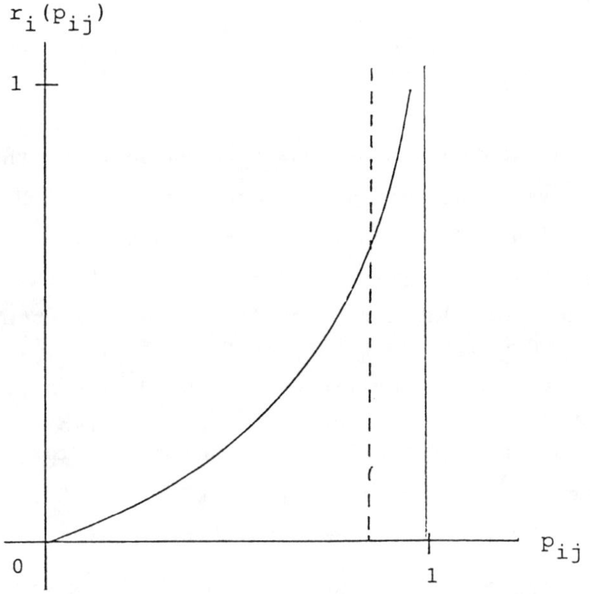

Figure 2. Representation of a risk function.

coefficient parameters, which shows the risk prone attitude. In short, the multiplicative form of the MRF holds a risk attitude consistent with the preference function to be assessed in an entire domain for both gains and losses, when the monotonicity of the singleattribute functions is presumed.

Assessment of the scaling constants k_i, $i = 1,\ldots,m$, on the SRF in (2.6) can be performed via the trade-off experiments between probabilities with which the values of the the risky attributes occur, as proved from the representation form (2.5) or (2.6). The indifference experiments are performed with answering the following questions in terms of the epistemic probabilities.

A preliminary work for determining the scaling constants is to derive an aversion order for the unwanted attributes in the reference set $x_j \subset \Omega$ in terms of their probabilities appearing in the risk profile p_j (2.3). Based on the derived ordering, relative values of the scaling constants are assessed via the trade-off experiment.

Question I (Assessment of the relative k_r values). Consider two qualitative probabilities $P_r[X_r]$ and $P_h[X_h]$ for unwanted attributes X_r and X_h, where X_h is the most important attribute to be taken into consideration. Suppose that a probability P_r for an occurrence of an unwanted attribute X_r will change from 1 (the worst situation) to 0 (the best situation). Then, how much degree of degradation for the most important attribute X_h can the decision maker (DM) allow in terms of its probability P_h ?

Question II (Assessment of the basic k_h value). On the one hand, consider a certain outcome in which the probability P_h for the most important attribute X_h takes 1 (the worst situation) and all other probabilities P_r, $r = 1,\ldots,m$, $r \neq h$, take 0 (the best situation). On the other hand, consider a probability mixture of uncertain consequences (lottery) in which all probabilities P_i, $i = 1,\ldots,m$, take 1 with a probability π or all probabilities P_i take 0 with a probability $1-\pi$. Then how much is the π-value which makes these two sets of consequences, the certain outcome and the lottery, indifferent for DM ?

An answer to Question I, via an assignment of the p_{hx}-value in quantitative terms, brings on an indifferent point such that

$$k_r = k_h r_h(p_{hx}) \tag{2.7}$$

on the $(p_r - p_h)$ plane. Then the relative k_r-value can be obtained from

$$\frac{k_r}{k_h} = r_h(p_{hx}), \; r \neq h, \; r \in I \triangleq (1,\ldots,m). \tag{2.8}$$

An answer to Question II determines the k_h-value by

$$k_h = \pi. \tag{2.9}$$

Then all numerical values of the scaling constants k_i, $i = 1,\ldots,m$, in (2.6) are uniquely determined.

It should be noted that, in the construction of MRF, the assessments for all constants and function forms of SRF are performed in "crisp" terms as a result of individual decision making although evaluation of the uncertain events are the matter of major concern.

In the next section, fuzzification of the MRF is introduced as a result of variation of the evaluation by multiple decision makers.

3. CONSTRUCTION OF THE FUZZY MULTIATTRIBUTE RISK FUNCTION FOR COLLECTIVE CHOICE

The risky choice problems usually involve linguistic ambiguousness in human decisions, which often is brought on by variations of risk evaluation due to the existence of multiple decision makers. These phenomena can be appropriately treated by the fuzzy set theory. Basic concepts of fuzzy ordering for human preferences have been provided by

Zadeh (1971). As an extension of this approach to preference theory, we have presented a device for including a fuzzification in constructing the multiattribute utility function (MUF) in the context of collective choice (Seo and Sakawa 1985). In the analogous discussion to this device, we are concerned with introducing fuzzy evaluation of the multiattribute risk function (MRF) which accrues from the collective choice. We will reformulate the representation form of MRF (2.6) as follows.

$$\tilde{R}(p_{1j},\ldots,p_{mj}) = \frac{1}{K}\left[\prod_{i=1}^{m}(K\tilde{k}_i\tilde{r}_i(p_{ij}) + 1) - 1\right], \quad (3.1)$$

where

$$\sum_{i=1}^{m}\tilde{k}_i \tilde{>} \tilde{1} \qquad \text{for } -1 < K < 0 \quad (3.2)$$

when the multiattribute risk averse attitude is embodied and

$$\sum_{i=1}^{m}\tilde{k}_i \tilde{<} \tilde{1} \qquad \text{for } K > 0 \quad (3.3)$$

when the multiattribute risk seeking attitude is embodied. We call the representation from (3.1) the fuzzy multiattribute risk function (FMRF). In FMRF (3.1), while the fuzzification is included in the scaling constant $\tilde{k}_i$ and the SRF $\tilde{r}_i$, the probability measure is still held for the assessment of the risk profile p_j.

It should be noted that the representation check by (3.2) and (3.3) for the multiplicative form of FMRF (3.1) uses the fuzzy comparison (Zadeh 1965) between the fuzzy numbers $\sum_{i=1}^{m}\tilde{k}_i$ and $\tilde{1}$.

3.1. Derivation of the fuzzy scaling constants

The fuzzification of the scaling constants $\tilde{k}_i$ is performed via the construction of the fuzzy aversion order for the probability P_{ij}, $i = 1,\ldots,m$, in the multiattribute risk profile p_j (2.3) before and after the trade-off experiment for assessing the relative values of the scaling constants. We will show it in two steps.

Step 1. Assessment of the fuzzy aversion order before the trade-off experiment.

(i) Making the score sheet for the multiple assessors, which describes the aversion order for the qualitative probabilities in the risk profile p_j (2.3) for a state j of a risky event A.

(ii) The membership function $\mu_R(P_r, P_s)$ for the fuzzy aversion ordering associates with an ordered pair of P_r, and P_s appearing in the risk profile P_j its "grade of membership" μ_R where R is a set of fuzzy binary relations for the aversion. The value of $\mu_R(P_r, P_s)$ takes a range of [0,1] and represents a strength of a binary relation R between P_r and P_s. According to a selected aggregation rule for the collective choice, the membership function $\mu_R(P_r, P_s)$ is assessed with a relative frequency of an aversion order appearing in the original score sheet in (i).

For example, a stranght-of-preference rule for collective choice is defined by

$$\mu_R(P_r, P_s) \triangleq \frac{1}{m} \sum_k w^k N^k(0_{rs}) \tag{3.4}$$

where $N^k(0_{rs})$ denotes a total number of the score (e.g. the number of votes) for an aversion ordering 0_{rs} represented as $P_r \succ P_s$, $P_r \prec P_s$ or $P_r \sim P_s$. The weighting coefficient w^k represents a degree of strength of an aversion for each ordered pair (P_r, P_s) of the qualitative probabilities. The m is the number of assessors (e.g. voters) multiplied by a maximum value of w^k, viz. $m \triangleq \max (w^k n)$ where n is the number of assessors. Note that the aversion order in 0_{rs} corresponds positively to the aversion order of the attributes, X_r and X_s.

(iii) The values of the membership function $\mu_R(P_r, P_s)$ are arrayed in the relation matrix:

$$[\mu_R] = \begin{array}{c} \\ P_1 \\ \vdots \\ P_m \end{array} \begin{array}{c} P_1 \quad \cdots\cdots\cdots \quad P_m \\ \begin{bmatrix} 0 & \ddots & \{\beta_{rs}\} \\ & \ddots & \\ \{\beta_{rs}\} & \ddots & 0 \end{bmatrix} \end{array}, \qquad r,s = 1,\ldots,m. \tag{3.5}$$

where $\beta_{rs} \triangleq \mu_R(P_r, P_s)$, e.g. defined as (3.4), and which represents the degree of strength of the binary relations in the collective choice. $1 \geqslant \beta_{rs} \geqslant 0$ when $r \neq s$ and $\beta_{rr} = 0$.

(iv) Transitivity check. The aversion order which is represented in the relation matrix (3.5) should hold the property of the fuzzy weak order for preserving the consistency of evaluation in the collective choice.

Definition 1 (Fuzzy weak ordering). Fuzzy weak ordering is defined with connectivity and transitivity as follows.

1) Connectivity:

$$P_r \neq P_s \Rightarrow \mu_R(P_r, P_s) > 0 \quad \text{or} \quad \mu_R(P_s, P_r) > 0. \tag{3.6}$$

2) Transitivity:

$$\mu_R(p_r, P_s) \geqslant \bigvee_t \mu_R(P_r, P_t) \wedge \mu_R(P_t, P_s), \tag{3.7}$$

where $\mu_R \vee \mu'_R$ denotes $\max(\mu_R, \mu'_R)$ and $\mu_R \wedge \mu'_R$ denotes $\min(\mu_R, \mu'_R)$.

When the connectivity is replaced by

3) Asymmetry:

$$\mu_R(P_r, P_s) > 0 \Rightarrow \mu_R(P_s, P_r) \not> 0, \tag{3.8}$$

the preference ordering is called the strong ordering.

The transitivity in (3,5) is not always held while the connectivity is usually held without difficulty. Thus the transitivity check should be performed by examining the numerical values appeared in the relation matrix (3.5) on the definition (3.7), which leads to construct the revised relation matrix embodying the transitivity as well as the connectivity.

(v) Derivation of a nonfuzzy ordering in the collective choice.

A nonfuzzy ordering is derived from the modified, weak-ordered relation matrix by using an α-level decomposition of the fuzzy relation R.

For a number α in [0.1], an α-level set R_α of the fuzzy binary relation R is defined by

$$R_\alpha = \{(P_r, P_s) \mid \mu_R(P_r, P_s) \geqslant \alpha\}. \tag{3.9}$$

The R_α is a nonfuzzy set of the binary relations and forms a nested sequence of nonfuzzy relations with $\alpha_r \geqslant \alpha_s \Rightarrow R_{\alpha_r} \subset R_{\alpha_s}$. The set of fuzzy binary relations R is decomposed into a union of several nonfuzzy relation sets R_α, which is nested sequentially in a descendant order into a hierarchical structure. The α is interpreted as an agreement level of a derived weak-ordered set of the binary relations in the collective choice.

A class C^j, j = 1,...,K, of the nonfuzzy weak-ordered relation set is constructed corresponding to the α-level decomposion by (3.9). The class C^j is defined as a set of all binary relations compatible with R_{α_j}. The intersection of these classes C^j forms a nonfuzzy weak ordering with a minimum α-level consistent with it, which shows an agreement level in the collective choice. The value α^* of the minimum α-level or the value of its membership function $\mu_{R_{\alpha^*}}$ indicates numerically "the degree of agreement for the derived nonfuzzy collective ordering." The nonfuzzy collective ordering at the α^*-level is derived on the qualitative probabilities for a state j as

$$P_r \succ \cdots \succ P_t \succ \cdots \succ P_s. \tag{3.10}$$

The trade-off experiment for assessing the relative values (2.8) of the scaling constants can be performed in terms of Question I on the qualitative probabilities for a risk profile (2.3). This work is performed as a crisp evaluation. The aversion order for the risk profile, however, has been constructed with an agreement level as a result of the fuzzy collective decisions.

Step 2. Assessment of the basic scaling constant as a fuzzy number.
The second step of the fuzzification is to assess the basic scaling constant $\tilde{k}_h$ as a fyzzy number. This experiment is performed on Question II and derives the probability $\tilde{\pi}$ for the prescribed lottery as a fuzzy number such as the L-R type (Dubois and Prade 1978, 1980):

$$\tilde{\pi} \triangleq (\bar{\pi}, \gamma, \delta) \tag{3.11}$$

where $\bar{\pi}$ is the "mean" value and γ and δ are the left and right spreads and represent a dispersion of the assessment. The basic scaling constant $\tilde{k}_h$ is assessed by

$$\tilde{K}_h = \tilde{\pi} \tag{3.12}$$

as the fuzzy number. Then all other scaling constants are assessed as the fuzzy numbers:

$$\tilde{k}_i = \tilde{k}_h \tilde{r}_h(p_{hj}), \quad i = 1,\ldots,m, \quad i \neq h, \tag{3.13}$$

where the SRF $\tilde{r}_h(p_{hj})$ can also be assessed as the fuzzy number with a dispersion as a result of the collective decision making. The forms of the membership function of the fuzzy numbers $\mu_{\tilde{k}_h}$ and $\mu_{\tilde{r}_h}$ are derived respectively with the values of the spreads for both sides, γ and δ, in their assessment.

3.2. assessment of the fuzzy singleattribute risk functions

The singleattribute risk function (SRF) $r_i(p_{ij})$, $i = 1,\ldots,m$, in FMRF (3.1) can also be assessed as the fuzzy number $\tilde{r}_i(p_{ij})$ such as the L-R type. We call $\tilde{r}_i(p_{ij})$ the fuzzy singleattribute risk function (FSRF). The construction of the membership function $\mu_{\tilde{r}_i}$, $i = 1,\ldots,m$, of FSRF is performed as a result of the varied assessments for risk evaluation on an assigned value of a probability p_{ij} in the collective choice (Figure 3).

Finally, the fuzzy multiattribute risk function (FMRF) (3.1) is derived with the fuzzy number operations for the multiplication (Dubois and Prade 1978, 1980).

It should be noted that, in this device, a probability $p_{ij}, i = 1,\ldots,m$, in the multiattribute risk profile p_j, $j = 1,\ldots,n$, is

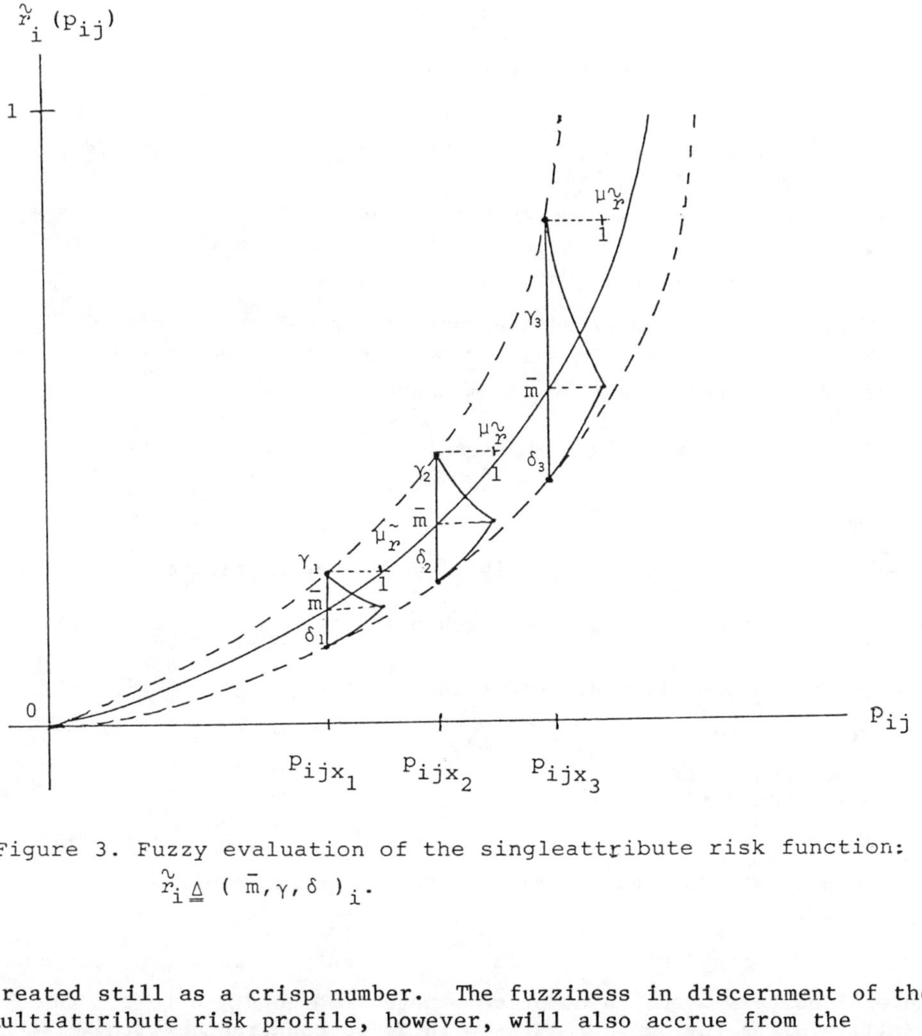

Figure 3. Fuzzy evaluation of the singleattribute risk function: $\tilde{r}_i \triangleq (\bar{m}, \gamma, \delta)_i$.

treated still as a crisp number. The fuzziness in discernment of the multiattribute risk profile, however, will also accrue from the variation in evaluation due to the collective choice. first, a fuzzy (= ill-defined) event can be described more properly with linguistic ambiguity, which is treated with possibilistic representations for the "degree of belief" rather than the probabilistic representations for "chance." Thus an epistemic probability p_{ij} in the multiattribute risk profile (2.3) can more naturally be recognized with the possibility measure in the fuzzy set theory (Zadeh 1978, Dubois & Prade 1985) than with the probability measure defined on the sigma field. In addition, one major source of the fuzziness is in dispersed recognition of the risk profile by multiple decision makers. The assessors have their own degree of sensitiveness to the fuzzy event. Thus the assessment of the multiattribute risk profile is not uniquely performed in crisp terms.

In the following discussion, the fuzziness in discernment of a "risky" event is treated in terms of the varied evaluation for a possibility ρ_i for an attribute x_i in the reference set Ω.

4. POSSIBILISTIC ESTIMATION FOR THE FMRF: DERIVATION OF THE FUZZY MULTIATTRIBUTE AVERSION FUNCTION

Now we will reconsider the FMRF in terms of the possibility theory.

A quantity x_i of an attribute X_i, $i = 1,\ldots,m$, in the reference set Ω for an event A is treated as a fuzzy variable $\tilde{x}_i$ in a fuzzy restriction set $\tilde{\Omega}$ which defined the fuzzy numerical set compatible with a proposition usually having the modifier in the description. The fuzzy set $\tilde{\Omega}$ is characterized by the membership function $\mu_{\tilde{\Omega}_i}(\tilde{x}_i)$ that is interpreted as the degree to which the fuzzy variable $\tilde{x}_i$ is compatible with the fuzzy restriction set $\tilde{\Omega}_i$ and takes a value in the interval [0,1], The fuzzy variable $\tilde{x}_i$ is associated with a possibility distribution Π_{X_i} in which the possibility that an attribute X_i takes a value $\tilde{x}_i$ is described with the membership function $\mu_{\tilde{\Omega}_i}(\tilde{x}_i)$ and is defined by the possibility distribution function $\pi(\tilde{x}_i)$ of Π_{X_i} associated with $\tilde{x}_i \in \tilde{\Omega}_i$. Naturally

$$\pi(\tilde{x}_i) = \mu_{\tilde{\Omega}_i}(\tilde{x}_i). \tag{4.1}$$

In these terms, the attribute X_i, $i = 1,\ldots,m$, can be enunciated in a form of a proposition. For example, suppose the attribute X_i be "the number of the fatalities due to the explosion", and be stated as a fuzzy description "is many" which defines a fuzzy restriction set $\tilde{\Omega}_i$ composed of several numbers within a numerical range from worst to best. The possibility distribution associated with X_i is constructed with:

$$\Pi_{X_i} \triangleq \{\tilde{x}_{ij}, \pi(\tilde{x}_{ij})\}, \quad j = 1,\ldots,n, \tag{4.2}$$

which can be denoted, for example, by

$$\Pi_{X_i} = 0.1/100 + 0.05/300 + 0.01/500 + 0.0001/1000 \tag{4.3}$$

(+ denotes a union).

In the case of collective choice,

$$\tilde{x}_i \triangleq (\tilde{x}_{i1},\ldots,\tilde{x}_{in}), \tag{4.4}$$

or in (4.3)

$$\tilde{x}_i = (100,\ 300,\ 500,\ 1000), \qquad (4.5)$$

is obtained as a result of multiperson decision making and represents a variation in the estimation of $\tilde{x}_i$.

It should be noted that the values of the possibility distribution function $\pi(\tilde{x}_{ij})$, $j = 1,\ldots,n$, does not obey the additivity axiom differing from the quantitative probability. The possibility measure is defined as $\pi(\tilde{x}_i) \triangleq \sup \pi(\tilde{x}_{ij})$, and $\pi(\phi) = 0$ and $\pi(\tilde{\Omega}_i) = 1$, where ϕ denotes an empty set while the probability measure p is defined with the additivity axiom: $p(\Omega_i) \triangleq p(\bigcup_{j=1}^{n} x_{ij}) = \sum_{j=1}^{n} p_j(x_{ij})$ and $\sum_{j=1}^{n} p_j(x_{ij}) = 1$ for $x_{ij} \in \Omega_i$, where $p_j(x_{ij}) \triangleq p_{ij}$, for the disjoint states.

An element P_{ij} of the multidimentional risk profile p_j (2.3) can be restated in terms of the possibility distribution function $\pi(\tilde{x}_i)$ which is defined to be equal to the membership function $\mu_{\tilde{\Omega}_i}(\tilde{x}_i)$. The possibility distribution Π_{X_i} associated with X_i describes the possibility of a proposition $\mathscr{P}_i$;

$$\mathscr{P}_i\text{: "The attribute } X_i \text{ takes their values in } \tilde{\Omega}_i\text{."} \qquad (4.6)$$

where

$$\tilde{\Omega}_i \ni (\tilde{x}_{i1},\ldots,\tilde{x}_{in}). \qquad (4.7)$$

The possibility distribution function $\pi(\tilde{x}_i)$ of Π_{X_i} takes their corresponding values:

$$\rho_i \triangleq (\pi(\tilde{x}_{i1}),\ldots,\pi(\tilde{x}_{in})\}, \qquad i = 1,\ldots,m. \qquad (4.8)$$

The set ρ_i (4.6) is called a possibility set for the attribute $\tilde{X}_i$. A set of values of the possibility distribution function $\pi(\tilde{x}_{ij})$,

$$\rho_j \triangleq (\pi(\tilde{x}_{ij}),\ldots,\pi(\tilde{x}_{mj})), \qquad j = 1,\ldots,n, \qquad (4.9)$$

is called a multiattribute possibility profile for a situation j discerned in a proposition for a prospect of an event. The representation (4.9) stresses the epistemic property of the information conveyed by a linguistic proposition for a hazardous event.

Using $\rho_{1j} \triangleq \pi(\tilde{x}_{1j}),\ldots,\ \rho_{mj} \triangleq \pi(\tilde{x}_{mj})$, the multiattribute possibility profile (4.9) is rewrited as

$$\rho_j \triangleq (\rho_{1j},\ldots,\rho_{mj}). \tag{4.10}$$

On the multiattribute possibility profile ρ_j (4.9), the fuzzy multiattribute aversion function (FMAF) is defined as

$$R(\rho_j) \triangleq R(\pi(\tilde{x}_{1j}),\ldots,\pi(\tilde{x}_{mj})). \tag{4.11}$$

or on (4.10) in an analogous term to MRF (2.4),

$$R(\rho_j) \triangleq R(\rho_{1j},\ldots,\rho_{mj}), \tag{4.12}$$

where $R(\rho_j)$ is normalized in [0,1]. We can derive the fuzzy multiattribute aversion function (FMAF) (4.12) in the representation form analogous to the multiattribute risk function (MRF) in the multiplicative form (2.6).

Assumption 1' (Noninteraction). The elements in the fuzzy restriction set $\tilde{\Omega}$ are decomposable. In other words, a set of the possibility distributions,

$$\Pi_X \triangleq (\Pi_{X_1},\ldots,\Pi_{X_m}), \tag{4.13}$$

associated with an attribute set $X \triangleq (X_1,\ldots,X_m)$ is represented as the Cartesian product of the possibility distributions Π_{X_i}, $i = 1,\ldots,m$,

$$\Pi_X \triangleq (\Pi_{X_1} \times \ldots \times \Pi_{X_m}), \tag{4.14}$$

where $X \triangleq (X_1,\ldots,X_m)$ is a set of the attributes taking their values $\tilde{x} \triangleq (\tilde{x}_1,\ldots,\tilde{x}_m)$ in $\tilde{\Omega} \triangleq \bigcup_i \tilde{\Omega}_i$. Using the possibility distribution function,

$$\pi(\tilde{x}_j) = \pi(\tilde{x}_{1j}) \wedge \ldots \wedge \pi(\tilde{x}_{mj}). \tag{4.15}$$

Or alternatively,

$$\rho_j = (\rho_{1j} \wedge \ldots \wedge \rho_{mj}). \tag{4.16}$$

(4.14)-(4.16) shows the noninteractivity of the fuzzy attributes $X \triangleq (X_1,\ldots,X_m)$. The noninteractivity is interpreted as the noncompensation in which a variation (increase or decrease) in an element of a possibility distribution (4.13) cannot be compensated by variations in the complementary elements.

The fuzzy multiattribute aversion function (FMAF) is represented, under the Assumptions 1' and 2, in the form of

$$R(\rho_j) = R(\rho_{1j} \wedge \ldots \wedge \rho_{mj}), \tag{4.17}$$

or using the fuzzy singleattribute aversion function (FSAF) $r_i(\rho_{ij})$,

$$R(\rho_j) = R(r_1(\rho_{1j}),\ldots, r_m(\rho_{mj})) \tag{4.18}$$

$$= R(r_s(\min_i \rho_{ij})). \tag{4.19}$$

The representation form is derived from (2.5) as

$$R(\rho_j) = \omega_s r_s(\min_i \rho_{ij}), \tag{4.20}$$

where ω_s is the scaling constant and $1 > \omega_s > 0$. $r_s(\min_i \rho_{ij})$ taking a value in [0,1] is a fuzzy singleattribute aversion function defined on

$$\min_i \rho_{ij} \triangleq (\rho_{1j} \wedge \cdots \wedge \rho_{mj}), \tag{4.21}$$

and $\frac{\partial r_i}{\partial \rho_{ij}} > 0$.

Under the Assumption 1' for the noninteraction, the assessment of FMAF is reduced to that of FSAF in term of (4.20). The assessment of the scaling constant ω_s, however, is still effective because the assessment of FSAF r_s is based on the evaluation for the multiattribute possibility profile ρ_j (4.10). A particular weight for the possibility, $\min_i \rho_{ij} \triangleq \rho_{sj}$, should be assigned according to the degree of relative importance of a possibility ρ_{sj} in the possibility profile (4.10) for a situation j. The ω_s-value is taken as a crisp number by an analogy to the scaling constant k_s for SRF which is assessed on the probability p_{sj} in (2.6), and thus the procedure for assessing ω_s is simply to answer the Questions I' and II' under the Assumption 3 in Section III in terms of the possibility values instead of the probability values. The trade-off experiments among ρ_{ij}, $i = 1,\ldots,m$, in a situation j where an attribute x_i takes a x_{ij} -value are performed as a deterministic (crisp) experiment by the decision maker (DM).

<u>Question I'</u> (Assessment of the relative ω_s value). Consider two possibilities ρ_s and ρ_h for unwanted fuzzy attributes $\tilde{x}_s$ and $\tilde{x}_h$ where $\tilde{x}_h$ is the most important attribute to be considered in the fuzzy restriction set $\tilde{\Omega}$. Suppose that a possibility ρ_s for an attribute $\tilde{x}_s$ to occur will vary from 1 (worst) to 0 (best). Then, how much degree of degradation for the most important attribute $\tilde{x}_h$ can the decision maker (DM) allow in terms of its possibility ρ_h?

Question II' (Assessment of the basic ω_h value). Consider a crisp outcome in which the possibility ρ_s for the fuzzy attribute $\tilde{x}_h$ takes 1 (worst) and the possibility ρ_s for the fuzzy attribute $\tilde{x}_s$ takes 0 (best). On the other hand, consider a possibilistic outcome where the possibilities ρ_h and ρ_s take 1 with a possibility ρ^*. Then how much is the ρ^*-value which makes these two outcomes, crisp and possibilistic, indifferent for DM ?

The Question I' derives

$$\frac{\omega_s}{\omega_h} = r_h(\rho_{hjx}), \tag{4.22}$$

via the assignment of ρ_{hjx}.

The Question II' determines the ω_h-value with

$$\omega_h = \rho^*. \tag{4.23}$$

Notice that $R(\rho_j)$ in (4.20) is assessed as a crisp number and the calculation of (4.20) does not require a fuzzy calculus. It is also possible, however, that the scaling constants $\tilde{\omega}_h$ and $\tilde{\omega}_s$ are assessed with the fuzzy numbers $\tilde{r}_h$ and $\tilde{\rho}^*$. In this case, the fuzzy calculus such as the scalar multiplication should be introduced. We will recommend to treat all of fuzzy numbers as the L-R type due to the easiness of fast computations.

5. REEXAMINATION OF THE EXPECTATION HYPOTHESIS FOR FMAF

This section is devoted to a reexamination of the expectation hypothesis when introducing the FMAF.

The expected utility model, developed by von Neumann and Morgenstern (1944), is based on several postulates on individual choice between gambles, which is described in Savage's axioms (1954). These are (i) weak order: connectivity and transitivity; (ii) the sure-thing principle: the substitution axiom, which asserts that a choice among actions is not affected by the payoff from a consequence of an event producing the same payoff for both of the alternative actions; (iii) admissibility: inadmissibility of dominated actions; (iv) independence: a probability for a consequence is assessed independently of the payoff; and (v) nontriviality: there is at least one pair of consequences (a,b) such that a $\succ$ b in an event.

The subjective expected utility (SEU) model is derived under Savage's axioms (i)-(v):

$$u(\hat{x}) = u\left(\sum_{j=1}^{n} p_j x_j\right) \tag{5.1}$$

$$= \sum_{j=1}^{n} p_j u(x_j) \tag{5.2}$$

where the prospect $\hat{x} \triangleq (x_1, p_1; \ldots; x_n, p_n)$ shows an n-chance fork gamble, and x_j is a consequence and p_j is a probability combined with x_j, where $j = 1, \ldots, n$. The linearity of the SEU model (5.1)-(5.2) should be noticed.

The questions to the SEU hypothesis have been raised. Allais's paradox (1952) mainly results from the violations of the consistency in holding the axiomatic properties. First is the violation of the substitution axiom (ii), which shows a preference reversal due to the shift from a sure gain to a gain with the probability one in an evaluation. Second is the violation of the independence axiom (iv) (Ellsberg 1961), which shows the asymmetric properties in assessing preference functions for gains and losses. These violations of the axiomatic properties represent the certainty effects, including the "more probability" effect and the "pseudo certainty" effect. The "more probability" effect shows that more probability is preferred to less probability and thus larger gains for nonprobable consequences than those for probable ones are demanded. The "pseudo certainty" effect shows that reframing a choice problem with a disguised certain gain induces a reversion of the preference. These effects lead to nonlinearity in the probability evaluation and as a result the linear additivity in the representation form of the SEU model (5.1)-(5.2) is confronted with contradiction.

In extension of the SEU model for coping with Allais's paradox, the following directions corresponding to the components, $u(x_j)$ and p_j, in (5.2) are proposed.

(1) Introducing the asymmetric property of the risk attitudes in the preference function for coping with the preference reversal. A counterexample for the symmetric property has been considered in terms of the reflection effect with skewed symmetry (Kahneman & Tversky 1979).

(2) Explicit introduction of nonlinearity in the probability evaluation. This means to use the weighted estimation of the probability distribution, by which the certainty effects can be explained (Ellsberg 1961, Kahneman & Tversky 1979, Karmarkar 1978, Handa 1977).

Concerning (1), the reflection effect is proposed in the prospect theory (PT) which asserts that, putting the present situation as the reference point, the preference function is concave in the domain for gains and convex for losses, and the response to losses is steeper than that to gains. This finding has been followed by the experiments for the multiattribute risk attitude (Payne, Laughhunn & Crum 1984): the multiattribute risk aversion holds for gains, and the multiattribute risk prone holds for losses. This finding is inconsistent with the risk attitude revealed in the representation form of the multiattribute utility function for which the preference should consistently be assessed with unvaried parameters and should not include the reversal of the parameters within an entire domain.

In the serious games with the hazardous events, however, differing from the cases of the hobby games, the phenomenon of the preference reversal in a domain for both of gains and losses will seldom occur.

The experiments supporting the PT hypothesis have been performed only for the hobby games. The serious game having the probability for occurrence of catastrophic consequences includes the following characteristics: (i) possible occurrence of irreversible losses, (ii) existence of a threshold level in acceptability of losses, and (iii) a more risk averse attitude to the ruinous losses than to the gains.

On the consideration for these points, we will confirm the consistency of the multiattribute risk function (MRF) in the representation form (2.6), as a counterpart of the multiattribute utility function (MUF), with the preference defined in an entire domain for both of positive and negative gains.

It is known, however, that the expectation hypothesis (5.3) for MRF in the von Neumann-Morgenstern sense

$$R(p) = \sum_{j=1}^{n} q_j R(p_j), \tag{5.3}$$

where q_j is a probability assigned to the risk profile p_j of a state j in an event A and denotes the "degree of belief" for the occurrence of the p_j, is still inconsistent with the nonlinearity to appear in the probability assessment of q_j.

Now we proceed to the analogous discussion for the possibilistic approach. Using the fuzzy multiattribute aversion function (FMAF) (4.20) derives the "expectation" form with a possibility q_j as follows:

$$R(\rho) = \bigoplus_{j=1}^{n} q_j \, R(\rho_j), \tag{5.4}$$

$$= \bigoplus_{j=1}^{n} q_s \omega_s r_s (\min_i \rho_{ij}), \tag{5.5}$$

where $\rho \triangleq (\rho_1, \ldots, \rho_n)$, and $\bigoplus$ stands for the "sigma" operation in (5.4) and (5.5) by analogy to (5.3). $R(\rho_j)$ is defined in (4.18). Recall that $R(\rho_j)$ in (5.5) is derived from the decomposability assumption of the fuzzy restriction set $\tilde{\Omega} \ni \tilde{x}_i$, $i = 1, \ldots, m$, in Section IV, which means the noninteractivity and is interpreted as the noncompensation. The representation form (5.5) does indicate no additive form in a decomposition of a multiattribute function, and simply indicates a "fuzzy expectation" of a singleattribute function scalar-multiplied, which is compatible with the multiattribute risk prone attitude for $R(\rho_j)$ and also is compatible with the multiattribute risk averse attitude of the assessor in the preference function for an entire domain of a fuzzy environment.

The remaining problem is to consider the implication of the q_j in (5.5). By analogy with the concept of "crisp" expectation, q_j is the degree of belief for the appearance of the ρ_{sj}-value in the possibility set ρ_s, where $\rho_{sj} \triangleq \min_i \rho_{ij} \triangleq \min_i \pi(\tilde{x}_{ij})$ and which takes dispersed values represented by the possibility distribution function $\pi(\rho_{sj}) = \mu_{T_j}(\rho_{sj})$. T_j is a set in which ρ_{sj} takes a value. The operation of $\oplus$ in (5.4) and (5.5) stands for the union and is replaced by $\vee$ (max) when $R(\rho_j)$ takes a crisp number. Then $R(\rho)$ takes in the following "expectation" form.

$$R(\rho) = \bigvee_j \mu_{T_j}(\rho_{sj})\omega_s r_s(\rho_{sj}) \tag{5.6}$$

$$= \bigvee_j \mu_{T_j}(\pi(\tilde{x}_{sj}))\omega_s r_s(\pi(\tilde{x}_{sj})) \tag{5.7}$$

where $\rho_{sj} = \min_i \rho_{ij} \triangleq \min_i \pi(\tilde{x}_{ij})$. Because the additivity is abandoned for the possibility measure in contrast with the probability measure which is used for the expected utility hypothesis, the nonlinearity which appears in the probability evaluation and leads to the violation of the expectation hypothesis (5.3) does face no difficulty in preserving the "expected" part of (5.5). In other words, appearance of the nonlinear attitudes in the possibility evaluation does not violate the construction of the fuzzy multiattribute aversion function (FMAF) in the "expectation" form (5.5).

6. CONCLUDING REMARKS

Human decisions on risk assessment for hazardous events usually include linguistic ambiguity due to the noncrisp properties in the framing of discernment of the events, which accrue from incompleteness of information and the epistemic nature of its interpretation. These properties particularly form varied evaluation in multiperson decision making. Thus the fuzzy extensions of risk assessment will be expected not only from the methodological points of view but also from the practical points of view. In the present paper, discussions on the multiattribute risk function embodying the consistencies with the SEU model and the MUF representation are extended, via the possibilistic approach, to fuzzy evaluation. The present study is based on the corresponding treatment of the possibility evaluation in the fuzzy multiattribute aversion function with the probability evaluation in the multiattribute risk function. There are some further elaborated relationships between both evaluation, however, which have been suggested in the theories of evidence and belief (Dempster 1967, Shafer 1976), for example. The elaboration is not taken into account in the present paper and remains for future investigations.

REFERENCES.

Allais, M., The foundations of a positive theory of choice involving risk and a criticism of the postulates and axioms of the American School (1952), in ed. by Allais, M. and O. Hagen, Expected Utility Hypothesis and the Allais Paradox, D. Reidel, 1979.

Dempster, A.P., Upper and lower probabilities induced by a multivalued mapping, Annals of Mathematical Statistics, **38**, 325-329, 1967.

Dubois, D. and H. Prade, Operations on fuzzy numbers, International Journal of Systems Science, **9(6)**, 613-626, 1978.

Dubois, D. and H. Prade, Fuzzy Sets and Systems, Theory and Applications, Academic Press, 1980.

Dubois, D. and H. Prade, Possibility Theory, Trans. by E.F. Harding, Plenum Press, 1988. (original edition in French, Masson 1985).

Ellsberg, D., Risk, ambiguity, and the savage axioms, Quarterly Journal of Economics, **75**, 1961.

Handa, J., Risk, probabilities, and a new theory of cardinal utility, Journal of Political Economy, **85(1)**, 1977, 97-122.

Kahneman, D. and A. Tversky, Prospect theory: an analysis of decision under risk, Econometrica, **47(2)**, 263-291, 1979.

Karmarkar, U.S., Subjectively weighted utility: a descriptive extension of the expected utility model, Organizational Behavior and Human Performance, **21**, 1978, 61-72.

Keeney, R.L. and H. Raiffa, Decision with Multiple Objectives: Preference and Value Tradeoffs, John Wiley & Sons, 1976.

Payne, J.W., D.J. Laughhunn, and R. Crum, Multiattribute risky choice behavior: the editing of complex prospects, Management Science, **30(11)**, 1350-1361, 1984.

Richard, S.F., Multivariate risk aversion, utility independence and separable utility functions, Management Science, **22(1)**, 12-21, 1975.

Savage, L.J., The Foundation of Statistics, Wiley, 1954.

Seo, F. and M. Sakawa, Fuzzy multiattribute utility analysis for collective choice, IEEE Trans. on Systems, Man and Cybernetics, Vol. SMC-15, No. 1, 1985 45-53.

Seo, F. and M. Sakawa, Multiple Criteria Decision Analysis in Regional Planning: Concepts, Methods and Applications, D. Reidel Publishing Co., 1988.

Shafer, G., A Mathematical Theory of Evidence, Princeton University Press, 1976.

von Neumann, J. and O. Morgenstern, Theory of Games and Economic Behavior, Wiley, 1944, 3rd ed. 1953.

Zadeh, L.A., Fuzzy set, Information and control, **8**, 338-353, 1965.

Zadeh, L.A., Similarity relation and fuzzy orderings, Information Sciences, **3**, 177-200, 1971.

Zadeh, L.A., Fuzzy set as a basis for a theory of possibility, Fuzzy Sets and Systems, **1**, 3-28, 1978.

CONSENSUS MEASURES FOR QUALITATIVE ORDER RELATIONS *

Piera Mazzoleni
University of Verona, via dell'Artigliere, 19, Italy.

Abstract. A new order relation is analyzed by applying consensus procedures and is developed in a fuzzy environment.
Such an order is then generalized to stochastic fuzzy decisions for economic and financial alternatives.
Some remarks on fuzzy generalized concavity are finally added.
Keywords. Convexity, dominance relation, fuzzy set, order relations, utility theory, consensus.

1. INTRODUCTION

The development of fuzzy set theory has emphasized the importance of qualitative properties for both numbers and orders.

Up to now the decision among available alternatives with multiple criteria in a fuzzy environment has usually been reduced to the definition of suitable weights for any choice alternative / criterion so that the final judgment is done in a crisp way.

Our aim in this paper is to give a constructive justification of fuzziness within the utility theory in both the univariate and multivariate case. In Section 2 we analyze a new fuzzy order relation and construct a utility criterion based on agreement procedures.

In Section 3 we generalize this new kind of comparison to random alternatives.

The study suggests also a new interpretation of concavity and certainty equivalent for the utility theory. Finally in Section 4 a new vector property of concavity is stated.

2. OPINIONS ON ORDER RELATIONS AND CONSENSUS MEASURES

A comparison between two quantities implies a reference level for economic agents. Therefore instead of verifying inequality $a \leq b$, it is more realistic to declare an "ideal" value v belonging to a suitable set V and test the implication $(a \leq \upsilon \rightarrow b \leq \upsilon)$.

(*)The paper has been partially supported by the National Research Council and the Italian Ministery of Education.

J. Kacprzyk and M. Fedrizzi (eds.), Multiperson Decision Making Using Fuzzy Sets and Possibility Theory, 219–230.

Let us consider a preference order $R_{\upsilon,a}$ on a finite set $X = \{x_i : i \in I\}$ of available alternatives, such that

$$x \, R_{\upsilon,a} \, y \Leftrightarrow (\mu\,(x,\upsilon) \geq a \Rightarrow \mu\,(y,\upsilon) \geq a) \qquad (1)$$

where μ is the membership function of a fuzzy binary relation on X xV, $v \in V$. Such an implication substitutes the direct comparison

$$\mu(x,y) \geq \mu(y,x) \qquad (2)$$

and represents the implicit reasoning of the economic agent who always thinks in relative terms.
$R_{\upsilon,a}$ satisfies the usual properties:

- reflexivity,

$$\mu(x,\upsilon) \geq a \Rightarrow \mu(x,\upsilon) \geq a$$

- antisymmetry

$$\{(\mu(x,\upsilon) \geq a \Rightarrow \mu(y,\upsilon) \geq a) \wedge (\mu(x,\upsilon) \geq a \Rightarrow \mu(y,\upsilon) < a)\} \Rightarrow$$

$$\Rightarrow \mu(x,\upsilon) = \mu(y,\upsilon) = a \; .$$

- transitivity

$$\{(\mu(x,\upsilon) \geq a \Rightarrow \mu(y,\upsilon) \geq a) \wedge (\mu(y,\upsilon) \geq a \Rightarrow \mu(z,\upsilon) \geq a)\} \Rightarrow$$

$$\Rightarrow (\mu(x,\upsilon) \geq a \Rightarrow \mu(z,\upsilon) \geq a)$$

and it is a fuzzy order relation.

The number $\mu(x,\upsilon)$ can be taken as the "cardinal" utility of x for the economic agent who weighs the "ideal" υ with the membership function μ.

But it is well known that opinions on "ideals" are imprecise and mainly given through a qualitative description. Let us consider υ as a normal fuzzy number $\tilde{\upsilon}$ with membership function $\mu_{\tilde{\upsilon}}$, support $V_0^+ = \{x \in \mathbb{R} : \mu_{\tilde{\upsilon}}(x) > 0\} =$

(w_0, w_1) and mode $V_1 = \{x \in \mathbb{R} : \mu_{\tilde{\upsilon}}(x) = 1\}\;) = [\upsilon_0, \upsilon_1]$

If we are comparing crisp alternatives, but we treat the inequality as a qualitative one, then, by applying the theory developed by Delgado, Verdegay and Vila (1988) we can define a distribution function

$$h_{\tilde{\upsilon}}\,(x) = \{0 \text{ if } x \leq w_0 ;\; \mu_{\tilde{\upsilon}}(x) \text{ if } x \in [w_0, w_1];\; 1 \text{ if } x \geq \upsilon_1\} \qquad (3)$$

and a corresponding comparison function

$$f_{\tilde{\upsilon},\lambda}(x) = ((\lambda+1)\, h_{\tilde{\upsilon}}\,(x)) \,/\, (1+\lambda h_{\tilde{\upsilon}}\,(x)),\; \lambda \in [-1,\, 0] \qquad (4)$$

which represents the membership function of the set of alternatives "believably $x \tilde{\geq} \upsilon$". Therefore we can set $u(x) = f_{\tilde{\upsilon},\lambda}(x)$ as a crisp utility function.

Only when f itself is described imprecisely and by means of qualitative properties, we truly need a fuzzy representation for the utility function (Chen, Lee and Yu (1984)).

Assume the economy under examination has a leading group G stating the rules of the comparison,

$$(\mu(x,\upsilon_g) \geq a,\ g\in G) \Rightarrow (\mu(y,\upsilon_g) \geq a,\ g\in G) \qquad (5)$$

$\geq$ denoting a suitable vector order relation, balanced between the two extreme cases of "all" and "at least one" component.

Then, by suitably modifying the "soft" measure of consensus introduced by Kacprzyk and Fedrizzi (1988a), we can describe a progressive agreement towards a suitable "ideal" reference point to be used by all the agents belonging to that economy, in order to define their personal utility function $u_i(x) = \mu_i(x,\upsilon^{ideal})$

Let us briefly summarize the construction of the agreement. Assume that the degree of preference of x over $\{\upsilon_g : g\in G\}$ can be represented by a vector r_{xg}.

The importance of any proposed "ideal" is given by its membership function $\mu(\upsilon_g)$ and the degree of strict agreement with respect to any pair of "ideals" is measured by

$$\mu(x;g1,g2) = \begin{cases} 1 \text{ if } r_{x,g1} = r_{x,g2} \\ 0 \quad \text{otherwise} \end{cases} \qquad (6)$$

The mutual importance of any pair of "ideals" is given by

$$b_{g1,g2} = \rho_1 \min\{\mu(\upsilon_{g1}),\ \mu(\upsilon_{g2})\} + \rho_2 \max\{\mu(\upsilon_{g1}),\ \mu(\upsilon_{g2})\} \qquad (7)$$

with $\rho_1 \geq 0$, $\rho_2 \geq 0$, $\rho_1 + \rho_2 = 1$ and the degree of agreement with respect to all the important pairs of ideals becomes

$$\mu_I(x) = \sum_{g1}\sum_{g2} \min\{\mu(x;\ g1,\ g2),\ b_{g1,\ g2}\} \Big/ \sum_{g1}\sum_{g2} b_{g1,\ g2}$$

Finally if the economic agents decide to refer their choices only to a subset of Q important "ideals" we can give a further degree of consensus by means of the membership function μ_Q

$$\mu_{QI}(x) = \mu_Q(\mu_I(x)) = \mu(x,\ \upsilon^{ideal}) \qquad (8)$$

Such a theory can be generalized to fuzzy alternatives

through the function $\beta_T(\tilde{x},\tilde{\upsilon}_g)=1-\sup_{Z\in\mathbb{R}} T(f_{\tilde{\upsilon}_g}(Z),\ \mu_{\tilde{x}}(Z))$ where T is a triangular norm.

Let us assume now that $\underline{x}\in\mathbb{R}^n$ is a basket of alternatives. Then we can compare $\underline{x}$ and the "ideal" reference by using one of the classical order relations: componentwise, lexicografical and noninferior.

We restrict our attention to fuzzy vectors such that $M^+ = \{\underline{x}\in\mathbb{R}^n:\ \mu(\underline{x}) > 0\}$ and $M_1 = \{\underline{x}\in\mathbb{R}^n:\ \mu(\underline{x}) = 1\}$ are intervals in $\mathbb{R}^n$ and set $\underline{M}_0 = \inf M^+$.

A joint distribution function can be associated with the fuzzy vector $\underline{x}$,

$$h_{\tilde{x}}(\underline{x}) = \{0 \text{ if } \underline{x} \ngeq \underline{M}_0;\ \mu(\underline{x}) \text{ if } \underline{x}\in M^+;\ 1 \text{ if } \underline{x} \geq \underline{M}_0,\ \underline{x}\notin M^+\}$$

and the basic implication (1) has to be applied to the measure

$$f_{\tilde{x},\lambda}(\underline{x}) = ((\lambda+1)\ h_{\tilde{x}}(\underline{x}))\ \Big/\ (1+\lambda h_{\tilde{x}}(\underline{x})) \tag{9}$$

More generally, assume that both the alternatives and the "ideal" reference on one hand and the order relation on the other hand, are fuzzy. Then for any realization $(\underline{x},\ \mu_{\tilde{x}}(\underline{x}))$, $(\underline{\upsilon},\ \mu_{\tilde{\upsilon}}(\underline{\upsilon}))$ we set

$$\mu_g(\underline{x},\ \underline{\upsilon}_g) = \begin{cases} 1 & \text{if } x_i \geq \upsilon_{ig} \quad \text{for any } i \\ c\in(0.5,\ 1) & \text{if there is } Q \subseteq \mathfrak{I} = \{1,..,n\}:\ x_Q \geq \upsilon_{Qg} \\ 0.5 & \text{if } x_i = \upsilon_{ig} \quad \text{for any } i \\ c\in(0,\ 0.5) & \text{if } x_{\bar{Q}} < \upsilon_{\bar{Q}g},\ \bar{Q} = \mathfrak{I} \setminus Q \\ 0 & \text{if } x_i < \upsilon_{ig} \quad \text{for any } i \end{cases} \tag{10}$$

with the further condition $\mu_g(\underline{x},\ \underline{\upsilon}_g) \leq \max\ \{\mu_{\tilde{x}}(\underline{x}),\ \mu_{\tilde{\upsilon}}(\underline{\upsilon})\}$

A suggestive way to represent this fuzzification procedure refers to the coordinate axes and to the corresponding order relation. If we represent the fuzziness by means of a suitable transformation we can visualize the procedure with a simple figure in $\mathbb{R}^2$ (Fig 1).

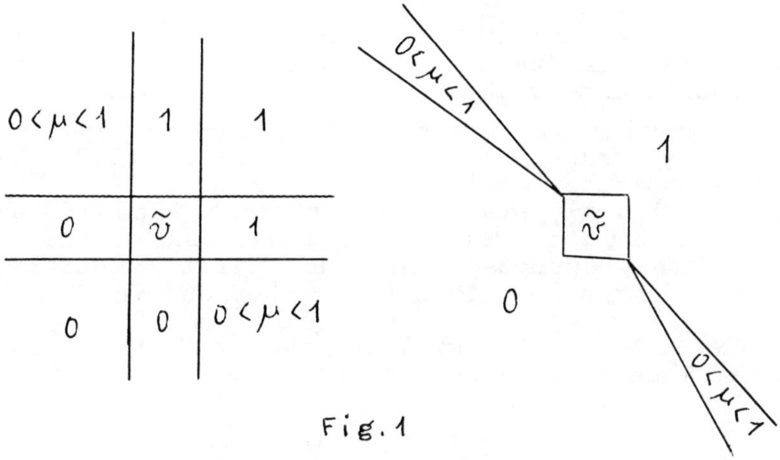

Fig. 1

3. A PARTICULAR PROPERTY OF FUZZY STOCHASTIC DOMINANCE

Fuzzy stochastic dominance has been studied by using a crisp utility function (Czogala and Disney (1988)) and a fuzzy utility function (Slyadz and Borisov (1988)). Let us briefly review these approaches.

Let us denote (Ω, B, P) a parameter space, $(\Omega_c, B_c) = ([0,1],$ Borel sets$)$ a characteristic space, $M = \{\mu : \Omega \to \Omega_c, (B, B_c)$-measurable$\}$ a family of characteristic variables.

A probabilistic set D on a set A of possible alternatives is defined by its membership function μ_D, $\mu_D : A \times \Omega \to \Omega_c$ where $\mu_D(x, .)$ is a (B, B_c) -measurable function.

Then we can write $F_{D(x)}(z)$ for each $x \in A$ where F stands for the distribution function of the respective probabilistic set, D(x).For any pair of alternatives x_i, $x_j \in A$, the possible impact on decision side is represented by the probability distributions $F_{D(x_i)}$, $F_{D(x_j)}$, on the unit interval $I = [0,1]$.

Czogala and Disney (1988) assume the existence of a crisp utility function and they state the r - th degree stochastic dominance as $F_{D(x_i)} \geq_r F_{D(x_j)}$ if $E(u,D)\,(x_i) \geq$

$$E(u,D)\,(x_j), u \in U_r \text{ where } E(u,D)\,(x) = \int_0^1 u(z)\, d\, F_{D(x)}(z)$$

and U_r denotes a class of utility functions satisfying

suitable properties according to the degree of dominance. They also give explicit inequalities for distribution functions of a single variable.

An imprecise utility approach is studied by Slyadz and Borisov (1988) in terms of fuzzy granules of information, g_i = "a consequence with attribute value G_i has utility V_i".

They develop a suitable algorithm to construct the corresponding membership functions in terms of the fuzzy attributes and fuzzy utilities.

In this section we analyze a different kind of dominance property and try to justify the fuzziness of the decision set. For this purpose let us first generalize the probability dominance, given by Wrather and Yu (1982).

DEFINITION 1. Let X and Y be two real - valued random variables. Then X is said to dominate Y with probability $\beta \geq 0,5 (X \beta Y)$ if

$$\text{Prob } (X > Y) \geq \beta \tag{11}$$

Therefore between two alternatives a decision - maker will choose the one which gives a better result more likely. Such a criterion allows us to compare cases in which neither the classical mean - variance nor the stochastic dominance of various degrees succeed.

If u is a bounded nondecreasing utility function, normalized over the unit interval [0,1], then it can be treated as the probability distribution of a suitable random variable V, according to Castagnoli (1989).

The following theorem reminds us of a very useful equality.

THEOREM 1. Let X and Y be two independent random variables with distribution function F_X and F_Y, respectively. Then we have

$$\text{Prob } (X \geqq Y) = \int_{\mathbb{R}} F_Y(x) d\, F_X(x) \tag{12}$$

Therefore the mean value of the utility function can be read as the probability of a suitable inequality

$$E\,(u(x)) = \int_{\mathbb{R}} u(x)\, dF_X(x) = \text{Prob } (X \geqq V) \tag{13}$$

Analogously, let X and Y be two independent fuzzy numbers with membership function μ_X and μ_Y, respectively, and h_X, h_Y denote the corresponding distributions: then the equality

$$\mu_{X \geqq Y} = \int_{\mathbb{R}} h_Y(x) \; dh_X(x) \tag{14}$$

gives another way of considering the normalized nondecreasing utility function u, that is, as the membership function of a suitable fuzzy variable, so we are led back to the analysis developed in the previous section.

Let us now give a new definition of stochastic dominance

DEFINITION 2. A random variable X dominates a random variable Y with probability β with respect to V if the following implication holds

$$\text{Prob}(X > V) \geq \beta \Rightarrow \text{Prob}(Y > V) \geq \beta \tag{15}$$

We can now assume that X and Y are probabilistic sets while V is either a random variable or a fuzzy variable and develop a new property of fuzzy stochastic dominance.

It is worth noticing that any comparison implies an ideal reference and the inequality $F_{D(x_i)}(z) \geq F_{D(x_j)}(z)$ is better represented as an implication.

Indeed the following holds

$$\begin{aligned} E(u,D)\ (x) &= \int_0^1 u(z)\ d\ F_{D(x)}(z) = \\ &= \int_0^1 F_V(z)\ d\ F_{D(x)}(z) = \text{Prob}(\tilde{X} \geqq V) \end{aligned} \tag{16}$$

Assume now that u is a fuzzy utility function so that a suitable membership function can be defined for $\tilde{E}(u, D)\ (x)$, say $\mu_{\tilde{E}}$. Then

$$\tilde{E}\ (\tilde{u}, D)\ (x) = \int_0^1 \tilde{u}\ (z)\ d\ F_{D(x)}(z) = \text{Prob}\ (\tilde{X} \geqq \tilde{V}) \tag{17}$$

is a fuzzy function with membership function $\mu_{\tilde{E}}$ (Dubois and Prade (1988)), which can be calculated by applying the extension principle.

DEFINITION 3. A fuzzy random variable $\tilde{X}$ dominates a fuzzy random variable $\tilde{Y}$ with degree β with respect to $\tilde{V}$ if the following implication holds

$$\mu_{\tilde{E}}(\tilde{X}, \tilde{V}) \geq \beta \Rightarrow \mu_{\tilde{E}}\ (\tilde{Y}, \tilde{V}) \geq \beta. \tag{18}$$

But what is even more interesting, we can apply the consensus procedure to the "ideal" V as in the previous section. Indeed, we can either leave V free to vary within a preassigned set $\mathcal{V}$ or define it more precisely for a leading group of economic agents by applying some "consensus" procedure. Let us consider the second approach.

Suppose that the economic agents of the group under examination face the ideal probabilistic sets $\{\tilde{V}_g : g \in G\}$ that are proposed by the leading group G and try to reach a compromise decision.

Let us follow Kacprzyk and Fedrizzi (1988). Assume for simplicity that the preference relation between the random fuzzy variables $\tilde{X}$ and $\tilde{V}_g$ can be discretely represented by a matrix $[r^g_{x,w}]$ according to the realization of the random variables and to the g - th "ideal".

The importance of any proposed "ideal" is given by the distribution function of the probabilistic set $\tilde{v}_g(w)$, $\mu(v_g, w)$and the mutual importance of any pair of "ideals" is defined as

$$b_{g1,\, g2} = \rho_1 \min \{\mu\, (v_{g1},\, w),\, \mu(v_{g2},\, w)\} + \rho_2 \max \{\mu\, (v_{g_1},\, w),\, \mu(v_{g_2},\, w)\} \tag{19}$$

If we introduce the degree of strict agreement with respect to any pair of "ideals"

$$\mu(\, x,\, w;\, g1,\, g2) = \begin{cases} 1 & \text{if } r^{g1}_{x,\, w} = r^{g2}_{x,\, w} \\ 0 & \text{otherwise} \end{cases} \tag{20}$$

then we can balance μ and b in order to define a degree of agreement with respect to all the important pairs of "ideals"

$$\mu_I(x,w) = \sum_{g1} \sum_{g2} \min\{\mu(x,w;\, g1, g2),\, b_{g1,g2}\} \Big/ \sum_{g1} \sum_{g2} b_{g1,\, g2} \tag{21}$$

Finally, if the economic agents restrict their attention to a subset of Q important "ideals" we can further balance μ_I towards $\mu_{QI}(x,\, w) = \mu_Q(\mu_I(x,\, w)) = \mu(\tilde{X},\, \tilde{V}^{ideal})$.

Assume now that X, Y are random vectors in $\mathbb{R}^n$.

Then the expected utility criterion

$$M\, (u\, (X_1, \ldots\ldots, X_n)) \geqq M\, (u\, (Y_1 \ldots\ldots, Y_n)) \tag{22}$$

for any u componentwise increasing and such that the mean values exist, holds if and only if we have the probability inequality

$$\text{Prob}\ (X_1>Z_1,\ \ldots,\ X_n>Z_n) \geqq \text{Prob}\ (Y_1>Z_1, \ldots,\ Y_n > Z_n) \qquad (23)$$

for any $(Z_1, \ldots\ Z_n)^T \in \mathbb{R}^n$. In this case a suitable random vector V can be found so that

$$E\ (u\ (X_1, \ldots, X_n)) = \text{Prob}\ (X \geqq V) \qquad (24)$$

The reverse inequality

$$\text{Prob}\ (X_1 \leqq z_1, \ldots, X_n \leqq z_n) \leqq \text{Prob}\ (Y_1 \leqq z_1, \ldots, Y_n \leqq z_n)$$

for any $(z_1, \ldots, z_n)^T \in \mathbb{R}^n$, is equivalent to

$$E\ (u\ (X_1, \ldots, X_n)) \leqq M\ (u\ (Y_1, \ldots, Y_n)) \qquad (25)$$

for any survival function u (Castagnoli (1989)) and we can again define an "ideal" vector V so that $E\ (u\ (X_1, \ldots, X_n)) = 1 - \text{Prob}\ (X < V)$. But the equality of the scalar case $\text{Prob}\ (X \geqq V) = 1 - \text{Prob}\ (X < V)$ no longer holds for vector random variables. Indeed a survival function should be represented as $1 - \text{Prob}\ (X \geq V)$, so that all the alternatives are explored.

These remarks show how valuable is to soften the vector inequality. The introduction of a fuzzy utility function $\tilde{u}$ corresponds to both fuzzifying $\tilde{V}$ and $\tilde{\geq}$ according to a suitable membership function $\mu_{\tilde{u}}$: for any pair $(X(w), \mu_{\tilde{X}}(x,w))$, $(V(w), \mu_{\tilde{V}_g}(v,w))$ and for any proposed "ideal" $g \in G$, we set

$$\mu_g\ (X\ (w), V_g(w)) = \begin{cases} 1 \quad \text{if } x_i(w) \geqq v_{ig}(w) \quad \text{for any } i \\ c_g \in (0.5,\ 1) \text{ if there is } Q \subseteq \mathfrak{I} = \{1, \ldots, n\} \\ \qquad x_Q\ (w) \geqq v_{Qg}(w) \\ 0.5 \quad \text{if } x_i(w) \text{ and } v_{ig}(w) \text{ are} \\ \qquad \text{indifferent for any } i,\ w \\ c_g \in (0,\ 0.5) \text{ if } x_{\bar{Q}}(w) < v_{\bar{Q}g}(w),\ \bar{Q} = \mathfrak{I} \setminus Q \\ 0 \quad \text{if } x_i(w) < v_{ig}(w) \text{ for any } i \end{cases} \qquad (26)$$

with the further condition $\mu_g(X(w), V_g(w)) \leq \max\{\mu_{\tilde{X}}(x,w), \mu_{\tilde{V}_g}(v,w)\}$

We think that this is a more realistic way to describe opinions on preferences for random vectors, that is, for random variables observed at a finite number of times. After having applied the consensus procedure to the proposals of the leader group $\mathcal{G}$, we can justify the qualitative and imprecise behavior of the economic agents belonging to a given economy in order to choose among fuzzy investments, fuzzy income streams, qualitatively described in their market behavior.

4. FUZZY ORDER AND GENERALIZED CONCAVITY

One of the classical properties of a utility function $u(x)$ is related to a comparison between associative means.

THEOREM 2 (Hardy, Littlewood and Polya, 1952). If ϕ and ψ are continuous and strictly monotone on the unit interval I and ψ is increasing, then a necessary and sufficient condition that $M_\psi(X) \le M_\phi(X)$ is that $\chi = \psi[\phi^{-1}(x)]$ should be concave.

Here X denotes a random variable and M generalized mean.

The classical concavity of $u(x)$ requires for instance that the simple arithmetic mean, $\alpha x^0+\bar{\alpha}x^1 = \alpha\phi(\xi^0)+\bar{\alpha}\phi(\xi^1)$ is transformed into $\lambda y^0+\bar{\lambda}y^1 = \lambda\psi(\xi^0)+\bar{\lambda}\psi(\xi^1)$, where $\lambda = \lambda(\alpha)$ and the upper bar denotes the complement to 1, *i. e.* $\bar{\alpha}=1-\alpha, \bar{\lambda}=1-\lambda$.

If ϕ is the identity function then theorem 2 assures that for any random variable X, the certainty equivalent $M_u(X)$ for a given utility function u is not greater than the expected value E (X) if and only if $u(x)$ is a concave function.

But it is self evident that $M_u(X)$ is a qualitative quantity, which heavily depends on the properties taken into account, and should be represented as a fuzzy number $\tilde{CE} = \tilde{M}_u(X) = [c^0, c^1]$. In correspondence we find an interval of possible weights modifying the ari metic mean of the utility values (Fig. 2)

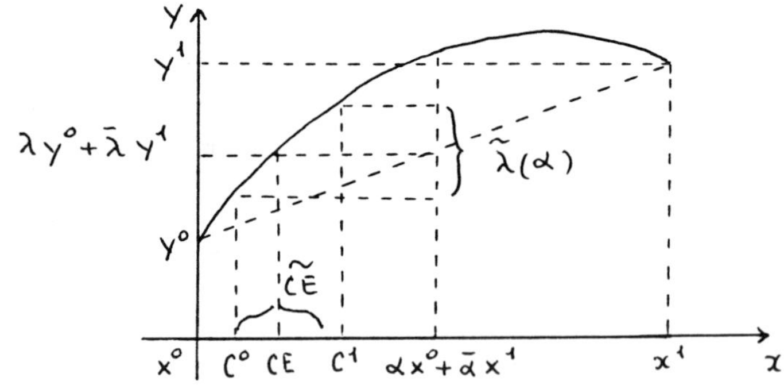

Fig. 2

DEFINITION 4. A function f: C → ℝ, defined on a convex subset C of $\mathbb{R}^n$ is said to be fuzzy -λ- concave if for any

$x^0 \in C$, $x^1 \in C$, $x^0 \neq x^1$ we have

$$f(\alpha x^0 + \bar{\alpha} x^1) \tilde{\geq} \tilde{\lambda}(\alpha) f(x^0) + \tilde{\bar{\lambda}}(\alpha) f(x^1) \tag{27}$$

with degree of membership $\mu_{\tilde{\lambda}}$.

Indeed we can weigh the risk/love aversion of a decision - maker and give intermediate cases between λ^0-concavity and λ^1- convexity

$$\lambda^0 f(x^0) + \bar{\lambda}^0 f(x^1) \leq f(\alpha x^0 + \bar{\alpha} x^1) \leq \lambda^1 f(x^0) + \bar{\lambda}^1 f(x^1) \tag{28}$$

by weighing the range of variability with a membership function

$$\mu_\lambda (f^0, f^1) = \begin{cases} 0 & \text{if } \lambda^0 = 1,\ \lambda^1 \notin [0, 1] \\ c \in (0, 0.5) & \text{if } \lambda^0 \in [0,1],\ \lambda \notin [0,1] \\ 0.5 & \text{if } \lambda^0 = \alpha,\ \lambda^1 = \alpha \\ c \in (0.5, 1) & \text{if } \lambda^0 \notin [0,1],\ \lambda^1 \in [0,1] \\ 1 & \text{if } \lambda^0 \notin [0,1],\ \lambda^1 = 1 \end{cases} \tag{29}$$

According to this definition, we can judge the behavior of the economic agent through its certainty equivalent.
Consider now a vector - valued function $f: C \to \mathbb{R}^K$ defined on a convex subset C of $\mathbb{R}^n$ and refer to the m-concavity property given by Castagnoli and Mazzoleni (1988):
- f is said to be m - concave if for any $x^0 \in X$, $x^1 \in X$, $\alpha \in I$ there are $\lambda(\alpha) \in I$, $i = 1, \ldots\ldots, K$ such that

$$f_i(\alpha x^0 + \bar{\alpha} x^1) \geq \lambda_i(\alpha) f_i(x^0) + \bar{\lambda}_i(\alpha) f_i(x^1) \tag{30}$$
$$i = 1, \ldots, K$$

We recognize two possible generalizations: first we can choose a fuzzy vector $\{\tilde{\lambda}_i(\alpha): i = 1,..,K\}$ with membership function $\mu_{\tilde{\lambda}_i}$, $i = 1,\ldots, K$ so that the second members of the inequalities (30) become fuzzy; second, we can weaken the kind of inequality towards the following membership function,

$$\mu(f(x(\alpha)), \lambda_f) = \begin{cases} 1 & \text{if } f_i > \lambda_{if} \text{ for any } i \\ c \in (0.5,1) & \text{if there is } Q \subseteq \mathfrak{I} = \{1,..,n\} \mid f_Q \geqq \lambda_{Qf} \\ 0.5 & \text{if } f_i = \lambda_{if} \text{ for any } i \\ c \in (0, 0.5) & \text{if } f_{\bar{Q}} < \lambda_{\bar{Q}g},\ \bar{Q} = \mathfrak{I} \setminus Q \\ 0 & \text{if } f_i < \lambda_{if} \end{cases} \tag{31}$$

Where $f(x(\alpha)) = f(\alpha x^0 + \bar{\alpha} x^1)$, $\lambda_f = \lambda(\alpha) f(x^0) + \bar{\lambda}(\alpha) f(x^1)$.

5. CONCLUDING REMARKS

The inequalities we have studied in Sections 2 - 3 have opened the way to new properties of concavity, which allow us to describe the preference towards risk in an imprecise and qualitative, and therefore presumably more realistic manner. We have a further reason to introduce fuzzy elements not only for the means used in the comparison but also for the orders and (α, λ) - concavity can be weighed within the known families and even across them.

REFERENCES

Castagnoli E. and Mazzoleni P.(1988) Scalar and vector generalized concavity, in Proceed, International School "Nonsmooth optimization and related topics" Erice (Tp) Plenum Press New York.

Castagnoli E. (1989) Qualche riflessione sull'utilità attesa, in Proceed. XIII Meeting of the Italian Association AMASES, Verona September 13 - 15.

Chen C. Q., Lee S. C. and Yu E. S. H. (1983)Application of fuzzy set theory to economics in Wang (1983) pp. 227 - 305.

Czogala E. and Disney P. L. (1988) Decision making in a probabilistic fuzzy environment, in Kacpryzk and Fedrizzi (1988b) pp. 215 - 226.

Delgado M., Verdegay J. L. and Vila M. A. (1988) A procedure for ranking fuzzy number using fuzzy relations, Fuzzy Sets and Systems 26, 49 - 62.

Dubois D. and Prade H. (1988) Decision evaluation methods under uncertainity and imprecision, in Kacprzyk and Fedrizzi (1988), 48 - 65.

Hardy C. H., Littlewood J. E. and Polya G. (1967) Inequalities, Cambridge Univ. Press.

Kacprzyk J. and Fedrizzi M. (1988a) A "soft" measure of consensus in the setting of partial "fuzzy" preferences, EJOR 34, 316 - 325.

Kacprzyk and Fedrizzi M. (Eds.) (1988b) Combining Fuzzy Imprecision with Probabilistic Uncertainity in Decision Making. Springer Verlag, Berlin.

Slyadz N. N. and Borisov A. N. (1988) Decision making based on fuzzy stochastic and statistical dominance, in Kacprzyk and Fedrizzi M. (1988b) 197 - 214.

Wang P. P., (ed.) (1983) Advanced in Fuzzy Sets, Possibility Theory and Applications, Plenum, New York.

Wrather C. and Yu P. L. (1982) Probability dominance in random outcomes, JOTA 36, 315 - 334.

ON A CONSENSUS MEASURE IN A GROUP MCDM PROBLEM

Michele Fedrizzi
Institute of Statistics and Operations Research
University of Trento, via Rosmini 42
38100 Trento, Italy

Abstract: A method for consensus measuring in a group decision problem is presented for the multiple criteria case. The decision process is supposed to be carried out according to Saaty's Analytic Hierarchy Process, and hence using pairwise comparison among the alternatives. Using a suitable distance between the experts' judgements, a scale transformation is proposed which allows a fuzzy interpretation of the problem and the definition of a consensus measure by means of fuzzy tools as linguistic quantifiers. Sufficient conditions on the expert's judgements are finally presented, which guarantee any a priori fixed consensus level to be reached.

Keywords: Group decision making, multiple criteria, degree of consensus, fuzzy preferences.

1. INTRODUCTION

An interesting issue within the group decision theory is that of measuring the consensus inside the group.

We consider the decision problem in which m experts have to express their judgements on n different alternatives on the basis of p criteria; from these judgements, priorities are then derived to be assigned to the alternatives.

In section 2 the structure of the problem and the resolution method are briefly described, both referring essentially to the Analytic Hierarchy Process (A.H.P.) introduced by T. Saaty in the 70's (Saaty, 1977,1980).

The aim of this paper is to analyze the consensus problem and to supply a suitable consensus measure in this particular framework.

The choice of studying the consensus measuring problem in this context is due to the wide interest the A.H.P. has produced both in scientific literature and its practical applications (several US - government agencies, consulting firms and corporations are currently using the A.H.P.).

In section 3 a dissimilarity measure between the opinions of the experts is introduced, coherent with Saaty's 1 to 9 ratio scale on which the opinions are expressed. This dissimilarity measure induces, in a natural way, a scale transformation, thus allowing a fuzzy formulation of the problem; a soft measure of the consensus is then defined for each criterion, according to the approach of Fedrizzi, Kacprzyk and Zadrozny (1988). More precisely, a fuzzy-logic-based calculus of linguistic quantified propositions is used to derive a measure of consensus which express the degree to which ,e.g., "almost all experts agree with the group's opinions concerning the most important alternatives".

To synthesize the degrees of consensus which refer to the different criteria, a parametrized operator due to Zimmermann and Zysno (1980,1983) is used, which allows the

J. Kacprzyk and M. Fedrizzi (eds.), Multiperson Decision Making Using Fuzzy Sets and Possibility Theory, 231–241.

choice of different values of compensation among the criteria.

In section 4 some conditions on the experts' judgements are supplied, which guarantee an a-priori fixed level of consensus.

2. THE DECISION PROBLEM AND THE RESOLUTION METHOD

This section briefly describes the decision problem and the method used to derive the priorities to be assigned to the alternatives.

The main features of the Analytic Hierarchy Process are supposed to be known, and only brief references will therefore be given.

Let $S = \{s_1, \ldots, s_n\}$ be a set of alternatives and $C = \{c_1, \ldots, c_p\}$ a set of criteria; each expert out of a group of m formulates his judgements on the alternatives by pairwise comparison on the basis of the given criteria.

It is required to calculate, by means of these judgements, the priority vector $\mathbf{w} = (w_1, \ldots, w_n)$, where w_i indicates the weight, or priority, the group assigns to s_i.

In the pairwise comparison matrices

$$A^{kh} = [a_{ij}^{kh}] \qquad k = 1, \ldots, m \quad ; \quad h = 1, \ldots, p$$

the element a_{ij}^{kh} represents the ratio between the priority of s_i and that of s_j, as subjectively judged by expert k, according to criterion c_h.

Saaty suggests measuring this intensity of preference a_{ij}^{kh} using a ratio scale, and precisely the 1 to 9 scale: $a_{ij}^{kh} = 1$ indicates indifference between s_i and s_j, $a_{ij}^{kh} = 9$ indicates that s_i is absolutely preferred to s_j, and $a_{ij}^{kh} \in \{2, 3, \ldots, 8\}$ indicates intermediate evaluations.

A^{kh} is completed by putting, for the remaining elements,

$$a_{ij}^{kh} = \frac{1}{a_{ji}^{kh}}.$$

Matrices A^{kh} are therefore called positive reciprocal matrices.

If the following equality holds

$$a_{ij}^{kh} \cdot a_{jl}^{kh} = a_{il}^{kh} \qquad \forall i, j, l = 1, \ldots, n, \tag{1}$$

i.e., if expert k is perfectly coherent in his judgements, matrix A^{kh} is said to be consistent.

Being (1) unrealistic for subjective evaluations, consistency is not required for the matrices A^{kh} (the measure of inconsistency of the matrices is nevertheless an important aspect of the theory).

The experts will also perform similar pairwise comparison in set C of criteria, thus providing m additional matrices $B^k = [b_{ij}^k]$, $k = 1, \ldots, m$. The quantities b_{ij}^k will therefore indicate the ratio between the relevance of criterion c_i and that of criterion c_j according to the opinion of expert k.

Matrices A^{kh} and B^k (for $k = 1,\ldots,m$) are aggregated by means of the geometric mean in order to derive the following matrices, which express the opinions of the group:

$$A^h = [a_{ij}^h], \quad h = 1, \ldots, p; \quad B = [b_{ij}],$$

where

$$a_{ij}^{h} = \left(\prod_{k=1}^{m} a_{ij}^{kh} \right)^{\frac{1}{m}} \qquad \text{and} \qquad b_{ij} = \left(\prod_{k=1}^{m} b_{ij}^{k} \right)^{\frac{1}{m}}. \tag{2}$$

About the functional properties of the geometric mean and the suitability of this kind of aggregation operator for the problem under examination, see the interesting paper by Aczél and Saaty (1983). As an example, it is easy to verify that the geometric mean preserves the reciprocity :

$$a_{ji}^{h} = \frac{1}{a_{ij}^{h}}, \quad b_{ji} = \frac{1}{b_{ij}}, \quad h = 1,\dots,p\,; \quad i,j = 1,\dots,n.$$

By means of group matrices A^h and B the problem can be hierarchically structured, and the A.H.P. method can be applied to calculate priority vector $\mathbf{w}$.

More precisely, let us consider the 3-levels group hierarchy where the third level contains the alternatives and the second one the criteria, the first being, as usual, simply the vertex of the hierarchy.

Following the A.H.P., the normalized eigenvector (say $\mathbf{w}^h$) corresponding to the maximum eigenvalue of A^h is calculated for $h = 1,\dots,p$, thus obtaining the so-called local priority vectors.

Analogously, the normalized eigenvector $\mathbf{b} = (b_1,\dots,b_p)$ is calculated, which corresponds to the maximum eigenvalue of B.

The global priority vector $\mathbf{w}$ is calculated according to the principle of the hierarchical composition:

$$\mathbf{w} = \sum_{h=1}^{p} b_h \mathbf{w}^h.$$

3. EVALUATION OF THE EXPERTS' AGREEMENT

In comparing alternatives it is a crucial point to determine not simply whether different opinions agree or not, but, also, how close the judgements are.

A suitable definition of the dissimilarity among experts' opinions is therefore a prerequisite to derive a consistent measure of consensus in the group.

3.1. A dissimilarity measure among experts' opinions

Taking into account the meaning of the subjective estimates $a_{ij}^{kh} \in [1/9, 9]$, it is evident that the euclidean distance $| a_{ij}^{k_1h} - a_{ij}^{k_2h} |$, for instance, is not a suitable dissimilarity measure between the judgement of the two experts k_1 and k_2 : it is obvious, for example, that the two estimates $a_{ij}^{k_1h} = 1$ and $a_{ij}^{k_2h} = 2$ are much more dissimilar than the estimates $a_{ij}^{k_1h} = 8$ and $a_{ij}^{k_2h} = 9$.

Having indicated by $d\left(a_{ij}^{k_1h}, a_{ij}^{k_2h}\right)$ the dissimilarity measure we are looking for, let us consider the following set of conditions that function d must satisfy. Let us, for simplicity, indicate by x and y the arguments of d, assuming, in the following, that x and y are positive real numbers.

(I) $d(x,y)$ is continous for $x,y>0$

(II) d is a distance: $\left.\begin{array}{l} d(x,y) = d(y,x) \\ d(x,y) = 0 \Leftrightarrow x = y \\ d(x,z) \leq d(x,y) + d(y,z) \end{array}\right\} \forall x,y,z>0$

(III) $d(x,y) = d(kx,ky) \quad \forall k > 0, \forall x,y> 0$

(IV) $d(1/x,y) = d(1/x,1) + d(1,y) \quad \forall x,y \geq 1$

(V) $d(1/9,9) = 1$

A short explanation is needed for conditions (III) $\rightarrow$ (V). Condition (III) states that equal distance is assigned to pairs of judgements with equal ratio; (IV) states that if two judgements are disagreeing, in the sense that the first (say $1/x$) prefers s_j to s_i and the second one (say y) prefers s_i to s_j, then the distance between them is the sum of the distance between the first and the indifference (=1) and the distance between the indifference and the second one. Finally (V) is a normalization condition.

By the following theorem a dissimilarity measure is supplied, which will be widely used in the rest of the paper.

Theorem 1

The only function d satisfying conditions (I) $\rightarrow$ (V) is the following:

$$d(x,y) = \frac{1}{2} \mid \log_9 x - \log_9 y \mid . \tag{3}$$

Proof

From (III), taking $k = 1/y$, is $d(x,y) = d(x/y,1)$. This means that, in order to calculate the distance between two points, it is sufficient to have the expression of the distance of an arbitrary point from point 1. Let us therefore indicate by $f(z)$ the distance of z from 1:

$$f(z) = d(z,1).$$

From (III), (II) and (IV) it is then, for $x,y \geq 1$

$$d(xy,1) = d(x,1) + d(y,1),$$

or

$$f(xy) = f(x) + f(y). \tag{4}$$

Taking into account (I), the general solution of functional equation (4) in $[1,+\infty[$ is (see Aczél, 1966):

$$f(z) = k_1 \log z \,, \quad k_1 \in \Re,$$

where $\log z$ is the natural logarithm of z.

Analogously, for $x, y \geq 1$, we obtain

$$f\left(\frac{1}{x} \cdot \frac{1}{y}\right) = f\left(\frac{1}{x}\right) + f\left(\frac{1}{y}\right),$$

or

$$f(xy) = f(x) + f(y) \qquad \text{for} \quad 0 < x, y \leq 1.$$

Solving the previous equation in $]0,1]$ we obtain again

$$f(z) = k_2 \log z \,, \quad k_2 \in \Re.$$

For $z \geq 1$, condition $f(z) \geq 0$ implies $k_1 > 0$ ($k_1 = 0$ violates (II)). For the same reason from $0 < z < 1$ it follows that $k_2 < 0$.
It is easy to verify that $k_2 = -k_1$; in fact, from (III) it is $d(1/z,1) = d(1,z)$, and hence $f(1/z) = f(z)$. By choosing, e.g., $z \geq 1$, it is therefore

$$k_2 \log\left(\frac{1}{z}\right) = k_1 \log z,$$

and then

$$-k_2 = k_1.$$

It is now possible to express f more synthetically (k stays for k_1):

$$f(z) = k \mid \log z \mid, \quad k > 0, z > 0.$$

Function d therefore takes the following expression:

$$d(x,y) = d\left(\frac{x}{y},1\right) = f\left(\frac{x}{y}\right) = k\left| \log\left(\frac{x}{y}\right)\right|,$$

where constant k is determined by (V):

$$1 = d\left(\frac{1}{9},9\right) = d\left(\frac{1}{9},1\right) + d(1,9) = 2d(1,9) = 2f(9) = 2k \mid \log 9 \mid,$$

and then

$$k = \frac{1}{2\log 9}.$$

Expression (3) is finally obtained:

$$d(x,y) = \frac{1}{2}\left| \frac{\log\frac{x}{y}}{\log 9} \right| = \frac{1}{2} \mid \log_9 x - \log_9 y \mid.$$

It is immediate to verify that conditions (II) are all satisfied:

$$d(x,y) = \frac{1}{2}\mid \log_9 x - \log_9 z\mid = \frac{1}{2}\mid \log_9 x - \log_9 y + \log_9 y - \log_9 z\mid \leq d(x,y) + d(y,z).$$

\\\

3.2. Obtaining fuzzy preference relations

Expression (3) suggests to interpret the distance d between two points x and y of the interval [1/9,9] as the usual euclidean distance between $g(x)$ and $g(y)$ where g represents a transformation of logarithmic type . More precisely, it is possible to rewrite (3) as

$$d(x,y) = \mid g(x) - g(y) \mid,$$

where

$$g(x) = \frac{1}{2}(1 + \log_9 x).$$

Function g maps [1/9,9] into [0,1] with some interesting properties. Putting

$$r_{ij}^{kh} = \mu_{R^{kh}}(s_i, s_j) = g\,(a_{ij}^{kh}),$$

a fuzzy preference relation $R^{kh} = [r_{ij}^{kh}]$ is defined for each expert k and each criterion c_h.

The following properties point out how matrices A^{kh} and R^{kh} relates:

$a_{ij}^{kh} = 1/9$	$\Leftrightarrow$	$r_{ij}^{kh} = 0$	stating that s_j is absolutely preferred to s_i.
$a_{ij}^{kh} = 9$	$\Leftrightarrow$	$r_{ij}^{kh} = 1$	stating that s_i is absolutely preferred to s_j.
$a_{ij}^{kh} = 1$	$\Leftrightarrow$	$r_{ij}^{kh} = 0.5$	stating indifference between s_i and s_j.
$a_{ij}^{kh} \cdot a_{ji}^{kh} = 1$	$\Leftrightarrow$	$r_{ij}^{kh} + r_{ji}^{kh} = 1.$	(5)

The last property is rather interesting, as it shows that g transforms the "multiplicative" reciprocity of Saaty's matrices in the "additive" reciprocity usually requested for the fuzzy preference relations (Tanino, 1988).

Another property of function g is pointed out by applying it also to the elements a_{ij}^h of the group matrices A^h, thus defining the group fuzzy preference relations $R^h = [r_{ij}^h]$:

$$r_{ij}^h = \mu_{R^h}(s_i, s_j) = g\,(a_{ij}^h) = \frac{1}{2}(1 + \log a_{ij}^h).$$

It is now easy to verify that r_{ij}^h is just the arithmetic mean of r_{ij}^{kh} for k=1,...m :

$$r_{ij}^h = \frac{\sum_{k=1}^{m} r_{ij}^{kh}}{m}$$

(remember that a_{ij}^h was obtained as the geometric mean of a_{ij}^{kh} for k=1,...m).

The following diagram synthesizes the previous results

$$\begin{array}{ccc} a_{ij}^{kh} & \xrightarrow{\quad g \quad} & r_{ij}^{kh} \\ \text{geometric mean} \downarrow & & \downarrow \text{arithmetic mean} \\ a_{ij}^{h} & \xrightarrow[\quad g \quad]{} & r_{ij}^{h} \end{array}$$

To conclude, it can be said that by means of function g it is possible to transform, in a certain sense, a "multiplicative" formulation of the problem into an "additive" one.

3.3. A consensus measure

In this subsection a fuzzy-logic-based calculus of linguistically quantified propositions is used, for which we refer to Fedrizzi et al. (1988).

Being $d(\cdot,\cdot) \in [0, 1]$ a dissimilarity measure (as defined in subsection 3.1), expression

$$v_{ij}^h(k) = 1 - d(a_{ij}^{kh}, a_{ij}^h), \qquad h = 1,\ldots,p; \quad k = 1,\ldots,m; \quad i,j = 1,\ldots,n$$

will therefore express the degree of agreement between expert k and the group as to their preferences between alternatives s_i and s_j on the basis of criterion c_h.

In order to measure the consensus on vector $\mathbf{w}^h$, let us define the relevance of the pair of alternatives (s_i, s_j) as

$$\beta_{ij}^h = \frac{w_i^h + w_j^h}{2} \qquad h = 1,\ldots,p \quad i,j = 1,\ldots,n.$$

The degree of agreement between expert k and the group as to their preferences between all the relevant pairs of alternatives can then be expressed by

$$V^h(k) = \frac{\sum_{i=1}^{n-1} \sum_{j=i+1}^{n} v_{ij}^h(k) * \beta_{ij}^h}{\sum_{i=1}^{n-1} \sum_{j=i+1}^{n} \beta_{ij}^h} \qquad h = 1,\ldots,p \tag{6}$$

where * is a t-norm. Note that denominator of (6) adds to (n-1)/2. Let Q_1 be a linguistic quantifier, i.e. a fuzzy set in [0,1], and let us consider, for instance, Q_1 = "most".

It is therefore possible to define the degree of agreement between expert k and the group as to their preferences between Q_1 relevant pairs of alternatives:

$$V_{Q_1}^h(k) = \mu_{Q_1}(V^h(k)) \quad , \qquad h = 1,\ldots,p; \quad k = 1,\ldots,m$$

The degree of agreement of all experts with the group as to their preferences between Q_1 relevant pairs of alternatives is

$$V_{Q_1}^h = \frac{\sum_{k=1}^{m} V_{Q_1}^h(k)}{m} \qquad h = 1,\ldots,p.$$

Let Q_2 be another linguistic quantifier similar to Q_1 (e.g. Q_2 = "almost all"); the degree of agreement of Q_2 experts with the group as to their preferences between Q_1 relevant pairs of alternatives is

$$E_h = \mu_{Q_2}\left(V_{Q_1}^h\right) \qquad h = 1,\ldots,p. \tag{7}$$

In order to obtain a global measure (say F) of the consensus which synthesizes the degrees of agreement (7), let us aggregate them by means of a parametrized operator due to Zimmermann and Zysno (1983):

$$F = \left(\prod_{h=1}^{p} E_h^{b_h}\right)^{1-\gamma} \left(1 - \prod_{h=1}^{p} (1 - E_h)^{b_h}\right)^{\gamma}. \tag{8}$$

As the compensation parameter γ varies from 0 to 1, the operator describes the whole class of operators between "AND" and "OR". It is therefore possible to choose the desired compensation among the degrees of agreement corresponding to the different criteria.

Note that (8) takes into account the relevance b_h of the criteria as determined by the A.H.P.. Since the Zimmermann-Zysno operator requires that $b_1 + \ldots + b_p = p$, every b_h must therefore be multiplied by p. For simplicity the same notation b_h is maintained in (8).

Finally F can be interpreted as the degree of agreement of almost all (Q_2) experts, on the basis of the p criteria, as to their preferences between most (Q_1) pairs of alternatives which have turned out to be relevant through the A.H.P. (remember that the degrees of agreement were weighted by the w_i^h's). It can then be said that F is a consensus measure on the priority vector **w**.

4. SUFFICIENT CONDITIONS FOR THE CONSENSUS

Some conditions on the matrices A^{kh} are presented, which guarantee an a-priori fixed level of consensus.

4.1. The one-criterion case

Let us consider, for a fixed criterion c_h, the following condition on the elements of matrices A^{kh} :

$$\exists \delta > 1 \quad : \quad \frac{a_{ij}^{k_1 h}}{a_{ij}^{k_2 h}} \leq \delta \qquad \begin{array}{l} \forall k_1, k_2 \in \{1, \ldots, m\} \\ \forall i, j \in \{1, \ldots, n\}. \end{array} \tag{9}$$

Condition (9) states that, for each pair of alternatives, no estimate can be so different from any other, to be more than δ times greater.

The next theorem shows that for each desired consensus level $z^* \in [0, 1]$ it is possible to find a suitable value of δ which guarantees, under assumption (9), the consensus E_h to be not less than z^*.

Before formulating the theorem, let us assume that membership functions μ_{Q_1} and μ_{Q_2} have the following expressions:

$$\mu_{Q_1}(x) = \begin{cases} 0 & \text{for} \quad 0 \leq x \leq \alpha \\ \dfrac{x-\alpha}{\beta-\alpha} & \text{for} \quad \alpha < x < \beta, \\ 1 & \text{for} \quad \beta \leq x \leq 1 \end{cases} \qquad \mu_{Q_2}(x) = \begin{cases} 0 & \text{for} \quad 0 \leq x \leq \alpha' \\ \dfrac{x-\alpha'}{\beta'-\alpha'} & \text{for} \quad \alpha' < x < \beta' \\ 1 & \text{for} \quad \beta' \leq x \leq 1. \end{cases} \tag{10}$$

Theorem 2

Under the following assumptions

(i) functions μ_{Q_1} and μ_{Q_2} are defined by (10),

(ii) the product is chosen as t-norm in (6),

(iii) for any $z^* \in [0,1]$, (9) is satisfied by taking

$$\delta = 9^{\sigma(z^*)}, \tag{11}$$

where

$$\sigma(z^*) = 2(1-\alpha-(\beta-\alpha)[\alpha' + z^*(\beta'-\alpha')]),$$

the following inequality holds

$$E_h \geq z^*.$$

Proof

From (9) it follows

$$\frac{1}{\delta} \leq \frac{a_{ij}^{k_1 h}}{a_{ij}^{k_2 h}} \leq \delta \qquad \forall k_1, k_2 \in \{1, \dots, m\} \quad \forall i,j \in \{1, \dots, n\}$$

and thus, being a_{ij}^h a mean,

$$\frac{1}{\delta} \leq \frac{a_{ij}^{kh}}{a_{ij}^{h}} \leq \delta. \qquad \forall k \in \{1, \dots, m\} \quad \forall i,j \in \{1, \dots, n\}$$

Taking the logarithms we have

$$-\log_9 \delta \leq \log_9 a_{ij}^{kh} - \log_9 a_{ij}^h \leq \log_9 \delta,$$

and then

$$d(a_{ij}^{kh}, a_{ij}^h) = \frac{1}{2} | \log_9 a_{ij}^{kh} - \log_9 a_{ij}^h | \leq \frac{1}{2} \log_9 \delta.$$

We therefore obtain

$$v_{ij}^h(k) \geq 1 - \frac{1}{2}\log_9 \delta,$$

and after some calculations, from (11) we have

$$V_{Q_1}^h(k) \geq \mu_{Q_1}\left(1 - \frac{1}{2}\log_9 \delta\right) = \mu_{Q_1}(\alpha + (\beta-\alpha)[\alpha' + z^*(\beta'-\alpha')]).$$

Taking into account (10) it is therefore

$$V_{Q_1}^h \geq \alpha' + z^*(\beta' - \alpha');$$

from (7) and (10) we finally obtain

$$E_h = \mu_{Q_2}\left(V_{Q_1}^h\right) \geq \mu_{Q_2}(\alpha' + z^*(\beta'-\alpha')) \geq z^*.$$

\\\

4.2. Global consensus

Let us now take into account all the criteria $c_1, \ldots, c_p$; the following theorem, which is analogous to theorem 2, supplies conditions which guarantee a fixed level t^* for the global consensus F given by (8).

Theorem 3

Under assumptions (i) and (ii) of theorem 2, for any $t^* \in [0,1]$, if (9) is satisfied by taking

$$\delta = 9^{\phi(t^*)}, \tag{12}$$

where

$$\phi(t^*) = 2(1 - \alpha - (\beta - \alpha)[\alpha' + (t^*)^{1/p}(\beta' - \alpha')]),$$

then

$$F \geq t^*.$$

Proof

Being

$$\phi(t^*) = \sigma((t^*)^{1/p}),$$

it follows from theorem 2

$$E_h \geq (t^*)^{1/p} \qquad h = 1, \ldots, p.$$

From (8) it is then (remember that $b_1 + \ldots + b_p = p$)

$$F \geq (t^*)^{1-\gamma}\left(1 - (1 - (t^*)^{1/p})^p\right)^\gamma.$$

It is now sufficient to prove that for any $t \in [0,1], \gamma \in [0,1]$ and $p \in N$ is

$$(t^*)^{1-\gamma}\left(1 - (1 - (t^*)^{1/p})^p\right)^\gamma \geq t^*. \tag{13}$$

Taking the logarithms in (13) we have, after some calculations,

$$t^* + (1 - (t^*)^{1/p})^p \leq 1,$$

which can be written as

$$((t^*)^{1/p})^p + (1 - (t^*)^{1/p})^p \leq 1.$$

The last inequality is easily verified to be true:

$$((t^*)^{1/p})^p + (1 - (t^*)^{1/p})^p \leq (t^*)^{1/p} + 1 - (t^*)^{1/p} = 1.$$

Inequality (13) and consequently theorem 3 are therefore proved.

\\\

It can be noted that inversion of (12) allows an alternative formulation of theorem 3. Instead of fixing a level t^* of the consensus and consequently derive sufficient conditions which guarantee the level t^* to be reached, it is possible to determine, by direct examination of matrices A^{kh}, the minimum value of δ, say δ^*, which satisfies (9).

A lower bound for the global consensus F is then obtained:

$$F \geq \left(\frac{\left(1 - \alpha - \frac{1}{2}\log_9 \delta^*\right)/(\beta - \alpha) - \alpha'}{\beta' - \alpha'} \right)^p .$$

5. CONCLUDING REMARKS

(a) Assumption (ii) of theorems 2 and 3 can be modified by choosing "min" as t-norm; more generally, any t-norm * s.t. $x*y \geq xy \quad \forall x, y \in [0,1]$ can be chosen. Statements of the theorems still hold, since the degree of agreement $V^h(k)$ given by (6) does not decrease by this substitution.

(b) If number 9, which Saaty proposes as the maximum value of the ratio scale in the A.H.P., is substituted with any other value $n \in \{2, 3, \ldots\}$, all the results of the previous paragraphs still hold, with obvious changes in the formulas.

REFERENCES

Aczél J.(1966), *Lectures on functional equations and their applications*, Academic Press, New York / London.

Aczél J., Saaty T.L.(1983), Procedures for synthesizing ratio judgements, *Journal of Mathematical Psychology*, **27**, 93-102.

Barzilai J., Cook W.D., Golany B.(1987), Consistent weights for judgements matrices of the relative importance of alternatives, *Operations Research Letters*, **6**, 131-134.

Crawford G., Williams C.(1985), A note on the analysis of subjective judgment matrices, *Journal of Mathematical Psychology*, **29**, 387-405.

Fedrizzi M., Kacprzyk J., Zadrozny S. (1988), An interactive multi-user decision support system for consensus reaching processes using fuzzy logic with linguistic quantifiers, *Decision Support Systems*, **4**, 313-327.

Harker P., Vargas L.G. (1987), The theory of ratio scale estimation: Saaty's analytic hierarchy process, *Management Science*, **33**, 1383-1402.

Kemeny J.G., Snell J.L. (1962), *Mathematical Models in the Social Sciences*, Blaisdell Publishing Company, Waltham, Massachusetts.

Saaty T. L. (1977), A scaling method for priorities in hierarchical structures, *Journal of Mathematical Psychology*, **15**, 234-281.

Saaty T.L.(1980), *The analytic hierarchy process*, Mc.Grow-Hill, New York.

Saaty T.L.(1986), Axiomatic foundation of the analytic hierarchy process, *Management Science*, **32**, 841-855.

Tanino T.(1988), Fuzzy preference relations in group decision making, in Kacprzyk J., Roubens M. (Eds.), *Non-Conventional Preference Relations in Decision Making*, Springer-Verlag, Berlin.

Zahedi F.(1986), The Analytic Hierarchy Process - a Survey of the Method and its Applications, *Interfaces*, **16**, 96-108.

Zimmermann H.J., Zysno P.(1980), Latent connectives in human decision making, *Fuzzy Sets and Systems*, **4**, 37-51.

Zimmermann H.J., Zysno P.(1983), Decisions and evaluations by hierarchical aggregation of information, *Fuzzy Sets and Systems*, **10**, 243-260.

VOTING PROCEDURES WITH A PRIORI INCOMPLETE INDIVIDUAL PROFILES

Jacek. W. Mercik
Institute of Production Engineering and Management
Technical University of Wrocław
25, Smoluchowski Str., 50-370 Wrocław, Poland

Abstract: A short review of voting procedures based on incomplete individual profiles is given. Two groups of such procedures: the approval and related Borda based ones are analysed via simulation with spatial models. Positional characteristics and efficiency of these procedures are presented.
Keywords: incomplete information, group decision making, voting.

1. INTRODUCTION

Voting is a commonly used way of group decision making. The solution of how to choose one from two alternatives via voting is essentialy simple: the alternative which is prefered by majority should be chosen. If one has to choose among three (or even more) alternatives, the answer is not that simple. There are many voting procedures leading to the solution. The results which one may obtain using different procedures may differ to a large extent.

Let $A=\{ a_1, a_2,...,a_n\}$ denote the set of alternatives; one of them should be chosen (via voting) as the winner; $card(A)=n\geq 3$.

We also assume that any member of an electorate is,

J. Kacprzyk and M. Fedrizzi (eds.), Multiperson Decision Making Using Fuzzy Sets and Possibility Theory, 242–251.

confronted with the given set of alternatives, the decisive voter, i.e. for any pair of the alternatives (a_i and a_j) from A she or he may decide that:

(1) $a_i > a_j$, when a_i is strictly preferred by this voter, or

(2) $a_i \approx a_j$, when there is indifference between a_i and a_j.

Usually, it is assumed that relation of strict preference is transitive.

When voter k may describe strict preference between any two alternatives then she or he may also generate an order over the set of alternatives, i.e. $a_{i1} >_k a_{i2} >_k \dots >_k a_{in}$, where "$>_k$" is the strict preference of voter k, and i1,i2,...,in is a permutation of the alternatives. Such an order may be called an individual profile of voter k; obviously that individual profiles of different voters are usually different.

The particular way how the voters vote and how the winnig alternative is chosen may be called a voting procedure.

In a majority of classical voting procedures it is assumed that every voter has their individual profile. Some of such voting procedures may be: simple majority (Condorcet, 1785), Borda count (Borda, 1781), average rank, plurality voting, plurality voting with run-off, Schwartz count (Schwartz, 1972), Dodgson count, Black count (Black, 1958), maximin method, Nanson voting, Hare voting (1859), Coombs voting (1954), Copeland, etc.

The use of these voting procedures is limited in that every voter is assumed to have a knownindividual profile. More realistically it should be assumed that relation between two chosen alternatives is rather indifferent or unknown. Such an individual profile is called a modified individual profile. Among the methods of finding one winning alternative we may show minimal variation method (Cook and Seiford, 1982) or Lehrer-Wagner Method (1981) based upon eigenvalue of matrix of alternative pair relations. Those methods are not, *de facto*, voting procedures. They are rather multicriterial solution findings methods. Also the consensus aproach methods (see for example Kacprzyk and Fedrizzi, 1986), methods via maximal value of subjective probability distribution (for example Badran,1988), or via fuzzy structure of alternatives (for example Rao *et al.* 1988).

In our paper we are going to present voting procedures

which do not need voters' individual profiles. We will expect at least only modified individual profiles. A certain characteristics of those voting procedures will be evaluated too.

2. SOME VOTING PROCEDURES

2.1. Voting procedures based upon approval voting.

The initial point of our analysis is the problem of capacity of information on voters' preferences. We assume that every voter may divide the set of alternatives A into three separated subsets: (1) approved alternatives, (2) disapproved alternatives, and (3) indifferent alternatives. It is obvious that maximum two of those subsets may be empty.

The basic voting procedure for such dividing of alternatives is approval voting (Brams,Fishburn;1978,1983). According to this procedure every voter gives one vote to their approved alternatives. The alternative with themaximum vote total is the winner. It seems to be important to say that approval voting is a method of group decision making with constant growing importance and acknowledgment (see: Brams,1988; Fishburn,Little,1988; Brams,Nagel, 1989).

The disapproval voting is another voting procedure. During this kind of voting any voter reject all alternatives he or she disapproves (i.e. that one vote is given to every alternative from the subsets of approved and indifferent alternatives). The alternative with maximal vote total is the winner. It is important to notice that disapproval voting is not a mirror image of approval voting, and it may lead toward election of one of indifferent alternatives (Mercik,1990).

When a voter gives one vote "for" his or her most preferred alternative or one vote "against" his or her most disapproved alternatives we call it negative voting. The alternative with maximal vote total is the winner. In some circumstances the negative voting is a reverse (negative) of approval voting. For example, a vote "against" one of the three alternatives means that two remaining alternatives receive "approbation" of a voter (approval voting gives them one vote for each).

Another procedure similar to the approval voting is

cumulative voting (Merrill,1981,1984). Every voter has at his or her disposal a given quantity of votes which he or she consequently distributes among his or her approved alternatives. The alternative with the most vote total is the winner. The cumulative voting may lead toward election of disapproved alternative.

2.2. Borda-count based voting procedures

According to Borda count (Borda voting) every voter uses his or her individual profile and gives points (ranks) as follows: the most preferred among n alternatives receives n-1 points, the second one n-2 points, and so on, till the last one with zero points. The alternative with the most vote total is the winner.

Let's change necessary information about voter's individual profile. We assume that every voter may divide the alternatives set A into m disjoint subsets (classes) $A_1, A_2, \ldots, A_m$ such that $\bigcup_{i=1}^{m} A_i = A$. The k-th class A_k includes alternatives which are indifferent to each other and in some aspects similar in the eyes of the voter. Similarily as in an individual profile we assume that the voter creates <u>class individual profile</u> when set of m classes may be ordered via a strict preferation relation, i.e. $A_{j1} \succ A_{j2} \succ \ldots \succ A_{jm}$.

On the basis of class individual profiles we may costruct a voting procedure which we call ε-Borda procedure. The alternatives belonging to a given class are ε-close – when they are distributed in matrics space then $\varepsilon > 0$ is a distance measure of sensitivity of voter(s),i.e. when two given alternatives are distanced less then ε (in the voter's eyes) they are indiferrent.

According to ε-Borda voting procedure all alternatives from certain class have the same rank which equals to arithmetic mean of their Borda rank. The alternative with the most vote total is the winner.

The same way is used for creating a jury procedure. Every voter builds his or her class individual profile (card $A_k \geq 1$, $k=1,2,\ldots,m$) and gives points (using common scale) to every alternative from the class. The alternative with the most total vote is the winner.

3.SIMULATION OF THE VOTING PROCEDURES.

Apart from analitical methods of research of voting procedures the simulation allows for a deeper account of uncertainty connected with voter behaviour. In simulation of voting procedures one uses two kinds of models: random models, and spatial models. In random models "directions" of preferation relation in individual profiles (also class type) are generated according to assumed probability distribution (usually uniform distribution - such electorate is called impartial cultures).

In spatial models the positions of alternatives and voters are mapped by points and distributed in r-dimensional space,called ideological dimension(s). All preferation relations are generated througth distance metrics (in most cases it is euclidean measure) between positions of alternatives and voters.

3.1.Positional characteristcs of the voting procedures

Let us assume that in one-dimensional spatial model the positions of alternatives and voters are distributed (for simplicity) along [0,1] interval. We are interested in answering to the following question: are the chances of the alternatives depending on their positions along [0,1] interval? To reject or accept such a hypothesis with different values of parameters (number of voters or/and alternatives) 1000 simulation runs were done for each combination of parameters (same as in Mercik,1988). The frequential probability (relation between number of runs where a given alternative is winning and total number of runs) were calculated too.

We assume that alternatives distributed along interval [0,1] are enumerated from left to right.

In fig.1 the results of the simulation for 3 alternatives and 460 voters are presented (for ε-Borda procedure ε=0.5). For comparison the result of simulation with plurality voting is also presented. One may notice that parts of the presented voting procedures are (on any accepted level of confidence) central procedures, i.e. the alternatives having central (relatively to the others) positions have more chances to win: ε-Borda, disapproval,

approval and cumulative procedures. Entered for compare plurality procedure is evidently extreme procedure - it is prefering alternatives displaced (relatively to the others) extremly.

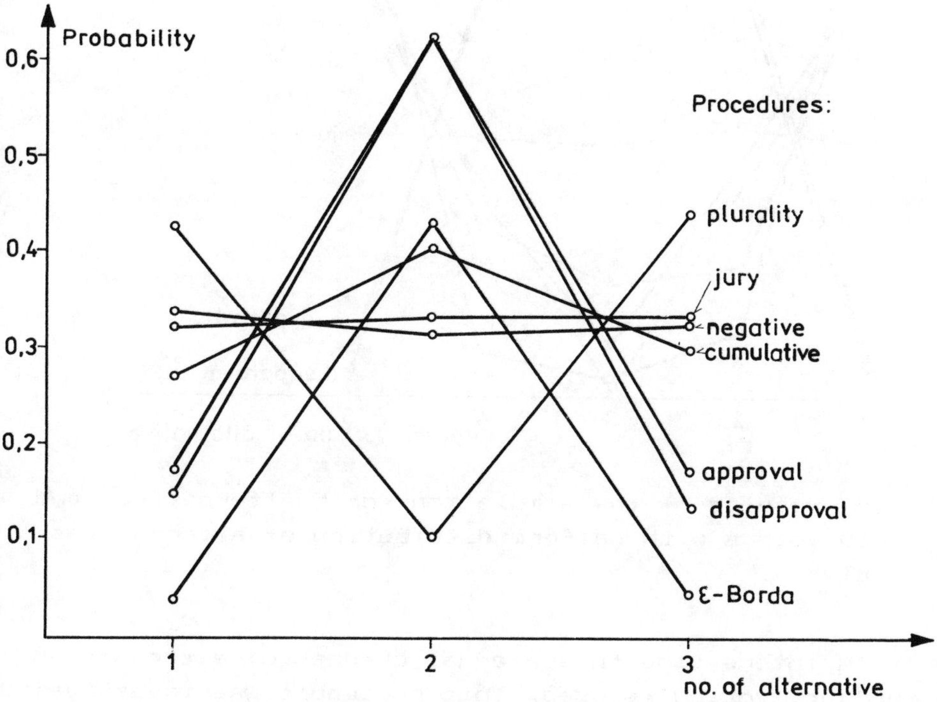

Fig.1. The results of the simulation for 3 alternatives and 460 voters distributed along [0,1] interval.

In fig.2 the results of the simulation are presented for 5 alternatives and 460 voters. During the runs it is assumed that only alternatives are uniformly distributed along [0,1] interval. The voters are distributed according to the two-modal probability distribution - expected chances of alternatives effecting from this distribution are presented in fig.2 by a dotted line. The procedures: approval (strong), disapproval (strong), cumulative (weak) and negative (weak) are central procedures. The plurality procedure (strong) is the extreme procedure.

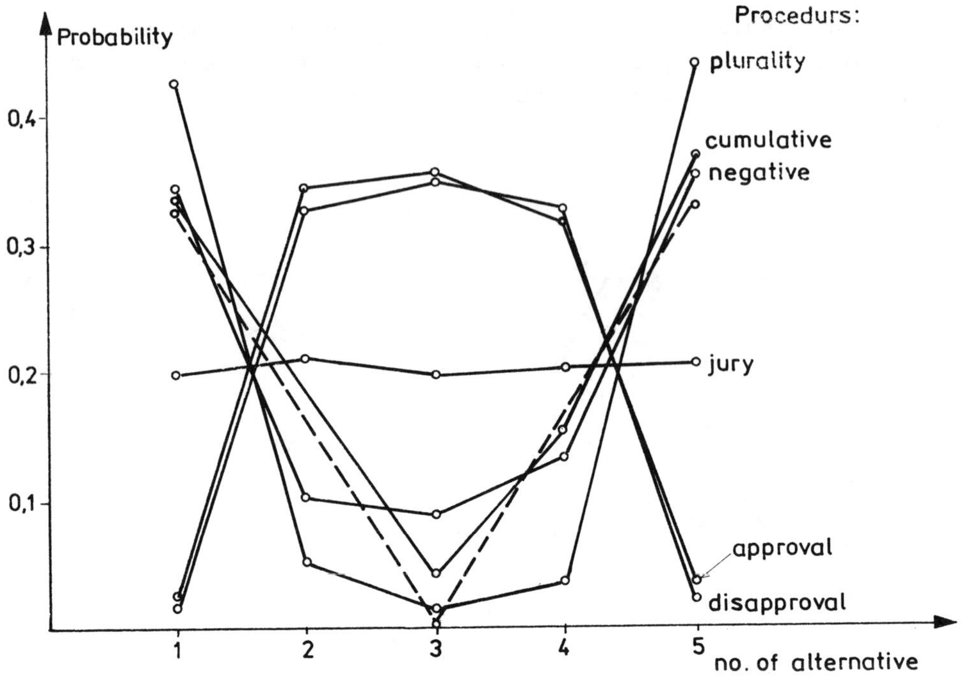

Fig.2. The results of the simulation for 5 alternatives and 460 voters with uniform ditribution of alternatives only.

It is interesting to see equal chances of alternatives when the jury voting is used. This procedure was investigated for different probability distributions of electorate and alternatives. When the alternatives are distributed uniformly, in all cases their chances do not depend on distribution of the electorate. One may interprete this result as deep indepedence of alternatives evaluation done by any voter.

3.2. Efficiency of the voting procedures

Let us re-call that we are interested in research of voting procedures with assumption of uncomplete individual profiles. Using a given procedure we are looking for the unique winning alternative.

The simulation efficiency of a given voting procedure is

understood as a pessimistic evaluation of a ratio of runs ended with one alternative choosing.

Among all voting procedures investigated with spatial model only ε-Borda and the cumulative one give sometimes no solution. In tab.1 the results of simulation for electorate of k=25,50,100 and 460 voters, uniformly distributed 3 alternatives and distribution of the lectorate same as on the fig.2 are given.

Tab.1.

Efficiency of ε-Borda and negative procedures.

Procedures:	Number of voters	Positions			No solution
		1	2	3	
ε-Borda	25 50 100 460	0,050 0,108 0,108 0,108	0,242 0,252 0,254 0,266	0,057 0,112 0,112 0,112	0,651 0,528 0,526 0,514
negative	25 50 100 460	0,282 0,337 0,334 0,333	0,104 0,106 0,130 0,134	0,300 0,376 0,377 0,377	0,314 0,181 0,159 0,154

3.3. Sensitivity of ε-Borda voting procedure to parameter chang

Up-to-now the results of the simulation of ε-Borda procedure have been obtained with assumption ε=0.5. In fig.3 the results with 7 alternatives and electorate distribution as on fig.2 are presented. A value of the ε parameter is changing: 0.1, 0.25 and 0.5.

Analysing the obtained for ε-Borda procedure results one may find that for certain number of alternatives there is optimal (from point of view of efficiency) value of ε. For the results presented in fig.3 the value of ε should be less then 0.1. The increasing of its value effects in strong decreasing of efficiency of ε-Borda procedure.

4.CONCLUSION.

A realistic evaluation of possibilities (and also

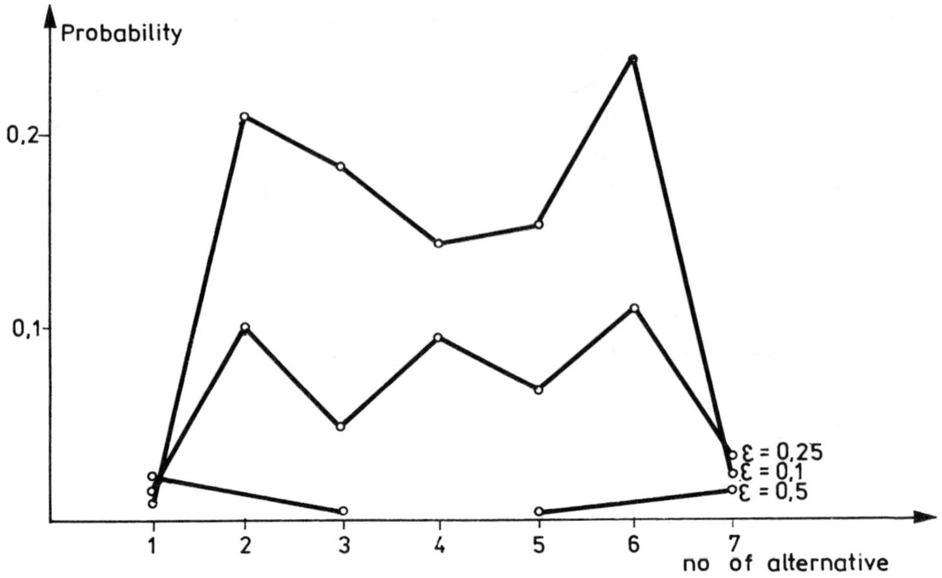

Fig.3. The results of the simulation for ε-Borda voting with different values of ε parameter.

wishfullness) of creation of individual profiles is the reason for voting procedures to use incomplete information on preference relation of voters. Such preferences, described partially or fuzzily, make choosing of certain voting procedures. Among them, the most important and useful one is the approval voting procedure. In the light of simulation research the use of approval voting in group decision making should be carefully analized when the entry problem occurs - central tendency of this voting procedure influences alternative chances for winning.

REFERENCES.

Black D. (1958), The Theory of Commiittees and Elections, Cambridge University Press, Cambridge.

Borda J.C. (1781), Sur la Forme des Elections. Mémoire sur les élections au scrutin, Histoire de L'Académie Royale de Science, Paris.

Brams S.J. (1988), Approval Voting in the 1987 MAA Elections, The Newsletter of the Mathematical Association of America, 8, no.3.

Brams S.J., Fishburn P.C. (1978), Approval Voting, The American Political Science Review, 172, 831-847.

Brams S.J, Fishburn P.C. (1983), Approval Voting, Birkhauser.

Brams S.J., Nagel J.H. (1989), Approval Voting in Practice, C. V. Starr Center for Applied Economics Report #89-07.

Badran Y. (1988), Preference ranking of discrete multiattribute instances, European J. Oprs. Res.,34, 308-315.

Condorcet, de Marquis (1785), Essai sur l'application de l'analyse a la probabilite des elections rendues a la pluralite de voix, Paris.

Cook W.D., Seiford L.M. (1982), On the Borda-Kendall consensus method for priority ranking problems, Management Science, 28, 621-637.

Coombs C.H. (1954), Theory of Data, J. Wiley, New York.

Fishburn P.C., Little J.D.C. (1988), An Experiment in Approval Voting, Management Science, 34, 555-568.

Hare T. (1859), The Election of Representatives, Parliamentary and Municipal: A Treatise, Longman, Green, London.

Kacprzyk J., Fedrizzi M. (1986), "Soft" concensus measures for monitoring real consensus reaching processes under fuzzy preferences, Control and Cybernetics, 15, 307-323.

Lehrer K., Wagner C. (1981), Rational Consensus in Science and Society, D. Reidel, Dordrecht.

Mercik J.W. (1988), Computer simulations of approval voting: positional in one dimension candidates' characteristics, Europ. Journal of Political Economy, 4, 517-527.

Mercik J.W. (1990), On some consistency of disapproval voting, Przegląd Statystyczny,R.XXXVII, zeszyt 1 (forthcoming).

Merrill S. (1981), Strategic decision under one-stage multi-candidate voting systems, Public Choice, 26, 115-134.

Merrill S.III (1984), A comparison of efficiency of multicandidate electoral systems, Amer. J. Pol. Sci., 28, 23-48.

Rao J.R., Tiwari R.N., Mohanty B.K. (1988), Preference structure on alternatives and judge in a group decision problem - a fuzzy aproach, Int. J. System Science, 19, 1795-1811.

Schwartz T. (1972), Rationality and the Myth of the Maximum, Nous, 6, 97-117.

4

TEAM DECISION MAKING UNDER FUZZINESS

A TEAM DECISION-MAKING MODEL FOR DISTRIBUTED PROBLEM SOLVING

Hideyuki Nojiri
Department of Commerce, Kumamoto University of Commerce,
2-5-1 Oe, Kumamoto 862, Japan

Abstract: Distributed decision making is a new research area for the design of information systems. To formulate a theoretical framework for the design of distributed information systems, we introduce the concept of dynamic team theory and fuzzy sets of type 2 to the problem of new product introduction decisions and we present a team decision-making model for distributed problem solving and also present the concept of an ultrafuzzy information system of a dynamic team and an evaluation function of ultrafuzzy information systems.

Keywords: Team model, Distributed problem solving, Fuzzy information. Fuzzy information system, Fuzzy strategy, Fuzzy expected utility.

1. INTRODUCTION

Distributed problem solving is a subfield of artificial intelligence that deals with the interaction of groups of intelligent agents attempting to cooperate to solve problems. Decker (1987) presented a taxonomy for distributed problem solving divided along two main lines such as the distribution of control or organizational features of a system, and the communications protocols observed. The goal of distributed problem solving is to create a team of coarse-grained cooperating agents that act together to solve a single task. Sage (1987) discussed many issues associated with the design and evaluation of information systems to aid real-time dynamic decision-making in environments where decision-making responsibility and knowledge bases are physically separated and potentially geographically distributed. One of the main problems in this area is the organization structure, which results in a network of communication channels, in which distributed multi-agent decision-making takes place. In the economic theory of teams, Marschak and Radner (1972) discussed the team problem as a problem of optimal networks. Ho and Chu (1972) studied the information structures of organizations and applied them to the problems of dynamic team decisions. Levis (1984) described an

J. Kacprzyk and M. Fedrizzi (eds.), Multiperson Decision Making Using Fuzzy Sets and Possibility Theory, 254–266.

analytical approach to modeling organizational structures for teams of decision-makers supported by command, control, and communication systems. In the past works given by Nojiri (1979, 1981a, 1981b, 1982, 1983, 1988), we studied fuzzy information structures of organization and presented various models of team decision-making under fuzziness.

In this paper we will study this problem of organizational structure that relates to the new product development at a conceptual level and make an extension of the previous static team decision model presented in Nojiri (1982). To formulate a theoretical framework for the design of distributed information systems, we introduce the concepts of dynamic team theory presented by Ho and Chu (1972) and fuzzy sets of type 2, i. e., ultrafuzzy sets (see Zadeh, 1975 ; Mizumoto, 1976 ; Nojiri, 1981b) to the problem of new product introduction decisions discussed in Boyd and Massy (1972), and Dalrymple and Parsons (1976). Then we formulate a sequential team decision model for distributed decision-making. Dynamic team theory is introduced to formulate the concepts of causal information systems and strategies. Fuzzy set theory seems appropriate for the presentation of the decision-making process by using linguistic variables instead of numbers because the human decision-making process is a vague and imprecise process.

Theoretical work in distributed problem-solving is relatively new (see Decker, 1987). In this work we present the concept of an ultrafuzzy information system of a sequential team and present an evaluation function of ultrafuzzy information systems.

2. SEQUENTIAL TEAM DECISION PROBLEM

To formulate a theoretical framework for the design of distributed information systems in new product development, the basic concept of dynamic team theory (Ho and Chu, 1972) is applied to the problem of new-product introduction decisions (Boyd and Massy, 1972 ; Dalrymple and Poasons, 1976) and we present a sequential team decision model.

Let us consider a hypothetical new product committee consisted of $i = 1, \ldots, n$ executives who represent each agency that is geographically distributed, and consider a situation in which it is being decided whether or not to continue the introduction of a new product. Assume that the order of decisions by each executive is fixed in advance and if the decision of executive j affects the information of executive i, then the decision d_i of executive i cannot affect the information of executive j.

We formalize this in the following definitions (Ho and Chu, 1972).

Definition 1 : We say j is related to i, jRi.

Definition 2 : We say j is a precedent of i, $j<i$ if and only if (a) jRi or (b) there exists distinct $r, s, t, \ldots, k \in I$ such that jRr and $rRs, \ldots, kRi$. Graphically, the idea of precedence can be presented in a precedence diagram. Each member of the committee is represented by a node so placed that the

node for j is above that for i if $j<i$. One then draws a directed segment from node j to node i if jRi and there exists no k such that $j<k$ and $k<i$. A path, consisting of a connected series of directed segments, exists between j and i if and only if $j<i$. An example of the precedence diagram of a sequential team is displayed in Fig. 2. 1.

Environmental conditions beyond the executive's control are termed sates of the environment. Let $S=\{s\}$ be the set of mutually exclusive states of the environment. The uncertainty about s is expressed by a probability distribution on S. Let us assume that the set S is finite and S does not contain any states that have zero probability. Denote the committee information by $y=(y_1, \ldots, y_n)$, where y_i, $i=1, \ldots, n$, is the part of the committee information received by the ith committee member. Let $Y_i=\{y_i\}$ be the set of alternative signals that committee member i can receive as information, and let $Y=\{y\}$ be the set of all alternative signals that the committee can receive as information, then $Y=Y_1 \times \cdots \times Y_n$.

A causal information system η_i for committee member i is a function from the Cartesian product of S and $\bar{D}_{i-1}$ to set Y_i such that $y_i=\eta_i(s, \bar{d}_{i-1})$, where $\bar{D}_0=\{\bar{d}_0\}$ is an empty set, $y_i \in Y_i$, $\eta_i \in \Pi_i$, $i=1, \ldots, n$, $d_j \in D_j$ for $j=1, \ldots, i-1$. $\bar{d}_{i-1}=(d_1, \ldots, d_{i-1}) \in \bar{D}_{i-1}=D_1 \times \cdots \times D_{i-1}$ where d_i is the decision variable of committee member i and $\bar{d}_{i-1}$ denotes the new product decisions of other precedent committee members of member i. It is assumed that the new product introduction decision of member j affect the information of member i, then d_i cannot affect the information of member j for all $i, j \in I=\{1, \ldots, n\}$. We note that y_j is embedded in y_i as components if $j<i$ and stress this fact by drawing a memory-communication line segment (dotted line) from j to i on the precedence diagram (see Fig. 2. 1).

A new product introduction decision made by a committee as a whole, called a new product introduction decision of the committee, also has n components; then $d=(d_1, \ldots, d_n)$, where d_i, $i=1, \ldots, n$ is a new product introduction decision, i. e., a go or no-go decision made by the ith committee member. Let $D_i=\{d_i\}$ be the set of alternative new product introduction decisions that committee member i can make, and let $D=\{d\}$ be the set of alternative new product introduction decisions that the committee can make; then $\{d\}=\{d_1\} \times \{d_2\} \times \cdots \times \{d_n\}$, $D=D_1 \times \cdots \times D_n$. The ith member of the committee makes his new product introduction decision d_i on the basis of his information signals y_i. A strategy δ_i for committe member i is a function from Y_i to D_i, *i. e.*, $d_i=\delta_i(y_i)$ is the new product introduction decision that committee member i makes if he receives the information signal y_i. Thus, given the causal information system η_i and the strategy δ_i, if the state of the environment s and the new product introduction decisions $\bar{d}_{i-1}$ of other members obtain, then committee member i will make the new product introduction decision $d_i=\delta_i[\eta_i(s, \bar{d}_{i-1})]=\delta_i \circ \eta_i[s, \bar{d}_{i-1}]$ for $i=1, \ldots, n$.

For any given set of strategies $\delta_1, \ldots, \delta_n$, and causal information systems $\eta_1, \ldots, \eta_n$, we define a set of composite decision function p_i from S to D_i such that $p_1(s) = \delta_1[\eta_1(s)]$, $p_i(s) = \delta_i[\eta_i(s, p_1(s), \ldots, p_{i-1}(s))]$, $i = 2, \ldots, n$, $d_i = \delta_i(y_i) = p_i(s)$, $i = 1, \ldots, n$. Schematically, p_i is a composite function which directly maps points of S space into D_i space (see Fig. 2. 2).

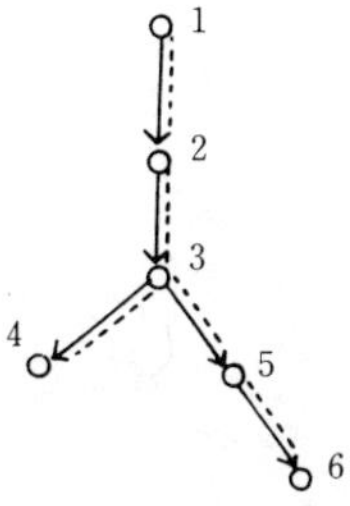

Fig. 2. 1

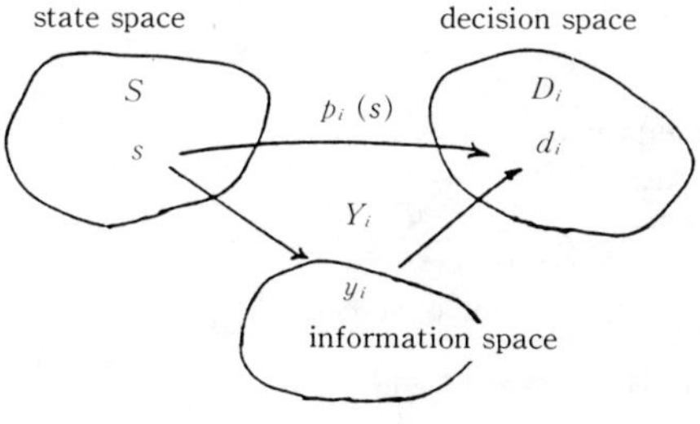

Fig. 2. 2

The n-tuple of information systems and strategies can be denoted by $\eta = (\eta_1, \ldots, \eta_n) \in \Pi = \Pi_1 \times \cdots \times \Pi_n$, and $\delta = (\delta_1, \ldots, \delta_n) \in \Delta = \Delta_1 \times \cdots \times \Delta_n$, respectively and called the information system and strategy of the new product committee.

The expected utility of the new-product committee resulting from a given pair (η, δ) is

$$\Phi(\eta, \delta; \omega, \xi) = \sum_{s \in S} \xi(s)\, \omega(s, \delta[\eta(s)]), \tag{2. 1}$$

where $\delta[\eta(s)] = [\delta_1[\eta_1(s)], \delta_2[\eta_2(s, p_1(s))], \ldots, \delta_n[\eta_n(s, p_1(s), \ldots, p_{n-1}(s))]]$. $\omega: S \times D \to R$ is a real-valued utility function of the committee and $\xi: S \to [0, 1]$ is a probability density function. Assume that all committee members have the same utility function and the same probability function, and further, the both functions are true ones.

With a given causal information system η^* and a known probability distribution over the set S of states s of the environment, the objective of the new product committee is assumed to be to choose the best strategies, that is the $\delta = \delta^\circ$ for which

$$\Psi(\eta^*) = \max_{\delta \in \Delta} \Phi(\eta^*, \delta) \tag{2. 2}$$

Let us now define that an information system η_1 is not more valuable than an information system η_2, given ω and ξ, if

$$\Psi(\eta_2) \geqq \Psi(\eta_1) \tag{2. 3}$$

After each information system is evaluated in terms of the best strategy that can be used with it, the obvious conclusion is to use the information system, η°, that maximizes $\Psi(\eta)$.

The situation of sequential team decision model is summarized as follows (see Fig. 2. 3).

For committee member 1,

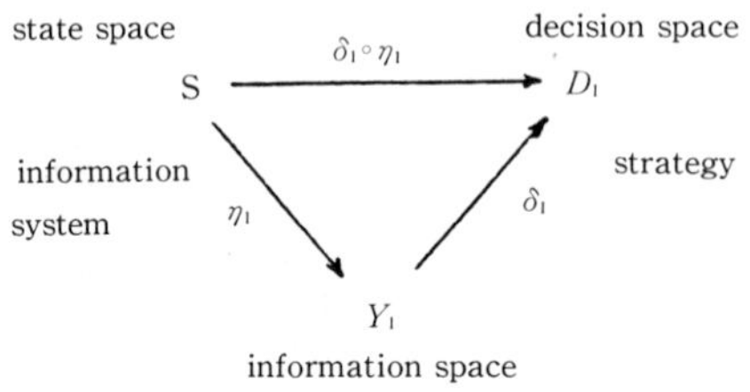

For the ith committee member, for $i = 2, \ldots, n$,

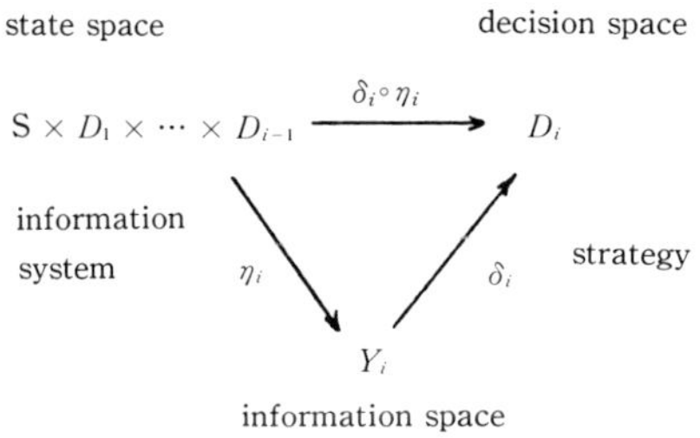

Fig. 2. 3 A diagram of sequential team decision model

3. SEQUENTIAL EXECUTIVE'S DECISION PROCESSES

In this section we address the sequential executive's decision processes that have incomplete information, and introduce the concept of fuzzy sets of type 2, i. e., ultrafuzzy sets given in Zadeh (1975), Mizumoto and Tanaka (1976), and Nojiri (1981b) to the sequential team decision problem defined in section 2 and formulate the sequential executive's decision processes using linguistic variables given in Zadeh (1975). It is assumed that the grades of fuzzy sets of type 2 are normal and convex. Normal convex fuzzy grades in $J \subseteq [0, 1]$ form a distributive lattice under $\sqcup$ and $\sqcap$, where the greatest element is $1/1$ $(= 1)$ and the least element is $1/0$ $(= 0)$.

We shall assume that each committee member i receives an ultrafuzzy information signals in the information space Y_i of the ith committee member. Then an ultrafuzzy set F_i in Y_i is characterized by a fuzzy membership function $h_{Fi}: Y_i \rightarrow [0, 1]^J$, for $i = 1, \ldots, n$. An information system η_i of the ith committee member is a function from the Cartesian product of the state space S and the decision space $\bar{D}_{i-1}$ to the information space Y_i. Then, for each committee member i, $i = 1, \ldots, n$, the inverse of F_i, written as $\eta_i^{-1}(F_i)$ is an ultrafuzzy set in $S \times \bar{D}_{i-1}$ whose fuzzy grade is defined by

$$h_{\eta_i^{-1}(F_i)}(s, \bar{d}_{i-1}) = h_{F_i}(\eta_i(s, \bar{d}_{i-1})), \qquad (3.1)$$
$$\forall (s, \bar{d}_{i-1}) \in S \times \bar{D}_{i-1},\ i = 1, \ldots, n.$$

Let us consider now a converse problem and for each member i, $i = 1, \ldots, n$. Let B_i be an ultrafuzzy set in S and $\bar{C}_{i-1} = C_1 \times \cdots \times C_{i-1}$ be an ultrafuzzy set in $\bar{D}_{i-1} = D_1 \times \cdots \times D_{i-1}$. Then $E_i = B_i \times \bar{C}_{i-1}$, is defined by

$$h_{E_i}(s, \bar{d}_{i-1}) = h_{B_i}(s) \sqcap h_{\bar{C}_{i-1}}(\bar{d}_{i-1}), \qquad (3.2)$$
$$\forall\ s \in S,\ \bar{d}_{i-1} \in \bar{D}_{i-1},\ i = 1, \ldots, n.$$

Let E_i be an ultrafuzzy set in $S \times \bar{D}_{i-1}$. Then, the information system η_i induces an ultrafuzzy set F_i of information signals in Y_i whose fuzzy grade is given by

$$h_{F_i}(y_i) = \begin{cases} \bigsqcup_{(s, \bar{d}_{i-1}) \in \eta_i^{-1}(y_i)} \{h_{E_i}(s, \bar{d}_{i-1})\} & if\ \eta_i^{-1}(y_i) \neq \phi \\ 0 & if\ \eta_i^{-1}(y_i) = \phi \end{cases} \qquad (3.3)$$

for all y_i in Y_i where the join $\sqcup$ is taken over the set of points $\eta^{-1}{}_i(y_i)$ in $S \times \bar{D}_{i-1}$ and $\eta^{-1}{}_i(y_i) = \{(s, \bar{d}_{i-1}) \mid \eta_i(s, \bar{d}_{i-1}) = y_i\}$.

Let an ultrafuzzy set E be the Cartesian product of ultrafuzzy sets $E_1, \ldots, E_n$. Then, an ultrafuzzy set $E = E_1 \times E_2 \times \cdots \times E_n$ in $S \times \bar{D}_{n-1}$ is defined by

$$h_E(s, \bar{d}_{n-1}) =$$
$$h_{E1}(s, \bar{d}_0) \sqcap h_{E2}(s, \bar{d}_1) \sqcap \cdots \sqcap h_{En}(s, \bar{d}_{n-1}),\ (s, \bar{d}_{i-1}) \in S \times \bar{D}_{i-1}, \qquad (3.4)$$

where the symbol $\sqcap$ means taking the meet of $h_{Ei}(s, \bar{d}_{i-1})$, $i = 1, \ldots, n$.

In order to define an ultrafuzzy set of information signals to the committee, let F be the Cartesian product of n ultrafuzzy sets $F_1, \ldots, F_n$. Then, $F = F_1 \times F_2 \times \cdots F_n$ is defined by

$$h_F(y) = h_F(y_1, \ldots, y_n),\ y_i \in Y_i,\ i = 1, \ldots, n \qquad (3.5)$$
$$= h_{F1}(y_1) \sqcap h_{F2}(y_2) \sqcap \cdots \sqcap h_{Fn}(y_n),$$

where the symbol $\sqcap$ means taking the meet of $h_{Fi}(y_i)$, $i = 1, \ldots, n$, and $\{y\} = \{y_1\} \times \cdots \times \{y_n\}$, $Y = Y_1 \times \cdots \times Y_n$.

F denotes an ultrafuzzy set of information signals $y = (y_1, \ldots, y_n)$ to the committee.

Let F_i be an ultrafuzzy set of information signals in the information space Y_i of the ith committee member. Then, for each committee member i, $i = 1, \ldots, n$, the strategy δ_i induces an ultrafuzzy set G_i of new product introduction decisions in the decision space D_i whose fuzzy grade is given by

$$h_{Gi}(d_i) = \begin{cases} \bigsqcup_{y_i \in \delta_i^{-1}(d_i)} \{h_{Fi}(y_i)\} & \text{if } \delta_i^{-1}(d_i) \neq \phi \\ 0 & \text{if } \delta_i^{-1}(d_i) = \phi \end{cases} \tag{3.6}$$

for all d_i in D_i, where the join $\bigsqcup$ is taken over the set of points $\delta^{-1}{}_i(d_i)$ in Y_i and $\delta^{-1}{}_i(d_i) = \{y_i | \delta_i(y_i) = d_i\}$.

Let G be the Cartesian product of n ultrafuzzy sets $G_1, \ldots, G_n$, then $G = G_1 \times G_2 \times, \ldots, G_n$ is defined by

$$\begin{aligned} h_G(d) &= h_G(d_1, \ldots, d_n) \qquad d_i \in D_i,\ i = 1, \ldots, n \\ &= h_{G1}(d_1) \sqcap h_{G2}(d_2) \sqcap \cdots \sqcap h_{Gn}(d_n), \end{aligned} \tag{3.7}$$

where the symbol $\sqcap$ means taking the meet of $h_{Gi}(d_i)$, $i = 1, \ldots, n$, and $\{d\} = \{d_1\} \times \{d_2\} \times \cdots \times \{d_n\}$, $D = D_1 \times D_2 \times \cdots \times D_n$.

G denotes an ultrafuzzy set of new product introduction decisions, $d = (d_1, \ldots, d_n)$, in the product space $D = D_1 \times D_2 \times \cdots \times D_n$ which are made by the new product committee as a whole.

Let us define a set of new product introduction decisions to the committee with fuzzy threshold r as follows:

$$G_r = \{d \mid h_G(d) \sqsupseteq r\}, \tag{3.8}$$

where r is assumed to be a normal convex fuzzy grade and $1/0 \sqsubseteq r \sqsubseteq 1/1$.

Then, the problem becomes one of choosing, for a given set G_r, the best decision, that is the $d = d^\circ$ for which

$$\begin{aligned} \Phi(d) &= \sum_{s \in S} \xi(s)\, \omega(s, d) \\ \Phi(d^\circ) &= \max_{d \in Gr} \Phi(d), \end{aligned} \tag{3.9}$$

where $\Phi(d)$ denotes the expected utility, $\omega: S \times D \to R$ is a real valued utility function of the committee, and $\xi: S \to [0, 1]$ is a probability density function. The expected utility $\Phi(d)$ is said to represent the preference ordering $\lesssim_G$ if for all decisions d_i and d_j for $i \neq j$ in G_r, $d_i \lesssim_G d_j$ if and only if $\Phi(d_i) \leq \Phi(d_j)$.

If the utility, the state of the environment, and the probability distribution are fuzzy and B is an ultrafuzzy set of states in S observed by the committee: then the fuzzy expected utility is defined as follows:

$$\Gamma (d) = \{\sum_{s \in S} \varepsilon (s)\ \nu (s, d)\ h_B (s)\} / \Xi, \tag{3. 10}$$

where $\Xi = \sum_{s \in S} \varepsilon (s)\ h_B (s)$. $\varepsilon : S \to [0, 1]^K$, $\nu : S \times D \to [0, 1]^U$, and $h_B : S \to [0, 1]^J$ denote a fuzzy probability function, a fuzzy utility function, and a fuzzy membership function, respectively, where $K = \{k_1, k_2, \ldots, k_g\}$, $U = \{u_1, u_2, \ldots, u_l\}$ and $J = \{w_1, w_2, \ldots, w_m\}$ are finite subsets of $[0, 1]$, respectively. Here $\varepsilon (s)$, $\nu (s, d)$ and $h_B (s)$ denote a fuzzy probability, a fuzzy utility, and a fuzzy grade which are fuzzy sets, i. e., normal convex fuzzy numbers, in K, U and J respectively. $\Gamma (d)$ is a fuzzy subset of the real line, and can be obtained by using the extension principle given by Zadeh (1975).

$\varepsilon (s)\ \nu (s, d)\ h_B (s)$ and the operation $\sum$ stand for the arithmetic product and arithmetic sum, respectively. The fuzzy expected utility $\Gamma (d)$ represents the preference ordering $\lesssim_D$ such that for any two new-product introduction decisions d and e in the finite decision space D, $d \lesssim e \Leftrightarrow \Gamma (d) \sqsubseteq \Gamma (e) \Leftrightarrow \Gamma (d) \sqcap \Gamma (e) = \Gamma (d)$ or $d \lesssim e \Leftrightarrow \Gamma (d) \dot{\sqsubseteq} \Gamma (e) \Leftrightarrow \Gamma (d) \sqcup \Gamma (e) = \Gamma (e)$.

For a single recommendation, a crisp decision can be made by the methods given in Baas and K wakernaak (1977), Yager (1980), Dubois and Prade (1982), and Watson and Donnell (1979). Suppose that the ultimate decision on whether to introduce a new product is dependent on the chief executive' s judgment.

If the new product committee takes an ultrafuzzy set G of new product introduction deisions and B is an ultrafuzzy set of states observed by the new product committee, then the maximal fuzzy expected utility of the new product committee, given its information system η^*, is the fuzzy expected utility maximized with respect to δ:

$$\Psi (\eta^*) = \max_{\delta \in \Delta} \Gamma (\eta^*, \delta) \tag{3. 11}$$

$$\Gamma (\eta^*, \delta) = \{\sum_{s \in S} \varepsilon (s)\ \nu (s, \delta [\eta^* (s)])\ h_{B \times G} (s, \delta [\eta^* (s)])\} / \Theta$$

where $\Theta = \sum_{s \in S} \varepsilon (s)\ h_{B \times G} (s, \delta [\eta^* (s)])$. $\Gamma (\eta^*, \delta)$ denotes the fuzzy expected utility over the Cartesian product of B and G. $\delta [\eta^* (s)] = \{\delta_1 [\eta_1^* (s)], \ldots, \delta_n [\eta_n (s, \delta_1 [\eta_1 (s)], \ldots, \delta_{n-1} [\eta_{n-1} (s)]\ )]\}$. $h_B (s) = h_{B1} (s) \sqcap \cdots \sqcap h_{Bn} (s)$. $h_G (d) = h_{G1} (d_1) \sqcap \cdots \sqcap h_{Gn} (d_n)$.

4. DISTRIBUTED FUZZY INFORMATION SYSTEMS

For the design of an information system that provides support for distributed multi-agent decision-making, we define the concept of a distributed ultrafuzzy information system of a sequential team as an extension of the concepts presented in Nojiri (1982).

4.1 Ultrafuzzy information system of a sequential team

We suppose that each executive i uses a given vague method of information gathering, namely, an ultrafuzzy information system η_i. Then, for each executive i of the committee member, $i = 1, \ldots, n$, an ultrafuzzy information system η_i is defined as follows.

Definition 4.1 : An ultrafuzzy information system η_i from $S \times \bar{D}_{i-1}$ to Y_i is an ultrafuzzy set on $S \times \bar{D}_{i-1} \times Y_i$ characterized by a fuzzy membership function $h_{\eta i} : S \times \bar{D}_{i-1} \times Y_i \rightarrow [0, 1]^J$, for $i = 1, 2, \ldots, n$.

An ultrafuzzy information $\eta_i\ (s, \bar{d}_{i-1})$ is an ultrafuzzy set on the information space Y_i of committee member i with fuzzy grade

$$h_{\eta i(s, \bar{d}i-1)}\ (y_i) = h_{\eta i}\ [(s,\ \bar{d}_{i-1}),\ y_i],\ i = 1, \ldots, n. \tag{4. 1}$$

Its inverse $\eta^{-1}{}_i\ (y_i)$ is an ultrafuzzy set on $S \times \bar{D}_{i-1}$ with fuzzy grade

$$h_{\eta \bar{i}^{1}(yi)}\ (s,\ \bar{d}_{i-1}) = h_{\eta i}\ [(s,\ \bar{d}_{i-1}),\ y_i],\ i = 1, \ldots, n. \tag{4. 2}$$

For each committee member i, an ultrafuzzy information system $\eta_i\ (s, \bar{d}_{i-1})$ is said to be finer than an ultrafuzzy information system $\theta_i\ (s, \bar{d}_{i-1})$ if

$$h_{\eta i(s, \bar{d}i-1)}\ (y_i) \sqsubseteq h_{\theta i(s, \bar{d}i-1)}\ (y_i)\ \forall\ (s,\ \bar{d}_{i-1}) \in S \times \bar{D}_{i-1},\quad y_i \in Y_i,\ i = 1, \ldots, n. \tag{4. 3}$$

For each committee member i, $i = 1, \ldots, n$, let E_i be an ultrafuzzy set on $S \times \bar{D}_{i-1}$ with fuzzy grade $h_{Ei}\ (s,\ \bar{d}_{i-1})$. Then the ultrafuzzy information system η_i induces an ultrafuzzy set F_i of information signals in Y_i whose fuzzy grade is defined by

$$h_{Fi}\ (y_i) = \bigsqcup_{(s, \bar{d}i-1) \in S \times \bar{D}_{i-1}} (h_{Ei}(s,\ \bar{d}_{i-1}) \sqcap h_{\eta i}[(s,\ \bar{d}_{i-1}),\ y_i]),\ y_i \in Yi,\ i = 1, \ldots, n. \tag{4. 4}$$

Let F be the Cartesian product of n ultrafuzzy sets $F_1, \ldots, F_n$. Then, $F = F_1 \times F_2 \times \cdots \times F_n$ is defined by

$$\begin{aligned} h_F\ (y) &= h_F\ (y_1, \ldots, y_n),\ y_i \in Y_i,\ i = 1, \ldots, n. \\ &= h_{F1}\ (y_1) \sqcap h_{F2}\ (y_2) \sqcap \cdots \sqcap h_{Fn}\ (y_n) \end{aligned} \tag{4. 5}$$

where $h_{Fi}\ (y_i)$, $i = 1, \ldots, n$, are given by (4. 4).

F denotes an ultrafuzzy set of information signals, $y = (y_1, \ldots, y_n)$, to the committee.

4. 2 Ultrafuzzy strategy

For each committee member i, $i = 1, \dots, n$, an ultrafuzzy strategy δ_i is defined as follows.

Definition 4. 2 : An ultrafuzzy strategy δ_i from the information space Y_i to the decision space D_i is an ultrafuzzy set on $Y_i \times D_i$ characterized by a fuzzy membership function $h_{\delta i} : Y_i \times D_i \rightarrow [0, 1]'$.

An ultrafuzzy new product introduction decision $\delta_i\ (y_i)$ is an ultrafuzzy set on D_i with fuzzy grade

$$h_{\delta i(yi)}\ (d_i) = h_{\delta i}\ (y_i,\ d_i),\ i = 1, \dots, n. \tag{4. 6}$$

Its inverse $\delta^{-1}{}_i\ (d_i)$ is an ultrafuzzy set on Y_i with fuzzy grade

$$h_{\delta i^{-1}(di)}\ (y_i) = h_{\delta i}\ (y_i,\ d_i),\ i = 1, \dots, n. \tag{4. 7}$$

An ultrafuzzy strategy $\delta_i\ (y_i)$ is said to be finer than an ultrafuzzy strategy $\sigma_i\ (y_i)$ if

$$h_{\delta i(yi)}\ (d_i) \sqsubseteq h_{\sigma i(yi)}\ (d_i),\ \forall\ y_i,\ d_i \in Y_i,\ D_i,\ i = 1, \dots, n. \tag{4. 8}$$

Let F_i be an ultrafuzzy set of information signals in the information space Y_i with fuzzy grade $h_{Fi}\ (y_i)$. Then the ultrafuzzy strategy δ_i induces an ultrafuzzy set G_i of new product introduction decisions in the decision space D_i whose fuzzy grade is defined by

$$\begin{aligned} h_{Gi}\ (d_i) &= h_{\delta i(Fi)}\ (d_i) \\ &= \bigsqcup_{yi \in Yi} (h_{Fi}\ (y_i) \sqcap h_{\delta i}\ (y_i,\ d_i)),\ d_i \in D_i,\ i = 1, \dots, n. \end{aligned} \tag{4. 9}$$

Let G be the Cartesian product of n ultrafuzzy sets $G_1, \dots, G_n$, then $G = G_1 \times G_2 \times \cdots \times G_n$ is defined by

$$\begin{aligned} h_G\ (d) &= h_G\ (d_1, \dots, d_n) \\ &= h_{G1}\ (d_1) \sqcap h_{G2}\ (d_2) \sqcap \cdots \\ &\quad \sqcap h_{Gn}\ (d_n),\ d_i \in D_i,\ i = 1, \dots, n. \end{aligned} \tag{4. 10}$$

where $h_{Gi}\ (d_i)$, $i = 1, \dots, n$, are given by the eq. (4. 9).

G denotes an ultrafuzzy set of new product introduction decisions $d = (d_1, \dots, d_n)$, in the product space $D = D_1 \times D_2 \times \cdots \times D_n$ which are taken by the committee as a whole.

4.3 Fuzzy expected utility

Here we consider the committee decision problem under ultrafuzzy constraints as an extension of the definitions given in Zadeh (1975). Then the maximization of the fuzzy expected utility of the committee, subject to the ultrafuzzy constraints represented by an ultrafuzzy set B in the state space S and an ultrafuzzy set G in the decision space D, reduces to a maximization of the function

$$\Gamma^* (d) = \Gamma (d) [h_G (d)]^{\lambda} \tag{4. 11}$$

where $\Gamma (d) = \{\sum_{s \in S} \varepsilon (s) \nu (s, d) h_B (s)\}/(\sum_{s \in S} \varepsilon (s) h_B (s))$. $\varepsilon : S \to [0, 1]^K$ and $\nu : S \times D \to [0, 1]^U$ are a fuzzy probability function and a fuzzy utility function, respectively. $h_B : S \to [0, 1]^J$ and $h_G : D \to [0, 1]^J$ are fuzzy membership functions, λ is any positive number.
$\Gamma^* (d)$ is a fuzzy subset of the real line, and can be obtained by using the extension principle. $\varepsilon (s) \nu (s, d) h_B (s)$, $\Gamma (d) [h_G (\mathrm{d})]^{\lambda}$, and the operation $\sum$ stand for the arithmetic products and arithmetic sum, respectively. The fuzzy expected utility $\Gamma^* (d)$ represents the preference ordering $\lesssim_D$ such that for any two decisions k and l in the finite decision space D, $k \lesssim l \Leftrightarrow \Gamma^* (k) \underset{\bullet}{\sqsubseteq} \Gamma^*(l) \Leftrightarrow \Gamma^* (k) \sqcap \Gamma^* (l) = \Gamma^* (k)$ or $k \lesssim l \Leftrightarrow \Gamma^* (k) \overset{\bullet}{\sqsubseteq} \Gamma^* (l) \Leftrightarrow \Gamma^* (k) \sqcup \Gamma^* (l) = \Gamma^* (l)$.

The problem then becomes one of choosing the best decision that is $d = \hat{d}$ for which

$$\Gamma^* (\hat{d}) = \max_{d \in D} \Gamma^* (d) \tag{4. 12}$$

The fuzzy expected utility of the committee resulting from the use of an ultrafuzzy information system η and an ultrafuzzy strategy δ is defined by

$$\Gamma (\eta, \delta ; \phi, \varepsilon) = \sum_{s \in S} \varepsilon (s) \phi (s, \delta [\eta (s)]), \tag{4. 13}$$

where for $s \in S$ and $d \in D$

$$\phi (s, \delta [\eta (s)]) \equiv \bigsqcup_{d \in D} [h_{\delta \circ \eta} (s, d) \sqcap \nu (s, d)]$$

$$h_{\delta \circ \eta} (s, d) \equiv h_{\delta 1 \circ \eta 1} (s, d_1) \sqcap \cdots \sqcap h_{\delta n \circ \eta n} \{(s, d_1, \ldots, d_{n-1}), d_n\}$$

$$h_{\delta 1 \circ \eta 1} (s, d_1) \equiv \bigsqcup_{y 1 \in Y 1} [h_{\eta 1} (s, y_1) \sqcap h_{\delta 1} (y_1, d_1)],$$

$$h_{\delta i \circ \eta i} \{(s, d_1, \ldots, d_{i-1}), d_i\}$$
$$\equiv \bigsqcup_{y i \in Y i} [h_{\eta i} ((s, d_1, \ldots, d_{i-1}), y_i) \sqcap h_{\delta i} (y_i, d_i)], \ i = 2, \ldots, n.$$

The maximal fuzzy expected utility of the committee, given its ultrafuzzy information system $\tilde{\eta}$, is the fuzzy expected utility maximized with respect to δ:

$$\Pi(\tilde{\eta}) = \max_{\delta \in \Delta} \Gamma(\tilde{\eta}, \delta), \tag{4. 14}$$

where $\Gamma(\tilde{\eta}, \delta)$ is given by eq. (4. 13). Δ denotes a collection of ultrafuzzy strategies of the committee.

The situation of sequential team decision model proposed in this research is summarized as follows (see Fig. 4. 1).

For committee member $i = 1$,

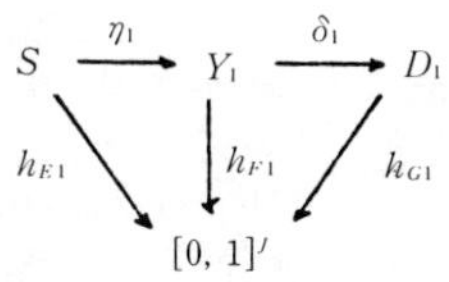

For the ith committee member, for $i = 2, \ldots, n$,

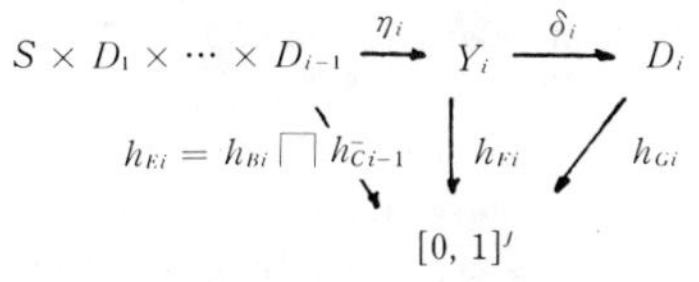

Fig. 4. 1 A diagram of sequential team decision model

5. CONCLUSIONS

There are still not many people working on formal theories of distributed problem solving systems (Decker, 1987). To formulate a theoretical framework for the design of distributed information systems, we have introduced the concepts of dynamic team theory and fuzzy sets of type 2 to the problem of new product introduction decisions, and have presented a sequential team decision model for distributed decision-making. The concepts of an ultrafuzzy information system and the fuzzy expected utility of a sequential team have been presented.

REFERENCES

Boyd H. W., and W. F. Massy (1972) Marketing Management. Harcourt Brace, New York.

Baas S. M., and H. K. Kwakernaak (1977) Rating and ranking of multiaspect alternatives using fuzzy sets. Automatica 13, 47-58.

Decker K. S. (1987) Distributed problem-solving techniques: A survey. IEEE Trans. Syst., Man, Cybern., 17, 729-740.

Dalrymple D. J., and L. J. Parsons (1976) Marketing Management, Text and Cases. Wiley, New York, 41-57.

Dubois D., and H. Prade (1982) A unifying view of comparison indices in a fuzzy set theoretic framework. Fuzzy Set and Possibility Theory, Pergamon Press, New York, 3-13.

Ho Y. C., and K. C. Chu (1972) Team decision theory and information structures in optimal control problems. Part I, Part II, IEEE Trans. Auto. Control AC-17, 1, 15-22, 22-28.

Levis A. H. (1984) Information processing and decision-making organizations: A mathematical description. Large Scale Systems 7, 151-163.

Marschak J., and R. Radner (1972) Economic Theory of Teams. Yale University Press, New Haven.

Mizumoto M., and K. Tanaka (1976) Some properties of fuzzy sets of type 2. Information and Control 31, 312-340.

Nojiri H. (1979) A model of fuzzy team decision. Fuzzy Sets and Systems 2, 201-212.

Nojiri H. (1981a) Fuzzy team decision problems in a sales organization. European journal of Operational Research 8, 256-263.

Nojiri H. (1981b) Ultrafuzzy team decisions in a dynamic environment. Information Sciences 23, 105-122.

Nojiri H. (1982) A model of the executive's decision processes in new product development. Fuzzy Sets and Systems 7, 227-241.

Nojiri H. (1983) A study on Fuzzy team decision making. A Monograph Series of the Institute of Economics and Business 9, Kumamoto University of Commerce, Kumamoto.

Nojiri H. (1988) Dynamic team decision problems with ultrafuzzy information structures, in M. M. Gupta and T. Yamakawa (Eds.): Fuzzy Logic in knowledge-Based Systems, Decision and Control. North-Holland, Amsterdam, pp 349-359.

Sage A. P. (1987) Information systems engineering for distributed decision making, IEEE Trans. Systems, Man, Cybern., 17, 921-936.

Watson S. R., Weiss J. J., and M. L. Donnell (1979) Fuzzy decision analysis. IEEE Trans. Syst., Man Cybern., 9, 1-9.

Yager R. R. (1980) On choosing between Fuzzy subsets. Kybernetes 9, 151-154.

Zadeh L. A. (1975) The concept of a linguistic variable and its application to approximate reasoning. Information Science 8, Part I, 199-249; 8, Part II, 301-357; 9, Part III, 43-80.

EVIDENTIAL TEAMS

T. Whalen and C. Brönn
Decision Sciences Department
Georgia State University
Atlanta, Georgia 30303-3083, USA

Abstract: The concepts of epistemic logics and the theory of evidence are combined with the theory of teams in a common notational system to more flexibly and realistically deal with the issues of communication and signalling between team members. A simple example of a two person dynamic team is analyzed to demonstrate the power of the evidential approach when one team member interprets the actions of another.
Keywords: Epistemic logic, evidence, teams

1. INTRODUCTION

A team is a collection of agents who act to maximize the expected utility of the team as a whole; team members have no goal conflict but they have information about different aspects of the unknown state of the world at the time each must act. Team problems were first defined and studied by the economist Jacob Marschak (1955); since that time the theory of teams has received attention both from economists (Marschak & Radner, 1972; McGuire, 1986; Radner, 1962) and control theorists (Ho, 1980; Ho & Chu, 1972).

Figure 1 shows a classification scheme for team problems; the ability to effectively solve a problem in this structure presupposes the ability to effectively solve all of the problems nested within it. Hence, our attention is primarily focused on the problems of decision and communication.

In the decision problem we must specify a decision rule for each agent on the basis of a fixed set of information channels. In the information system problem we must select information channels for each agent to maximize the benefit of an optimal solution to the decision problem, net of the cost of information. The communication problem

J. Kacprzyk and M. Fedrizzi (eds.), Multiperson Decision Making Using Fuzzy Sets and Possibility Theory, 267–285.

bridges information and decision systems since in dynamic teams one agent's action forms part of another agent's information. The casting problem is the selection of physical agents to implement the logical agents making up a team. The capacities of the physical agents chosen in casting partly determines what sorts of information systems, communication protocols, and decision rules are feasible. Finally, the scoping problem concerns how many agents will be used to achieve a given goal.

Figure 1: Nested Hierarchy of Team Design Problems

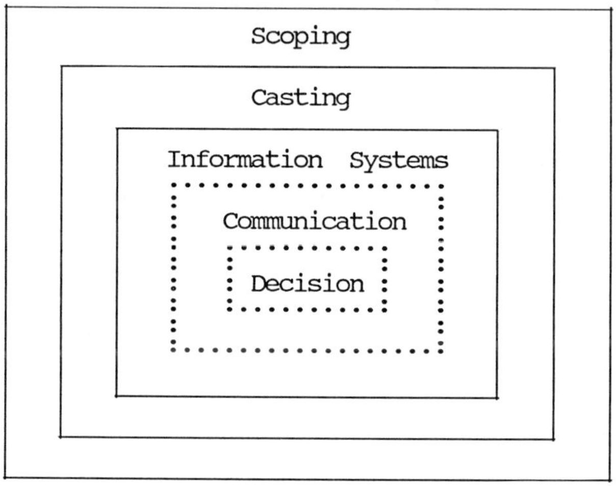

2. THE DECISION PROBLEM: BASIC SPECIFICATION

In a team decision problem the following mathematical structure is assumed to be given:

T = number of agents in the team.

$A_i = \{A_{i1}, A_{i2}, \ldots, A_{im}, \ldots, A_{i|Ai|}\}$, i = 1, 2, ..., T. A_i is the set of alternative actions available to the ith agent.

Ω = universe of discourse = $\{w_1, w_2, \ldots, w_x, \ldots, w_{|\Omega|}\}$, w_x is a possible world.

Y^W = utility function in terms of states of the world, mapping from Ω x A_1 x A_2 x ... x A_T to the real line. $Y^W(w_x, A_{1m}, A_{2m'}, \ldots, A_{Tm''})$ is the payoff to the team if each agent chooses the specified action and the exact state of the world is w_x.

$\theta_Y = \{\theta_{Y1}, \theta_{Y2}, \ldots, \theta_{Yj}, \ldots, \theta_{Y|\theta Y|}\}$, partition of Ω into equivalence classes with respect to Y^W. Changing the state of the world from one state to another within the same θ_{Yj} leaves the utility for any given set of actions unchanged. The elements of this partition are referred to as payoff-relevant sets.

Y = utility function for the payoff-relevant sets, mapping from $\theta_Y \times A_1 \times A_2 \times \ldots \times A_T$ to the real line.
$Y(\theta_{Yj}, A_{1m}, A_{2m'}, \ldots, A_{Tm''})$ is equal to
$Y^W(w_x, A_{1m}, A_{2m'}, \ldots, A_{Tm''})$ if w_x is an element of θ_{Yj}.

$Z_i = \{Z_{i1}, Z_{i2}, \ldots, Z_{ik}, \ldots, Z_{i|Zi|}\}$ = the set of possible messages agent i might receive. Each Z_{ik} specifies a complete set of information acquired by an agent up to the time of decision; agent i receives exactly one element of Z_i.

ν_i^W = agent i's information system for states of the world, mapping from Ω to Z_i. $\nu_i^W(w_x)$ is the particular message set agent i will receive whenever the state of the world is exactly w_x.

$\theta_i = \{\theta_{i1}, \theta_{i2}, \ldots, \theta_{ik}, \ldots, \theta_{i|Zi|}\}$ = partition of Ω into equivalence classes with respect to ν_i^W. All states w within a particular set θ_{ik} result in agent i receiving the same message Z_k. The elements of this partition are referred to as agent i's discernable sets. States within an equivalence class cannot be distinguished on the basis of information available to agent i.

ν_i = agent i's information system for discernable sets, mapping from θ_i to Z_i. $\nu_i^W(w_x)$ is equal to $\nu_i(\theta_{ik})$ if w_x is an element of θ_{ik}.

$\psi_i = \{\psi_{i1}, \psi_{i2}, \ldots, \psi_{i|\psi i|}\}$ = frame of reference for agent i, the partition of Ω into events to which agent i can assign probabilities on the basis of ν_i. Each element of ψ_i is referred to as an "aleatory set" for agent i.

$D_i = \{d_{i1}, d_{i2}, \ldots, d_{ip}, \ldots, d_{i|Di|}\}$ = set of feasible decision rules for agent i. $d_{ip} : Z_i \rightarrow A_i$ is a mapping which specifies what action agent i will perform upon receipt of each possible message.

Given a specification of this structure, the team decision problem is to assign a decision rule d_{ip} for each agent. To guide this assignment, we also need some kind of information about the relationships among the partitions θ_Y and the various θ_i. Most treatments of team theory have used a Bayesian approach, which assumes a degree of probabilistic knowledge equivalent to specifying the joint probability of every element of the Cartesian product:

$$\theta_Y \times \theta_1 \times \theta_2 \times \ldots \times \theta_T.$$

When such information is available, Bayesian methods are provably optimal. Our attention is directed toward a more general model of information, but including the Bayesian model as a special case.

Table 1: Subscript Conventions

$i, i', i'' \leq T$	denotes agents
$x, x', x'' \leq \lvert\Omega\rvert$	denotes states of the world
$m, m', m'' \leq \lvert A_i\rvert$	denotes actions
$j, j', j'' \leq \lvert\theta_Y\rvert$	denotes utility-relevant sets
$k, k', k'' \leq \lvert Z_{ik}\rvert$	denotes messages
$p, p', p'' \leq \lvert D_i\rvert$	denotes decision rules

3. THE DECISION PROBLEM: EPISTEMIC ALGEBRA

3.1 The Epistemic Universe

Ruspini's (1987) treatment of evidential reasoning on a logical foundation, drawing on Carnap's logic of probability (1956), Hintikka's logic of knowledge (1962), and Shafer's mathematics of evidence (1976), provides a language for specifying the relationships among the partitions of Ω. Ruspini's system involves objective states of the world, the knowledge about the states of the world possessed by various agents, and probability distributions defined over both of these. We extend this system to include actions taken by the agents and knowledge by one agent of the actions of another.

The elementary propositions of the logical system are the smallest units of analysis which carry truth values. Elementary propositions in our system are divided into two classes: ontological and praxeological. Ontological propositions correspond to Ruspini's "objective" propositions. The collection of elementary ontological propositions consists of one proposition for each element of the partition θ_Y plus one proposition for each element of each agent's partition of θ_i. The collection of elementary praxeological propositions consists of one proposition for each alternative action for each agent. Just as each onotological proposition corresponds to a set of states of the world for which the proposition is a true description, each praxeological proposition also corresponds to a set of states of the world; those states in which the indicated agent chooses the action in question.

A true proposition may be known to an agent at the time the agent selects an action, or the agent may have to act in ignorance of the proposition. If agent i knows the proposition ξ, this is indicated by the epistemic operator K_i applied to ξ: $K_i\xi$. If a proposition ξ is common knowledge (i.e. $K_i\xi$, $K_iK_{i'}\xi$, etc. for all i, i', etc.) this is indicated by $K\xi$ without the subscript.

The standard logical operators $\wedge$, $\vee$, ~ , -->, and <--> are defined as usual. Also, $\xi \Longrightarrow f$ means $\xi \dashrightarrow f$ in every possible world and similarly for $\xi \Longleftrightarrow f$. The system obeys the axioms of the standard modal logic S5, including $K_i\xi \Longrightarrow \xi$ (truth), $K_i\xi \Longrightarrow K_iK_i\xi$ (positive introspection), and $\sim K_i\xi \Longrightarrow K_i\sim K_i\xi$ (negative introspection). By abuse of language, we will use the same symbol to denote a proposition and the corresponding set of states of the world; thus, the compound proposition $\theta_{Yj}\wedge K_i\theta_{ik}\wedge\sim K_i\theta_{Yj}\wedge A_{im}$ asserts that the state of the world w is in the set:

$$(\theta_{Yj} \ \& \ K_i\theta_{ik} \ \& \ \overline{K_i\theta_{Yj}} \ \& \ A_{im}).$$

In addition to the axioms of S5, the system also contains as common knowledge necessary implications of the form $K(\xi \Longrightarrow f)$ which function as additional axioms. In particular, these include a set of implications defining each partition. For the partitions θ_Y, these are $\theta_{Y1}\vee\theta_{Y2}\vee \cdots \vee\theta_{Y|\theta_Y|}$ which asserts that the partition is exhaustive, and a collection of $_{|\theta_Y|}C_2$ implications of the form $\theta_{Yj} \Longrightarrow \sim\theta_{Yj'}$, $j \neq j'$ which collectively assert that the sets making up the partition are mutually exclusive. Similar axioms ensure that each θ_i is a partition, as is the collection A_i of actions A_{im} from which each agent must select exactly one.

Propositions of the form $K_iK_{i'}\xi$, asserting one agent's direct awareness of another agent's state of knowledge, are not permitted. However, if ν_i (the information system belonging to agent i) contains $\nu_{i'}$ (that of agent i'), in the sense of Marschak, it is possible for agent i to know which element of $\theta_{i'}$ contains w, hence which signal z_{ik} agent i' must have received. Even if ν_i does not contain $\nu_{i'}$, ν_i may give agent i information about $A_{i'm}$ from which useful information about $\theta_{i'}$ might be determined. (Another implicit exception is common knowledge, which "everyone knows everyone knows".)

Thus, three classes of elementary propositions have been defined: elementary ontological propositions θ_{Yj} or θ_{ik}; elementary praxeological propositions A_{im}; and elementary epistemic propositions $K\xi$ or $K_i\xi$, where ξ is an elementary proposition. An atomic proposition is a conjunction containing either the assertion or the negation of every elementary proposition in the system. Each atomic proposition which does not violate any necessary implication corresponds one-to-one with a possible world w_x in Ω.

The sentence space S of Ω consists of the collection of all logical formulas that can be constructed from the elementary propositions and logical operators. Each sentence ξ in S corresponds to a specific (but perhaps empty) set of worlds in Ω.

3.2 Probabilities in the Epistemic Universe

In addition to the things an agent knows for sure, each agent may also know the probability of various subsets of the epistemic universe conditional upon some particular other subsets. Any "objective" sentence ξ in S defines three important types of subsets of Ω: truth sets, epistemic sets, and support sets. The truth set $t(\xi)$ for an objective sentence ξ is the set of worlds in which ξ is true. The support set $k_i(\xi)$ for agent i is the set of worlds in which $K_i\xi$ is true; by the truth axiom $k_i(\xi)$ is a subset of $t(\xi)$. The epistemic set $e_i(\xi)$ is the set of worlds in which ξ is the most specific knowledge possessed by agent i, so that anything i knows can be derived from $K_i\xi$.

In the sequel, we assume that the effect on agent i of receiving signal Z_{ik} is, first, to add $K_i\theta_{ik}$ to the knowledge of agent i, and second, to induce a probability distribution over the epistemic sets of all sentences ξ in S. As Ruspini (1987, p.27) shows, the probability of an epistemic set is equivalent to the basic probability assignment (Shafer, 1976) of the corresponding sentence:

$$m_i(\xi) = P_i(e(\xi)) = P(e(\xi)|Z_{ik}).$$

3.2.1 Perfect Knowledge: The most desirable form for m_i to take is to assign a probability of 1.0 to a sentence which informs agent i of which utility-relevant state θ_{Yj} is the true one and what is the action $A_{i'm}$ of each agent other than agent i. In such a case, agent i can readily pick that action A_{im} which maximizes the payoff to the team.

3.2.2 Perfect Probabilistic Knowledge (Bayesian Teams): The form of knowledge most widely studied by team theorists is one in which each agent's information system provides just enough information to distribute a unit of probability over the full spectrum of sentences of the kind discussed in Section 3.2.1. This basic probability assignment reduces to an ordinary probability distribution defined over the Cartesian product of θ_Y and the various $A_{i'}$ partitions, and allows agent i to compute a point valued expected utility for each A_{im} and select the one for which the expected utility is greatest. Since Bayes theorem generally plays a significant role in calculating the probabilities, teams in which each member has an information system of this type will be referred to as Bayesian teams.

In any event, the notation ψ_i can be used to refer to a partition of Ω to which agent i can assign an ordinary probability distribution. The fineness of ψ_i can range from the full set of atomic sentences at one extreme to the trivial distribution $P(\Omega) = 1$ at the other.

3.2.3 General Basic Probability Assignments (Evidential Teams): The probability of sentences which specify θ_{Yj} and each $A_{i'm}$, for all $i' \neq i$, are the probabilities of direct interest to agent i. However, there is no guarantee in general that these are the only sentences to which ν_i gives nonzero probability. Historically (Ho & Chu, 1972; Marschak & Radner, 1972), this possibility has been ignored; sufficient conditional probabilities are assumed to convert ν_i into an information system of the form discussed in Section 3.2.2. Nevertheless, the most precise representation of the knowledge that is actually present is often the use of interval probabilities and associated basic probabilities.

This is especially true in dynamic teams: if one agent has a decision rule which calls for maximizing the expected value of some quantity, another agent who observes the first agent's action can use sensitivity analysis to infer an interval within which the first agent's probability assignment must have fallen.

Very similar to this is the case where one agent's action is to transmit standardized assessments of a probability or risk; for example, the military color code ranging from "condition green" for negligible threat through yellow and orange to a full alert status at "condition red."

Another way in which interval probabilities can arise naturally is when a team coordinator accepts inputs from specialized agents with disparate information systems and must translate to another frame of discernment to guide a final decision. This case is discussed in a subsequent section.

Teams whose members' knowledge obeys the laws of epistemic logic and evidential calculus (Ruspini, 1987) will be referred to as evidential teams. Evidential teams include Bayesian teams as a special case, but are more general.

4. OPTIMALITY IN STATIC TEAMS

4.1 Agent-by-Agent Optimality Versus Team Optimality

The two person team problem is that of finding feasible decision rules d_1^* and d_2^* (elements of the sets of feasible decision rules D_1 and D_2) such that, for expected payoff function $\underline{Y}$:
$\underline{Y}(d_1^*, d_2^*) \geq \underline{Y}(d_1, d_2)$, where $\underline{Y}(d_1, d_2)$ is the expected value of $Y(w_x, d_1(\nu_1(w_x)), d_2(\nu_2(w_x)))$, the expectation taken over all w_x in Ω.

In static teams a necessary condition of team optimality is called person by person satisfactoriness (pbps) (Radner, 1986). A team decision function is pbps if it is impossible to improve the expected payoff to the team by changing any one team member's decision function.

The pbps conditions for optimality of $\underline{Y}(d_1,d_2)$ are:

Find d_1^* and d_2^* such that:

$\underline{Y}(d_1^*,d_2^*) \geq \underline{Y}(d_1,d_2^*)$ for all d_1 in D

$\underline{Y}(d_1^*,d_2^*) \geq \underline{Y}(d_1^*,d_2)$ for all d_2 in D.

However, pbps is not a sufficient condition for team optimality, since it ignores the possibility of improving the team payoff by changing the actions of two or more agents. For example, consider a two agent team in which the combination of strategies (d_{11}, d_{21}) yields an expected utility of 2, (d_{11}, d_{22}) and (d_{12}, d_{21}) each yield an expected utility of 1 and (d_{12}, d_{22}) yields an expected utility of 3. In this case, both (d_{11}, d_{21}) and (d_{12}, d_{22}) satisfy pbps but only (d_{12}, d_{22}) is optimal.

If the team utility function is equal to the sum of $|T|$ functions, each of which depends only upon the state of the world and the action of a single agent, the utility function is "separable." If a team has a separable utility function and no agent knows any other agent's action (static team conditions), then pbps is a sufficient condition. But, if the team is dynamic, it can be optimal for one agent to sacrifice some utility from the agent's term of the separable utility in order to send a more informative signal to some other agent. This is the communication problem discussed below.

4.2 The Estimation Subproblem

The essential difference between estimation procedures and decision procedures lies in the fact that the latter includes the final actions and their specific utilitarian consequences. In the classical estimation model these considerations are included in an informal manner through the establishment of confidence intervals and significance levels (Luce & Raiffa, 1958).

The problem of estimation is to determine parameter values on the basis of prior information (if available) and by information gathered through experimentation and observation. In control theory teams, the emphasis has been on minimizing the mean squared deviation of the result of a control action from the state of nature, sometimes subject to a cost or power constraint. The utility of various outcomes is not explicitly incorporated in the analysis.

The most well-studied type of team is the linear quadratic Gaussian (LQG) team. In this formulation, the payoff is a quadratic function of the team action. In the special case where the coefficients of the second degree terms do not depend on the state of the world and the relevant random variables are normally distributed, the optimum decision variables are linear functions of the information variables.

The original static team solutions developed by Radner (1962) were based on this setup. The LQG team has also found application in problems of optimal distributed control where it is used to optimize a team's performance index (Ho & Chu, 1972).

4.3 General Decision Problem in Static Teams

The static team problem is a special form of individual decision making where the outcomes to the individual are coupled with the actions of the others. The static team consists of a group of agents who each must select an action on the basis of different information about the same true state of the world. Prior to the actual decision, the team members are free to communicate with each other. Consequently, each agent knows both the decision-relevant state partition θ_Y and the decision rules d_i of the other team members. Because there are no communications during the decision process, the agents do not know what observations their colleagues have made nor are the actions taken by other team members observed ($\sim K_i A_{i'm}$, $\sim K_i Z_{i'm}$).

The final action selected by each agent results from the signal received from the agents' information systems and the criteria of maximizing the expected value of the common utility function. The probabilities assigned to the discernable states can result from applying either Bayesian methods for point estimates or the interval estimation techniques of Shafer. There is no need for applying the epistemic algebra because knowledge of the actions of other intelligent agents is excluded from consideration by definition.

5. PARTIALLY NESTED (PN) DYNAMIC TEAMS

5.1 Dynamic Teams

The dynamic team introduces the influence of other agents on the actions of a particular team member. In dynamic team problems some team members are able to observe the actions of other team members before selecting their own actions. The frames of discernment of some agents include the action set A_i of other agents ($K_i A_{i'm}$ for some $i \neq i'$).

5.2 Reduction to Static Team Form

The analysis of dynamic teams is more complicated than static teams because the actions of previous team members affects the information and, consequently, the decisions of the subsequent decision makers. This results in two problems (Ho, Kastner, & Wong, 1978, p.307): (1) observations are no longer well defined random variables until the decision rule for the predecessors are specified, resulting in the determination of the probability measures needed for solution to the problem becoming solution dependent; and, (2) the payoff function becomes more complex because of the relationship between the decision rules of the agents.

In PN dynamic teams the team members who act later can always determine the information of a predecessor without observing their actions because that information is already contained within their own information systems. In other words, each element $\theta_{i'k'}$ of $\theta_{i'}$ (the partition of Ω induced by ν_i, the information system pertaining to the agent i' who acts later), is a subset of some element θ_{ik} of θ_{ik} (the partition induced by the information system pertaining to the agent i who acts earlier).

Under certain configurations the information system of a dynamic team may be reduced, through a process of backward induction, to an equivalent static team. The closed form solutions for static teams can then be applied to the dynamic problem. Information systems which permit this reduction include partially nested ISs, decomposable tree ISs, and combined ISs (Ho & Chu, 1974).

6. COMMUNICATION AND SIGNALLING PROBLEMS IN DYNAMIC TEAMS

6.1 Signalling Subproblem

The difficulty introduced by dynamic teams, that the information of subsequent decision makers becomes dependent upon the actions of their predecessors, can be bypassed where the team has partially nested information systems. Teams with non-partially nested (nonclassical) information systems do not allow such simplifications and emphasize the role of communication in achieving the team objectives.

Signalling is a characteristic of dynamic teams (Ho, 1980) and has been studied in an economic context by Spence (1974). By appropriate choice of actions, predecessor agent i can control what agent i' knows about the state of nature through the ability to modify the information signal agent i' receives after agent i has acted. The role of signalling in a team problem is to reduce the uncertainties of other team members through information sharing, thus allowing more informed decision making on the part of the subsequent agents ($K_iA_{i'm}$ for at least some $i \neq i'$).

6.2 An Example of a Dynamic Evidential Team

The simplest case of a dynamic team is a two agent team with an "expert" who makes direct observations of the state of nature and a "coordinator" who receives information signals from the expert. For added simplicity, we will assume that the the action of expert agent 1 is a pure signal. In other words, the team payoff depends only on the action of agent 2 and the state of the world, but the only source of data about the state of the world that is available to agent 2 is the action of agent 1. Our example is formulated in terms of three payoff-relevant sets of states of the world: $\theta_Y = \{\theta_{Y2}, \theta_{Y2}, \theta_{Y3}\}$, three alternative actions available to agent 1: $A_1 = \{A_{11}, A_{12}, A_{13}\}$, and three alternative actions available to agent 2: $A_2 = \{A_{21}, A_{22}, A_{23}\}$.

On the basis of this limited information, $Z_2 = \{A_{11}, A_{12}, A_{13}\}$, agent 2 must assess the probability of the three payoff-relevant sets and choose an action to maximize the expected value of the utility function shown in Table 2.

Table 2: Team Payoff Matrix

	θ_{Y1}	θ_{Y2}	θ_{Y3}
A_{21}	25	5	0
A_{22}	15	3	15
A_{23}	0	8	22

We assume that agent 1 has access to an information system ν_1 with a sufficient number of possible signals Z_{1k} to allow agent 1 to assess the probabilities over a different partition $\psi = \{\psi_1, \psi_2, \psi_3\}$ of the same universe Ω as θ_Y. ψ_1, ψ_2 and ψ_3 are not "discernable sets" or "epistemic sets" for agent 1 since there is no deterministic mapping from Z_1 to ψ, but ψ does carry useful probabilistic information. We will refer to them as "aleatory sets" for agent 1.

Figure 2 shows the relationship between the ψ partition and the θ_Y partition; each set in the θ_* partition is defined by the intersection of a set in ψ with a set in θ_Y. This relationship (in particular, the impossibility of ψ_1 & θ_1 or ψ_3 & θ_3) is referred to as "structural knowledge" and is assumed known to both agents on the team.

One very simple way agent 1 can choose an action is to use the three alternative actions to encode which of the three aleatory sets ψ_1, ψ_2, or ψ_3 is most likely. However, as we shall show by example, other strategies for the team can lead to expected payoff intervals that are preferable to those arising from this "maximum likelihood" approach.

6.2.1 Maximum Likelihood Rule for Agent 1: In this case we analyze the results of assigning to agent 1 the following decision rule: Do A_{11} if ψ_1 is the most likely of the three aleatory states in agent 1's frame of discernment; Do A_{12} if ψ_2 is the most likely; Do A_3 if ψ_3 is most likely.

Agent 2 learns which action agent 1 has performed, and must use this to make inferences about the probability distribution over ψ, then translate into terms of θ_Y. When agent 2 learns that agent 1 has selected A_{11}, this implies that $P(\psi_1) \geq 33\%$. In addition, it implies that $P(\psi_1 \cup \psi_2) \geq 50\%$ (since otherwise ψ_3 would be most likely) and $P(\psi_1 \cup \psi_3) \geq 50\%$.

These lower probabilities (one set for each of agent 1's alternative actions) can be converted into the basic probability assignments, b.p.a. (Shafer, 1976) as given in Table 3.

Table 3: Basic Probability Assignments for Agent 1's Frame of Discernment (Percent)

	A_{11}	A_{12}	A_{13}
$m(\psi_1) =$	33	0	0
$m(\psi_2) =$	0	33	0
$m(\psi_3) =$	0	0	33
$m(\psi_1 \cup \psi_2) =$	17	17	0
$m(\psi_1 \cup \psi_3) =$	17	0	17
$m(\psi_2 \cup \psi_3) =$	0	17	17
$m(\Omega) =$	33	33	33

Figure 2 reveals that the smallest element of the powerset of θ_Y which contains the set ψ_1 is the set $\theta_{Y2} \cup \theta_{Y3}$. The smallest element of the powerset of θ_Y which contains the set ψ_3 is the set $\theta_{Y1} \cup \theta_{Y2}$. No other elements of the powerset of ψ are contained in any elements of the powerset of θ_Y smaller than Ω. Thus, the following set of equations allow translation from any b.p.a. over ψ to the most specific b.p.a. over θ_Y:

$$m'(\theta_{Y1} \cup \theta_{Y2}) = m(\psi_3)$$

$$m'(\theta_{Y2} \cup \theta_{Y3}) = m(\psi_1)$$

$$m'(\Omega) = m(\psi_2) + m(\psi_1 \cup \psi_2) + m(\psi_1 \cup \psi_3) + m(\psi_2 \cup \psi_3) + m(\Omega)$$

$$m'(\theta_{Y1}) = m'(\theta_{Y2}) = m'(\theta_{Y3}) = m'(\theta_{Y1} \cup \theta_{Y3}) = 0$$

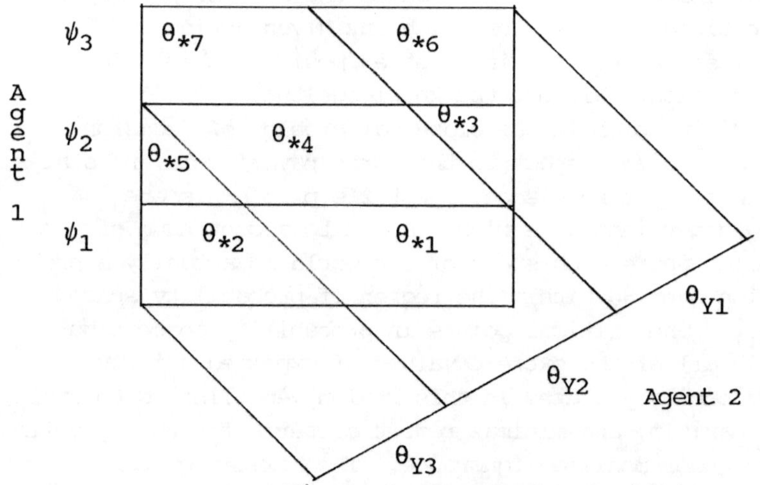

Figure 2: Team Member Partitions

When agent 1 selects A_{11}, agent 2's b.p.a. over θ_Y is $m'(\theta_{Y2} \cup \theta_{Y3}) = 33\%$ and $m'(\Omega) = 67\%$. Six critical points can now be identified in the space of probability distributions over θ_Y, by assigning the 33% to θ_{Y2} or to θ_{Y3}, then assigning the 67% to θ_{Y1}, θ_{Y2}, or θ_{Y3}. The minimum and maximum expected utility of each of agent 2's alternative actions (and of the differences between these expected utilities) can be found by examining these six critical points. Essentially, this is equivalent to solving a number of simple linear programming problems, in which the b.p.a. provides the constraints and the six critical points include the corner points of the simplex. (Note that all probabilities are expressed as percentages.)

Table 4: Given Observed Action A_{11}

PERCENT PROBABILITIES			EXPECTED UTILITIES			EU(A_{21}) minus	EU(A_{21}) minus	EU(A_{22}) minus
θ_{Y1}	θ_{Y2}	θ_{Y3}	A_{21}	A_{22}	A_{23}	EU(A_{22})	EU(A_{23})	EU(A_{23})
100	0	0	25.0	15.0	0	10.0	25.0	15.0
33	67	0	11.67	7.0	5.33	4.33	6.33	1.67
33	0	67	8.33	15.0	14.67	-6.67	-6.33	0.33
67	33	0	18.4	11.4	2.64	7.36	15.76	8.4
0	100	0	5.0	3.0	8.0	2.0	-3.0	-5.0
0	33	67	1.67	11.0	17.33	-9.33	-15.67	-6.33
Max Expected Regret:			15.67	10.0	25.0			

No action is dominated by any other, since there exist probability distributions compatible with A_{11} for which any given action is optimal. Nevertheless, agent 2 is in a better position than before receiving the information. Without the knowledge that agent 1 has chosen A_{11}, agent 2's problem is the classical problem of decision making under ignorance with respect to the three payoff-relevant states θ_{Y1} θ_{Y2} and θ_{Y3}. With this knowledge, agent 2's problem is one of decision making under ignorance with respect to a continuum of "states of the world" where each state of the world is actually a probability distribution over θ_Y within the region of probability space compatible with A_{11}. The critical points in probability space tabulated above contain all of the extreme values of expected utility of each action for probability states in this region, enabling us to evaluate the maximax, maximin, and minimax regret criteria for this problem of decision making under modified ignorance. The maximax criterion favors A_{21}, while the maximin and minimax regret criteria favor A_{22}.

When the action chosen by agent 1 is A_{12}, no restrictions are placed upon the probability distribution over θ_Y. Agent 2 faces the classical problem of decision making under ignorance. Action A_{22} is indicated by the Laplace, maximin, and minimax regret criteria while action A_{21} is indicated by the maximax criterion. (Note that the modified problem of ignorance with respect to points in probability space is identical to the classical problem of ignorance with respect to θ_Y when the probability space is unrestricted.)

Table 5: Given Observed Action A_{12}

PERCENT PROBABILITIES			EXPECTED UTILITIES			EU(A_{21}) minus EU(A_{22})	EU(A_{21}) minus EU(A_{23})	EU(A_{22}) minus EU(A_{23})
θ_{Y1}	θ_{Y2}	θ_{Y3}	A_{21}	A_{22}	A_{23}			
100	0	0	25.0	15.0	0	10.0	25.0	15.0
0	100	0	5.0	3.0	8.0	2.0	-3.0	-5.0
0	0	100	0	15.0	22.0	-15.0	-22.0	-7.0
Max Expected Regret:			22.0	10.0	25.0			

Finally, where the most likely state is ψ_3, action A_{23} satisfies the maximax criterion for decision making under ignorance with respect to points in probability space, while A_{22} satisfies the maximin and minimax regret criteria.

Table 6: Given Observed Action A_{13}

PERCENT PROBABILITIES			EXPECTED UTILITIES			EU(A_{21}) minus EU(A_{22})	EU(A_{21}) minus EU(A_{23})	EU(A_{22}) minus EU(A_{23})
θ_{Y1}	θ_{Y2}	θ_{Y3}	A_{21}	A_{22}	A_{23}			
67	0	33	16.75	15.0	7.26	1.75	9.49	7.74
0	67	33	1.67	11.0	17.33	-9.33	-15.67	-6.33
0	0	100	0	15.0	22.0	-15.0	-22.0	-7.0
67	33	0	18.4	11.4	2.64	7.36	15.76	8.4
0	100	0	5.0	3.0	8.0	2.0	-3.0	-5.0
0	33	67	3.33	7.0	12.67	-3.67	-9.33	-5.67
Max Expected Regret:			22.0	7.33	9.33			

6.2.2 A Better Decision Rule for Agent 1: In this case, the only change in assumptions involves agent 1's decision rule, which now is: Do A_{11} if $P(\psi_1) \geq 90\%$; Do A_{13} if $P(\psi_3) \geq 90\%$; Otherwise do A_{12}.

The information that agent 1 has selected a given act defines a collection of lower probabilities in the sense of Dempster (1967). These lower probabilities (one set for each of agent 1's alternative actions) can be converted into the basic probability assignments given in Table 7.

Table 7: Basic Probability Assignments for Agent 1's Frame of Discernment (Percent)

	A_{11}	A_{12}	A_{13}
$m(\psi_1) =$	90	0	0
$m(\psi_2) =$	0	0	0
$m(\psi_3) =$	0	0	90
$m(\psi_1 \cup \psi_2) =$	0	10	0
$m(\psi_1 \cup \psi_3) =$	0	0	0
$m(\psi_2 \cup \psi_3) =$	0	10	0
$m(\Omega) =$	10	80	10

When agent 1 has chosen action A_{11}, the corresponding basic probability assignment assigns 90% probability to ψ_1, with 10% probability assigned to the entire universe of discourse Ω. ψ_1 is equivalent to $\theta_{*1} \cup \theta_{*2}$ which is a subset of $\theta_{Y2} \cup \theta_{Y3}$ (see Figure 2).

Thus, by observing action A_{11}, but without any direct knowledge of agent 1's data or probability assessments, agent 2 can ascertain that $P(\theta_{Y2} \cup \theta_{Y3}) \geq 90\%$. Six critical points can be identified in the space of probability distributions over θ_Y, by assigning the 90% to θ_{Y2} or to θ_{Y3}, then assigning the 10% to θ_{Y1}, θ_{Y2}, or θ_{Y3}. The minimum and maximum expected utility of each of agent 2's alternative actions (and of the differences between these expected utilities) can be found by examining these six critical points.

Table 8: Given Observed Action A_{11}

PERCENT PROBABILITIES			EXPECTED UTILITIES			$EU(A_{21})$ minus $EU(A_{22})$	$EU(A_{21})$ minus $EU(A_{23})$	$EU(A_{22})$ minus $EU(A_{23})$
θ_{Y1}	θ_{Y2}	θ_{Y3}	A_{21}	A_{22}	A_{23}			
10	90	0	7.0	4.2	7.2	2.8	-0.2	-3.0
0	100	0	5.0	3.0	8.0	2.0	-3.0	-5.0
0	90	10	4.5	4.2	9.4	0.3	-4.9	-5.2
10	0	90	2.5	15.0	19.8	-12.5	-17.3	-4.8
0	10	90	0.5	13.8	20.6	-13.3	-20.1	-6.8
0	0	100	0	15.0	22.0	-15.0	-22.0	-7.0
Max Expected Regret:			22.0	7.0	0			

The expected utility of A_{23} is always 7.2 or higher, while that of A_{21} never rises above 7.0; thus the information that agent 1 has selected A_{11} clearly implies that A_{23} is preferable to A_{21}. Although the utility range for A_{23}, [7.2, 22.0], overlaps that of A_{22}, [4.2, 15.0], nevertheless A_{23} is strictly preferable to A_{22}. The reason is that, for any given probability distribution over θ_Y which is compatible with the optimality of A_{11}, $EU(A_{23}) \geq EU(A_{22})$.

When the selected action is A_{12}, no restrictions are placed upon the probability distribution over θ_Y. Agent 2 faces the classical problem of decision making under ignorance, action A_{22} is indicated by the Laplace, maximin, and minimax regret criteria while action A_{21} is indicated by the maximax criterion.

Finally, when the chosen action is A_{13}, action A_{22} is dominated by A_{11} since $EU(A_{11}) \geq EU(A_{12})$ for every particular probability in the region where A_{13} is optimal for agent 1. A_{21} is better than A_{23} under some possible probability distributions, but worse for others, so there is no absolutely dominant action indicated for agent 2 in this situation. Ideally, agent 2 should seek additional information

Table 9: Given Observed Action A_{12}

PERCENT PROBABILITIES			EXPECTED UTILITIES			$EU(A_{21})$ minus $EU(A_{22})$	$EU(A_{21})$ minus $EU(A_{23})$	$EU(A_{22})$ minus $EU(A_{23})$
θ_{Y1}	θ_{Y2}	θ_{Y3}	A_{21}	A_{22}	A_{23}			
100	0	0	25.0	15.0	0	10.0	25.0	15.0
0	100	0	5.0	3.0	8.0	2.0	-3.0	-5.0
0	0	100	0	15.0	22.0	-15.0	-22.0	-7.0
Max Expected Regret:			22.0	10.0	25.0			

to disambiguate the utility intervals; however, if a decision must be made on the basis of the information presented so far, action A_{21} is preferred by all three of the maximin, maximax, and minimax regret criteria.

Table 10: Given Observed Action A_{13}

PERCENT PROBABILITIES			EXPECTED UTILITIES			EU(A_{21}) minus EU(A_{22})	EU(A_{21}) minus EU(A_{23})	EU(A_{22}) minus EU(A_{23})
θ_{Y1}	θ_{Y2}	θ_{Y3}	A_{21}	A_{22}	A_{23}			
10	90	0	7.0	4.2	7.2	2.8	-0.2	-3.0
0	100	0	5.0	3.0	8.0	2.0	-0.3	-5.0
0	90	10	4.5	4.2	9.4	0.3	-4.9	-5.2
90	0	10	22.5	15.0	2.2	7.5	20.3	12.8
90	10	0	23.0	13.8	0.8	9.2	22.2	13.0
100	0	0	25.0	15.0	0	10.0	25.0	15.0
Max Expected Regret:			4.9	10.0	25.0			

7. DISCUSSION

In this paper we have presented a model for team decision making based on epistemic logic and the mathematical theory of evidence, and analyzed two simple decision rules in the context of an elementary two agent dynamic team problem. Nearly all past research in team theory has been devoted to finding necessary and/or sufficient conditions for an optimal solution to team problems subject to a great many very precise assumptions about conditional and joint probability distributions and utility functions. This prior research has also focused primarily on static teams and on special kinds of dynamic teams which are reducible to equivalent static team problems.

Our approach in this paper has been quite different. Our example problem is simple enough that it would be possible to find an optimal team strategy with the addition of a few more assumptions such as a uniform prior distribution over θ_*, assumptions of the sort which tend to be accepted without question in theoretical papers despite their limited reliability in application. Rather than seeking this sort of optimality, our approach has been to satisfice (Simon, 1960): we have presented a strategy for the team which provides nontrivial help to the key decision maker (agent 2) in those circumstances in which this help is most effective, without making too many extraneous assumptions.

Our results are only preliminary; a great deal of work remains to be done in this area. In terms of the decision problem, we intend to explore the use of other techniques such as the Laplace criterion or the use of an optimism/pessimism index to disambiguate the choice among nondominated options for decisions like that of agent 2.

In the communication problem, we intend to develop systematic ways of searching for good, robust decision rules for agents like agent 1 in the example, and to introduce direct costs and benefits from the actions of agents of this type in addition to the communication value of their actions. We also intend to develop a methodology for cost/benefit analysis of communication channel capacity. This means measuring the benefit of increasing the repertoire of actions available to agent 1 net of the cost of communicating the additional information to agent 2.

In the information system problem, we will investigate the effect of varying the quality and quantity of the information available to agents like agent 1, as well as the effect of supplying agents like agent 2 with direct as well as indirect probabilistic or possibilistic information about θ_Y or some other partition of Ω.

In the casting problem, we will investigate variations in the information-handling capacity of the agents making up the team. When agents are human beings, variation in capacity may refer to varying native ability, training, and experience; but in many important cases the most important source of variation in agent capacity is the amount of attention that a person filling the role of an agent in a particular team can devote to that role, especially its communication aspects, in competition with other demands upon the attention of the person. If some of the agents in the team are automatons, the question of agent capacity becomes one of applying classical engineering economics to artificial intelligence system implementation.

Finally, in the scoping problem we obviously need to extend our analysis beyond the dyadic team. Ultimately, we will apply the satisficing approach to finding a good systematic way of arriving, for a given problem, at a satisfactory number of agents with a good cost/benefit ratio for the agents' capacities, equipping each agent with a cost effective information system and communication channel, and providing each agent with a good set of rules for action in coordination with the rest of the team.

BIBLIOGRAPHY

Carnap, R. (1956) Meaning and Necessity. University of Chicago Press, Chicago, IL.

Dempster, A.P. (1967) Upper and lower probabilities induced by a multivalued mapping. Ann. Math. Stat. 38, 325-329.

Hintikka, J. (1962) Knowledge and Belief. Cornell University Press, Ithaca, NY.

Ho, Y-C. (1980) Team decision theory and information structures. Proc. IEEE 68, 644-654.

Ho, Y-C. and Chu, K.C. (1972) Team decision theory and information structure in optimal control problems (Parts I and II). IEEE Trans. on Auto. Cont. AC-17, 15-38.

Ho, Y-C. and Chu, K.C. (1974) Information structure in dynamic multi-person control problems. Automatica 10, 341-351.

Ho, Y-C., Kastner, M.P., and Wong, E. (1978) Teams, signalling, and information theory. IEEE Trans. Auto. Cont. AC-23, 305-311.

Luce, R.D. and Raiffa, H. (1958) Games and Decisions. John Wiley and Sons, New York, NY.

Marschak, J. (1955) Elements for a theory of teams. Mgmt. Sci. 1, 126-137.

Marschak, J. and Radner, R. (1972) Economic Theory of Teams. Yale University Press, New Haven, CT.

McGuire, C.B. (1986) Comparison of information structures, in C.B. McGuire and R. Radner (Eds.): Decision and Organization, 2nd ed., University of Minnesota Press, Minneapolis, MN.

Radner, R. (1986) Teams, in C.B. McGuire and R. Radner (Eds.): Decision and Organization, 2nd ed., University of Minnesota Press, Minneapolis, MN.

Radner, R. (1962) Team decision problems. Ann. Math. Stat. 33, 857-881.

Ruspini, E.H. (1987) The Logical Foundations of Evidential Reasoning. Technical Note 408, SRI International, Menlo Park, CA.

Shafer, G. (1976) A Mathematical Theory of Evidence. Princeton University Press, Princeton, NJ.

Simon, H.A. (1960) The New Science of Management Decision. Harper and Row, New York, NY.

Spence, M. (1974) Market Signalling: Information Transfer in Hiring and Related Screening Processes. Harvard University Press, Cambridge, MA.

5

Fuzzy Games

FUZZY GOALS AND SETS OF CHOICES IN TWO-PERSON GAMES

S. A. Orlovski
Computing Center of the USSR Academy of Sciences
Vavilova 40, 117333 Moscow, USSR

Abstract. The greatest attention of researchers in game theory is attracted by two principles of rational behavior: the principle of maximum guaranteed result due to Germeyer, and the principle based on the concept of equilibrium due to Nash. In this paper games are analyzed in which the players' goals and sets of choices are formulated as fuzzy subsets of the respective spaces, and the application of both the above principles is analyzed. The presentation is limited to two-person games, however, many of the results obtained can be extended to games with greater number of participants.
Keywords: two-person game, maximum guaranteed result, equilibrium situation, fuzzy set of strategies

1. INTRODUCTION

A game is a term commonly used to describe decision-making situations involving a number of decision-makers typical for many fields of human activities. The analysis of such situations aimed at the determination of rational behavior of a player should take into account possible actions of all other players.

Probably, the greatest attention of researchers in this field is attracted by two principles of rational behavior: the principle of maximum guaranteed result and the principle based on the concept of equilibrium. Both these principles are considered in this paper in the analysis of two-person games in a fuzzy environment and here we shall briefly outline them.

In the framework of the principle of maximum guaranteed result, a two-person game is viewed as a decision-making problem for one of the players (we shall refer to him as player

J. Kacprzyk and M. Fedrizzi (eds.), Multiperson Decision Making Using Fuzzy Sets and Possibility Theory, 288–297.

1). Choices of player 2 are considered as uncertain factors of this problem. This uncertainty can mean, for instance, that at the moment of making a choice player 1 may know not a concrete reaction of player 2, but only some set of his possible reactions (choices). The more player 1 knows about the goals and possibilities of player 2, the "narrower" is that set and, therefore, the less the uncertainty with regard to player's 2 behavior from the point of view of player 1.

This principle asserts that a rational way of the evaluation by player 1 of his choices consists in taking into account the worst for him reaction of player 2 from the set of possible reactions. Consequently, the best choice by player 1 is a choice having the best evaluation in the above sense.

Applying this principle it is important to take into consideration all information about player 2 that player 1 will have when making his concrete choice, since it is this information that enables him to judge about possible reactions of player 2. This flexibly understood principle of maximum guaranteed result and its applications to various types of games are described and analyzed in Germeyer (1976).

Another rationality principle considered in this paper is based on the concept of equilibrium due to Nash. This principle is applicable to cases when players can communicate with each other to choose a situation that is mutually acceptable. According to this principle, rational choices x_0, y_0 of players 1 and 2, respectively, should possess a stability property in the following sense: a unilateral change of this choice is not beneficial to any of the players. In other words, if for example, player 1 chose x instead of x_0, but player 2 sticked to y_0, player 1 would not obtain a better result. A pair (x_0, y_0) having this stability property is called the Nash equilibrium situation.

With all its attractiveness this principle has two drawbacks. First, equilibrium situations are not stable with respect to simultaneous changes of choices by both players (for instance, both players can obtain better results in some other, not necessarily equilibrium situation). Second, there can be a multitude of various equilibrium situations in a game, and different of them can be preffered by different players.

The most "preferable" are games in which there are situations that are rational with respect to both the above principles simultaneously. Many types of so the called zero-sum two-person games (antagonistic games) fall into this category. In this paper we show that in a certain sense this property is characteristic to some games with opposite preferences in a fuzzy environment.

In this paper we deal with games in which the players' goals and sets of choices are formulated as fuzzy subsets of respective spaces. And in our analyses of them we demonstrate the application of both the principles outlined above. We limit the presentation to two-person games only, however, many of the results obtained can be extended to games with a greater number of participants.

2. DESCRIPTION OF A GAME

We consider players 1 and 2 and assume that their feasible choices (strategies) are described by fuzzy subsets $\mu_1: Y \to [0,1]$ and $\mu_2: X \to [0,1]$ of some basic sets X and Y, respectively. This description reflects that from the players' viewpoints choices differ in their feasibility degrees, and only some of them can be considered as definitely (to degree 1) feasible.

We also assume that two pay-off functions $f_1, f_2: X \times Y \to R^1$ are specified with values interpreted as evaluations of situations $(x,y) \in X \times Y$ by the respective players.

Each of the players strives to attain his own fuzzily specified goal. We assume that the goal of player i is described as a fuzzy subset $\mu_g^i: R^1 \to [0,1]$ of the number axis R^1. Player 1 tries to obtain a value of $f_1(x,y)$ belonging to μ_g^1 to the highest possible degree. Interpretations of this type of goals can be found in many papers based on the use of the Bellman-Zadeh formulation of decision-making problems in a fuzzy environment. Rather than repeating those interpretations here, we should note that the goal considered by a player may happen to be poorly compatible or fully incompatible with his possibilities, i.e. with his set of feasible choices and/or information that he has about the other player. In the following this question is considered in greater detail.

To analyze the behavior of the players we introduce a more explicit description of the goals in the form of fuzzy subsets of X×Y:

$$\mu_g^i(x,y) = \mu_g^i(x,y)(f_i(x,y)), \quad i=1,2; \quad x,y \in X \times Y \tag{1}$$

and we also introduce the following fuzzy subsets D_1 and D_2 of X×Y:

$$\mu_D^1(x,y) = \min\{\mu_1(x), \mu_g^1(x,y)\}; \quad \mu_D^2(x,y) = \min\{\mu_2(x), \mu_g^2(x,y)\}.$$

As can be seen, fuzzy sets D_1 and D_2 are intersections of the respective fuzzy sets of choices and of the goals. The meaning of D_1 and D_2 can be explained in the following way. If, for example, player 1 knew a concrete choice $y \in Y$ by player 2, he would face a problem of attaining his fuzzy goal $\mu_D^1(x,y)$ by making choices from fuzzy set $\mu_1(x)$. Using the approach due to Bellman and Zadeh (1970) the fuzzy solution to that problem is defined as $\mu_D^1(x,y)$. Therefore, fuzzy set $\mu_D^1(x,y)$ can be considered as a family with respect to parameter y of fuzzy solutions to problems of attaining fuzzy goals $\mu_g^1(x,y)$. Similarly, $\mu_D^2(x,y)$ can be understood.

The value $\mu_D^1(x,y)$ is the degree to which the fuzzy goal μ_g^1 of player 1 is attained when the players make their respective choices x and y.

We shall assume in the following that for each fixed choice by one player the other player chooses a strategy that maximizes the respective function $\mu_D^i(x,y)$, i.e. a strategy belonging to set D_i to the maximum degree. With this assumption we can consider our game as a traditional (nonfuzzy) two-person game with X, Y being sets of choices of the players and with $\mu_D^1(x,y)$, $\mu_D^2(x,y)$ being their respective pay-off functions.

3. MAXIMUM GUARANTEED RESULTS

Let us assume that player 1 can make his choice and communicate it to player 2. Then the maximum guaranteed result of player 1 (maximum guaranteed degree of attaining the fuzzy goal of player 1)[1] is:

$$M_1 = \max_{x\in X} \min_{y\in Y(x)} \mu_D^1(x,y) = \max_{x\in X} \min \{\mu_1(x), \min_{y\in Y(x)} \mu_g^1(x,y)\}.$$

The set Y(x) in this equation that depends on x and is known to player 1, is the set of possible reactions of player 2 to the choice x made by player 1. In this sense the mapping Y(x) reflects the knowledge of player 1 about interests and possibilities of player 2. If for example, player 1 knows only set Y of player 2, then his fuzzy set of choices has the form:

$$\mu_1^1(x) = \min \{\mu_1(x), \min_{y\in Y} \mu_g^1(x,y)\}.$$

If, on the other hand, player 1 has more information about player 2 (i.e. in the case of $Y(x) \subset Y$, $x\in X$, then the fuzzy set of his choices becomes:

$$\mu_1^2(x) = \min \{\mu_1(x), \min_{y\in Y(x)} \mu_g^1(x,y)\}.$$

As can easily be seen, $\mu_1^2(x) \geq \mu_1^1(x)$, $x\in X$, which means that with more information player 1 has a wider fuzzy set of choices. In this sense we can say that having more information

---

[1] For simlicity, we assume in the following that max and min are always attained on the corresponding sets.

about his opponent, player 1 has more possibilities to guarantee the attainment of his goal.

If the value of M_1 is too small, then the goal that player 1 tries to attain is too high with respect to his possibilities. In this connection the following problem is of interest: *Given a formulation of a fuzzy goal of player 1 in the form of fuzzy subset μ_g^1, what should be his fuzzy set of choices that can guarantee the attainment of this goal to a prespecified degree α ?*

In other words, what should be the function $\mu_1(x)$ describing the set of strategies of player 1, to ensure:

$$M_1 = \max_{x \in X} \min \{\mu_1(x), \min_{y \in Y(x)} \mu_g^1(x, y)\} \geq \alpha \ ?$$

If we introduce a set

$$X(\alpha) = \{ \min_{y \in Y(x)} \mu_g^1(x, y) \geq \alpha\} \subseteq X,$$

then it can easily be seen that when $X(\alpha) = \emptyset$ we may have $M_1 < \alpha$, and therefore, in this case player 1 may not be able to guarantee the attainment of his goal to the desired degree whatever his set of strategies is. (The reason, for instance, can be the lack of information about player 2 which is determined by the mapping $Y(x)$.)

If $X(\alpha) \neq \emptyset$ then $M_1 \geq \alpha$ can be guaranteed iff $\mu_1(x) \geq \alpha$ for at least one $x \in X(\alpha)$.

We can also consider the inverse problem: *Given a fuzzy set of choices of player 1 the attainment of what goals can he guarantee to a degree not lower than α?*

To analyze this problem let us introduce a set:

$$X^*(\alpha) = \{x \mid \mu_1(x) \geq \alpha\}.$$

A goal $\mu_g^1(x, y)$ can be attained to a degree not lower than α, iff

$$\min_{y \in Y(x)} \mu_g^1(x^1, y) \geq \alpha \text{ for some } x \in X^*(\alpha),$$

or

$$\mu_g^1(x, y) \geq \alpha, \ y \in Y(x).$$

Let $\mu_g^1 : R^1 \rightarrow [0, 1]$ be some goal function of player 1. We introduce a set:

$$Z(\alpha) = \{z \mid \mu_g^1(z) \geq \alpha\} \subseteq R^1.$$

Now we can conclude that a goal μ_g^1 can be attained to a degree not lower than α by player 1, iff for at least one $x \in X^*(\alpha)$ we have

$$f_1(x, y) \in Z(\alpha) \text{ for any } y \in Y(x),$$

where $f_1(x, y)$ is the pay-off function introduced in sect. 2.

4. GAMES WITH OPPOSITE PREFERENCES OF PLAYERS

If X and Y are the respective sets of choices of the players, then the product set $X \times Y$ is called the set of situations of the game. A situation has the form $s = (x, y)$ with $x \in X$, $y \in Y$. If for player i situation s_1 is more preferable than situation s_2, we shall write $s_1 \succ^1 s_2$.

The difference between the players' preferences in a game can mean, for example, that for some situations s_1, s_2, s_3, s_4 we have:

$$s_1 \succ^1 s_2, \quad s_1 \succ^2 s_2, \quad s_3 \succ^1 s_4, \quad s_4 \succ^2 s_3,$$

i.e. preferences of the players can be the same for some pairs of situations (s_1 and s_2) and be opposite for other pairs (s_3 and s_4).

If preferences of the players are opposite *for all situations*, then we have a game with opposite preferences of the players. In other words, in such a game for every pair of situations (s, q) we have

$$s \succ^1 q \Rightarrow q \succ^2 s.$$

Similarly, by a game with opposite preferences in a fuzzy environment we shall mean a game in which the goal functions $\mu_g^i: R^1 \to [0, 1]$, $i = 1, 2$ and the pay-off functions $f_i: X \times Y \to R^1$ are such that

$$\mu_g^1(z_1) > \mu_g^1(z_2) \Rightarrow \mu_g^2(z_1) < \mu_g^2(z_2),$$

$$\mu_g^1(z_1) = \mu_g^1(z_2) \Rightarrow \mu_g^2(z_1) = \mu_g^2(z_2),$$

$$f_1(x, y) = f_2(x, y), \quad (x, y) \in X \times Y.$$

Using the above definitions (1) of functions $\mu_g^i(x, y)$ (cf. Section 2) we can formulate this definition of a game with opposite preferences in the following more explicit form for any (x_1, y_1) and (x_2, y_2):

$$\mu_g^1(x_1, y_1) > \mu_g^1(x_2, y_2) \Rightarrow \mu_g^2(x_1, y_1) < \mu_g^2(x_2, y_2),$$

$$\mu_g^1(x_1, y_1) = \mu_g^1(x_2, y_2) \Rightarrow \mu_g^2(x_1, y_1) = \mu_g^2(x_2, y_2).$$

Similarly to Section 2 we shall consider this game as the game with pay-off functions of the form:

$$\mu_D^1(x, y) = \min \{\mu_1(x), \mu_g^1(x, y)\}, \quad \mu_D^2(x, y) = \min \{\mu_2(x), \mu_g^2(x, y)\}$$

with X and Y being sets of choices of players 1 and 2 respectively.

A pair of choices (x_0, y_0) is called an equilibrium situation if for any $x \in X$ and $y \in Y$ we have:

$$\mu_D^1(x_0, y_0) \geq \mu_D^1(x, y_0), \quad \mu_D^2(x_0, y_0) \geq \mu_D^2(x_0, y). \tag{3}$$

To consider some interesting properties of this type of equilibrium situations we introduce two sets:

$$C_1 = \{(x, y) \mid (x, y) \in X \times Y, \ \mu_1(x) \geq \mu_g^1(x, y)\},$$

$$C_2 = \{(x, y) \mid (x, y) \in X \times Y, \ \mu_2(x) \geq \mu_g^2(x, y)\}.$$

In the case when an equilibrium situation $(x_0, y_0) \in C_1$ we can conclude that the set

$$C_1(y_0) = \{x \mid x \in X, \ (x, y_0) \in C_1\}$$

is not empty. Therefore, as follows from the definition of C_1 we have that $\mu_1(x) \geq \mu_g^1(x, y_0)$ for any $x \in C_1(y_0)$, hence

$$\min \{\mu_1(x), \mu_g^1(x, y_0)\} = \mu_g^1(x, y_0), \ \text{any } x \in C_1(y_0).$$

But then we have from (3) that $\mu_g^1(x_0, y_0) \geq \mu_g^1(x, y_0)$ for any $x \in C_1(y_0)$. From this and from definition (2) of the opposite preferences we have

$$\mu_g^2(x_0, y_0) \geq \mu_g^2(x_0, y_0) \ \text{for any } x \in C_1(y_0),$$

and therefore, $\mu_D^2(x_0, y_0) \geq \mu_D^2(x, y_0)$.
Using (3) we can also write

$$\mu_D^2(x_0, y_0) \geq \mu_D^2(x_0, y), \ y \in C_1(x_0),$$

where

$$C_1(x_0) = \{y \mid y \in Y (x_0, y) \subseteq C_1\}.$$

Finally we have

$$\mu_D^2(x_0, y_0) \geq \mu_D^2(x_0, y), \ y \in C_1(x_0),$$

$$-\mu_D^2(x_0, y_0) \geq -\mu_D^2(x, y_0), \ x \in C_1(y_0).$$

The last two inequalities mean that the equilibrium situation $(x_0, y_0) \in C_1$ considered is an equilibrium situation in a zero-sum game (μ_D^2 - pay-off function of player 1, and $-\mu_D^2$ - pay-off function of player 2: $\mu_D^2 + (-\mu_D^2) = 0$) in which only those situations are feasible which belong to set $C_1 \subseteq X \times Y$ (but not all situations in $X \times Y$). Games of this type are referred to as games with prohibited situations or games with constraints (see Orlovski 1973, 1975, 1978, 1981).

Similarly, it can be shown that in the case when an equilibrium situation $(x_0, y_0) \notin C_1 \cup C_2$, this situation is an equilibrium situation in the game where $\mu_1(x)$ is the pay-off function of player 1, $\mu_2(y)$ - pay-off situation of player 2, and the set of feasible situations is $C_1^* \cap C_2^*$ with $C_i^* = X \times Y \setminus C_i$, $i = 1, 2$.

5. MAXIMUM GUARANTEED RESULT AND EQUILIBRIUM SITUATIONS

In a traditional (nonfuzzy) game with opposite interests the players in any equilibrium situation receive their maximum guaranteed pay-offs. In view of this it is of interest to analyze the relation between maximumn guaranteed results of the players and their results in equilibrium situations in a fuzzy game with opposite preferences.

In calculating maximum guaranteed results we shall assume that each of the players knows only the basic set of choices of his opponent (X or Y respectively for players 1 and 2). In such a case the maximum guaranteed results have the following form:

$$M_1 = \max_{x \in X} \min_{y \in Y} \mu_D^1(x, y)$$

$$M_2 = \max_{y \in Y} \min_{x \in X} \mu_D^2(x, y).$$

Let us represent M_1 in the form:

$$M_1 = \max \{ \min_{x \in X_0^*} \mu_1(x), \max_{x \in X_0} \min_{y \in C_1(x)} \mu_g^1(x, y) \} \qquad (4)$$

with

$$X_0 = \{x \mid x \in X, \ C_1(x) \neq \emptyset\}, \ X_0^* = X \setminus X_0,$$

$$C_1(x) = \{y \mid y \in Y, \ (x, y) \in C_1\},$$

and C_1 as introduced in Section 3.

For simplicity we assume that $C_1 = C_2 = C$. Let (x_0, y_0) be an equilibrium situation of the game and $(x_0, y_0) \in C$. In the following we show that in this case the pay-offs of the players in the equilibrium situation (x_0, y_0) are their maximum guaranteed results.

1. To show first that $M_1 \leq \mu_D^1(x_0, y_0)$ we note that

$$\min_{y \in Y} \mu_D^1(x, y) \leq \mu_D^1(x, y_0),$$

therefore,

$$M_1 = \max_{x \in X} \min_{y \in Y} \mu_D^1(x, y) \leq \max_{x \in X} \mu_D^1(x, y_0) = \mu_D^1(x_0, y_0)$$

i.e. obtain the desired inequality.
2. To show that $M_1 \geq \mu_D^1(x_0, y_0)$ we note that

$$\mu_g^2(x_0, y_0) \geq \mu_g^2(x_0, y)$$

for any $y \in C(x_0) = \{y \mid (x_0, y) \in C\}$.

From this and using definition (2) of opposite preferences of the players (Section 2) we conclude that for any $y \in C(x_0)$:

$$\mu_g^1(x_0, y_0) \leq \mu_g^1(x_0, y),$$

hence

$$\min \{ \mu_1(x_0), \mu_g^1(x_0, y_0)\} \leq \min \{ \mu_1(x_0), \mu_g^1(x_0, y)\},$$

and therefore,

$$\mu_D^1(x_0, y_0) = \min_{y \in C(x_0)} \mu_D^1(x_0, y) = \min_{y \in C(x_0)} \mu_g^1(x_0, y).$$

As can easily be seen

$$\max_{x \in X_0} \min_{y \in C(x_0)} \mu_D^1(x, y) \geq \min_{y \in C(x_0)} \mu_D^1(x_0, y) = \mu_D^1(x_0, y_0).$$

From this and using (4) we have: $M_1 \geq \mu_D^1(x_0, y_0)$. Finally, we have $M_1 = \mu_D^1(x_0, y_0)$. Similarly, it can be shown that under the same assumptions we have $M_2 = \mu_D^2(x_0, y_0)$.

Now we can conclude that in a game with opposite preferences formulated here the equilibrium degrees of attainment of the players' goals corresponding to an equilibrium situation from set C are equal to maximum guaranteed degrees of their attainment. In particular, this means that all equilibrium situations from set C are equivalent for both the players since in any of them they obtain the same respective degrees M_1 and M_2 of their fuzzy goals.

6. CONCLUDING REMARKS

The formulation of fuzzy games considered in this paper is based on the approach to decision-making in a fuzzy environment due to Bellman and Zadeh. Our assumption is that using this approach the players make decisions on choices of their strategies. This assumption enabled us to reduce fuzzily formulated games to their corresponding nonfuzzy formulations and apply to their analyses the game-theoretic principles of maximum guaranteed result due to Germeyer and of equilibrium due to Nash. Some properties of the respective solutions and their interrelationships in games with opposite preferences of the players are also outlined in this paper.

REFERENCES

Bellman R., and Zadeh L. (1970). Decision-Making in a Fuzzy Environment. Management Sci. 17, 141-164.

Germeyer Y.B. (1978). Games with Nonantagonistic Interests. Nauka Publ. Moscow (in Russian).

Orlovski S.A. (1973). Infinite Games with Prohibited Situations. J. of Comput. Math. and Math. Phys. 13, 775-781 (in Russian).

Orlovski S.A. (1975). Equilibrium Situations in Non-Coalition Games with Constraints. J. of Comput. Math. and Math. Phys. 15, 1597-1601 (in Russian).

Orlovski S.A. (1981). Problems of Decision-Making with Fuzzy Information. Nauka Publ., Moscow, (in Russian).

PLAYING MATRIX GAMES DEFINED BY LINGUISTIC LABELS

M. Delgado, J.L. Verdegay, M.A. Vila

Departamento de Ciencias de la Computacion e Inteligencia Artificial

Universidad de Granada

18071 Granada (Spain)

ABSTRACT. Matrix (two person zero-sum) games are a particular case of interest conflict. In establishing a matrix game, it is usual to assume the rewards to be obtained by the players are perfectly known by them. But real situations are quite different from that. In fact, it is more frequent to hear from the players or decision makers statements like "the reward will be about x", "the reward may be essentially smaller than y", etc. than precise ones like "the reward will be z" ($x,y,z \in \mathbb{R}$). This is in correspondence with two different approaches: the one emerging from the fuzzy sets field and another one from the classical Theory of Games, respectively. This paper focuses on a non conventional case in which the players, rather than an exact knowledge about their rewards in playing the game, have an approximate knowledge which they express by means of linguistic labels (good, bad, high, etc.). For such a case we propose an approach for this kind of "ill defined games" which provides a solution consonant with the former features from all this context.

Keywords: Matrix Game, Approximate Reasoning, Linguistic Label

1. INTRODUCTION

The Mathematical Theory of Strategic Games has received and is receiving a very considerable attention because of its important implications into the economic field. Usually, the uncertainty assumed for the games is of a probabilistic kind. But, in real problems it may be considered another kind of uncertainty which is inherent with the way in which the players express their payoffs, if this is done by them in a linguistic way. In fact it is very easy for the players to express and exchange information in a linguistic way. From this point of view, it seems adequate to use labels (linguistic valuations) to represent the uncertainty (imprecise information) the players have about their payoffs.

Thus if one supposes the players give, rather than real

J. Kacprzyk and M. Fedrizzi (eds.), Multiperson Decision Making Using Fuzzy Sets and Possibility Theory, 298–310.

numbers, linguistic labels to define their payoffs, it appears the above referred linguistic uncertainty. That linguistic lack of precision may be modeled by fuzzy sets, and thus it arises a new kind of games: the Fuzzy Games.

In particular this paper focuses on Two-Person Zero-Sum Games in which the payoffs are defined by linguistic labels. Section 2 presents elements of the classical Theory of Matrix Games. In Section 3, the Matrix Games with non standard payoffs are introduced and after that, in Section 4, an approach for solving those fuzzy, or labelled, matrix games is proposed. Finally some examples illustrate the approach.

2. CONVENTIONAL MATRIX GAMES

A two-person zero sum game, or a matrix game, is a conflict of interests defined as follows. One supposes the players have stated a matrix

$$M = (a_{ij}),\ i \in I,\ j \in J \tag{1}$$

in which $a_{ij} \in \mathbb{R}$ and $I = \{1,2,...n\}$ and $J = \{1,2,...m\}$ are the sets of strategies for the first and second player, respectively. As it is well known, if player 1 chooses strategy $i \in I$ and player 2 chooses strategy $j \in J$, player 1 obtains a payoff $a_{ij} \in \mathbb{R}$ given by M.

The solution of the game, that is, the strategy that each player must choose to maximize his reward, is defined as a pair of strategies $(i^*,j^*) \in I x J$ such that the corresponding $a_{i^*j^*} \in M$ is both the maximum in its column and the minimum in its row. Such value is referred to as the value of the game and the pair (i^*,j^*) is a saddle point. Whether there is not such a pair of strategies, the Minimax Theorem (von Neumann and Morgenstern, 1947) guarantees there exists a pair (x^*,y^*) such that

$$v = \underset{x \in X}{\text{Max}}\ \underset{y \in Y}{\text{Min}}\ A(x,y) = \underset{y \in Y}{\text{Min}}\ \underset{x \in X}{\text{Max}}\ A(x,y) = A(x^*,y^*) \tag{2}$$

where

$$A(x,y) = \sum_{i \in I} \sum_{j \in J} x_i a_{ij} y_j \tag{3}$$

$$X = \{x \in \mathbb{R}^n / \ x_i \geq 0, \sum_{i \in I} x_i = 1\} \tag{4}$$

$$Y = \{y \in \mathbb{R}^m / \ y_j \geq 0, \sum_{j \in J} y_j = 1\} \tag{5}$$

X and Y are sets of mixed strategies for the players 1 and

respectively, and v is called the value of the game.
As is evident, if formerly the game has a saddle point, then

$$(i^*,j^*) \equiv (x^*,y^*)$$

From now on, if there is not ambiguity, the mixed strategies in X and Y will be simply referred to as strategies.

3. MATRIX GAMES WITH NON STANDARD PAYOFF

Now, a more general situation than the above one is considered (Campos and Verdegay, 1986). One supposes the matrix M is defined by fuzzy numbers, that is, it is assumed the players, perfectly knowing their possible strategies, are not able to exactly fix the corresponding payoffs or, from another point of view, they are more comfortable by specifying their payoffs as fuzzy numbers. Thus, the payoff matrix is now defined as

$$R = (a_{ij}),\ a_{ij} \in \mathbb{N}(\mathbb{R}),\ i \in I,\ j \in J \qquad (6)$$

where $\mathbb{N}(\mathbb{R})$ denotes the set of triangular fuzzy numbers on $\mathbb{R}$ [all of them represented by $(a, \underline{a}, \bar{a})$ as usual], and

$$\forall (i,j) \in I x J \Rightarrow \exists \mu_{ij} : \mathbb{R} \longrightarrow [0,1] \qquad (7)$$

A game defined according to these elements, (I,J,R), is called a Fuzzy Matrix (FM) game.

As it has been showed (Campos, 1989), the pair of optimal strategies and the value for that FM game can be found, similarly as in the classical case, from the solutions of the following pair of Fuzzy Linear Programming (FLP) problems, for player 1 and player 2, respectively

$$\begin{aligned} &\text{Min: } \sum_{i \in I} u_i \\ &\text{s.t.:} \\ &\quad \sum_{i \in I} a_{ij} u_i \ \textcircled{\geq}\ 1 - p_j(1-\alpha),\ j \in J \\ &\qquad\quad u_i \geq 0,\ i \in I,\ \alpha \in (0,1] \end{aligned} \qquad (8)$$

$$\begin{aligned} &\text{Max: } \sum_{j \in J} v_j \\ &\text{s.t.:} \\ &\quad \sum_{j \in J} a_{ij} v_j \ \textcircled{\leq}\ 1 + q_i(1-\alpha),\ i \in I \\ &\qquad\quad v_j \geq 0,\ j \in J,\ \alpha \in (0,1] \end{aligned} \qquad (9)$$

where:

a) $u_i = x_i/v$, $i \in I$ and $v_j = y_j/w$, $j \in J$.

b) v and w are defined, similarly to the classical case, from the following FLP problems,

$$\text{Max } \{v \,/\, \sum_{i \in I} a_{ij} x_i \gtrsim v,\ j \in J;\ x_i \geq 0,\ \sum_{i \in I} x_i = 1\} \tag{10}$$

$$\text{Min } \{w \,/\, \sum_{j \in J} a_{ij} y_j \lesssim w,\ i \in I;\ y_j \geq 0,\ \sum_{j \in J} y_j = 1\} \tag{11}$$

c) p_j and q_i, $\forall i,j$, express the maximum violations that the players permit in the accomplishment of the respective fuzzy constraints from (10) and (11), and

d) $\textcircled{\leq}$ is some relation for ranking fuzzy numbers.

In the following, this approach will be generalized. First, one assumes the players are not able to express their rewards in an exact way, that is, as real numbers. But, they may state those payoffs as linguistic labels chosen from some set of such possible labels. Each chosen label will reflect information that the players have about the reward they hope to obtain if the corresponding strategies are followed.

Thus, if the set of labels is denoted by

$$\Lambda = \{\lambda_1, \lambda_2, \ldots, \lambda_k\} \tag{12}$$

and one supposes there exits on Λ an order relation such that

$$\lambda_1 \leq \lambda_2 \leq \ldots \leq \lambda_k \tag{13}$$

then the game is defined by the triple (I,J,L), where now L is a matrix of labels such that $\forall (i,j) \in I \times J \Rightarrow \lambda_{ij} \in \Lambda$. The elements λ_{ij} of L express the reward that player 1 (2) will gain (lose) if he chooses strategy $i \in I$ ($j \in J$) and player 2 (1) strategy $j \in J$ ($i \in I$). This kind of games will be called Labelled Matrix (LM) games.

For example, let Λ be defined by Λ = {bad, indifferent, good, excellent}. Hence, one can for instance think on LM games (I,J,L) in which I = J = {1,2,3}, and

$$L = \begin{pmatrix} \text{excellent} & \text{indifferent} & \text{good} \\ \text{bad} & \text{bad} & \text{excellent} \\ \text{good} & \text{bad} & \text{indifferent} \end{pmatrix} \tag{14}$$

As is evident, every FM game may be expressed like an LM one.

4. SOLVING LABELLED MATRIX GAMES

Consider an LM game (I,J,L) with its rewards on L verifying (13). Player 1 is interested in choosing a strategy to obtain a higher possible payoff, whereas player 2 is interested in a strategy to lose as little as possible. Thus, both the players are interested in looking for two strategies (i^*,j^*) such that

$$\lambda_{i^*j} = \underset{i\in I}{\text{Max}}\ \underset{j\in J}{\text{Min}}\ \lambda_{ij} \tag{15}$$

and,

$$\lambda_{ij^*} = \underset{j\in J}{\text{Min}}\ \underset{i\in I}{\text{Max}}\ \lambda_{ij} \tag{16}$$

From (13) it is evident there always exists such labels, and moreover one can see that

$$\lambda_{i^*j} \leq \lambda_{ij^*} \tag{17}$$

To obtain a solution of this LM game, two possibilities should be considered:

a) The game has a saddle point. In such a case, a solution will be given by a pair of strategies $(i^*,j^*)\in I\text{x}J$ such that

$$\lambda_{i^*j^*} = \underset{i\in I}{\text{Max}}\ \underset{j\in J}{\text{Min}}\ \lambda_{ij} = \underset{j\in J}{\text{Min}}\ \underset{i\in I}{\text{Max}}\ \lambda_{ij} \tag{18}$$

i.e., $\lambda_{i^*j^*} \in \Lambda$ is the value of the game which, as it is evident, has the same nature than the former payoffs. This is exactly the case of the above LM game (14) for which

$$\lambda_{12} = \underset{i\in I}{\text{Max}}\ \underset{j\in J}{\text{Min}}\ \lambda_{ij} = \underset{j\in J}{\text{Min}}\ \underset{i\in I}{\text{Max}}\ \lambda_{ij}$$

that is, (1,2) constitutes an optimal pair of strategies for both the players, and correspondingly the value of the game is the saddle point λ_{12} = Indifferent.

b) The matrix L has no saddle point. In that case, it arises a new problem. In fact, now because of the payoff nature, it makes no sense to introduce, as in the classical case, a new payoff function similar to (3). Observe that if a conventional matrix game has no saddle point, the Minimax Theorem guarantees the existence of a pair of optimal mixed strategies and a value v given by (2). But v does not coincides with any entry of R (it will be some real number). For the LM games with no saddle point one reproduces this situation, but now if we would be able to find some value to that LM game (obviously in terms of a label), such a value would not coincide with any former label in Λ. Hence one has a double problem: 1) to obtain such labelled value, and 2) to find some label, or labels, in Λ to be associated with that labelled value.

For solving this last case, in the following we propose an approach for obtaining a label in Λ as a value of the LM game. Let (I,J,L) be an LM game, with no saddle point. Suppose that with each label λ_{ij} in L, one associates a fuzzy number $a_{ij} \in \mathbb{N}(\mathbb{R})$ which expresses, from the point of view of the players, a reward defined by λ_{ij}, and such that

$$\forall\ \lambda_{i_1j_1}, \lambda_{i_2j_2} \in \Lambda : \lambda_{i_1j_1} \le \lambda_{i_2j_2} \Rightarrow a_{i_1j_1} \circledleq a_{i_2j_2} \qquad (19)$$

where $\le$ is the order relation defined by (13) and $\circledleq$ is some transitive-relation for ranking fuzzy numbers. The set of those fuzzy numbers will be called a valuation of Λ.

Consider now an FM game (X,Y,A) defined from the above LM game (I,J,L) by taking X and Y as (4) and (5) and as payoff function

$$A(x,y) = \sum_{i\in I} \sum_{j\in J} x_i a_{ij} y_j \qquad (20)$$

where, in accordance with (19), $a_{ij} \in \mathbb{N}(\mathbb{R})$.

As it is evident, this new game is an FM one. Hence, as said before its solution can be found from the pair of FLP problems (8) and (9), and therefore from the point of view of this approach, two problems are to be solved:

1) If (x^*,y^*), $v^* \in \mathbb{N}(\mathbb{R})$ is a solution obtained from (8) and (9) for that FM game (strategies and value), what label $\lambda^* \in \Lambda$, definitely defining the value of the former LM game, is to be associated with v^*?

2) Does the label $\lambda^* \in \Lambda$, that is $(x^*,y^*) \in X\times Y$ and $v^* \in \mathbb{N}(\mathbb{R})$, depend on the fuzzy numbers a_{ij} associated with each $\lambda_{ij} \in \Lambda$?.

In the following these two problems will be solved. Now, the next result gives an answer to the second question.

Theorem 4.1. Let $\Lambda = \{\lambda_{ij}, i\in I, j\in J\}$ be a set of labels in which a relation $\leq$ of total order is defined. Suppose that for each $\lambda_{ij}\in\Lambda$ there exist $a_{ij}, b_{ij}\in\mathbb{N}(\mathbb{R})$ $(a_{ij}\neq b_{ij})$ such that

$$\forall\lambda_{i_1j_1}, \lambda_{i_2j_2}\in\Lambda: \lambda_{i_1j_1} \leq \lambda_{i_2j_2} \Rightarrow a_{i_1j_1} \circledleq a_{i_2j_2}, \ b_{i_1j_1} \circledleq b_{i_2j_2}$$

where, as usual, $\circledleq$ is some transitive relation for ranking fuzzy numbers. Denote $\mathfrak{A} = \{a_{ij}\}$ and $\mathfrak{B} = \{b_{ij}\}$ and consider A: $X\text{x}Y \longrightarrow \mathbb{N}(\mathbb{R})$, and B: $X\text{x}Y \longrightarrow \mathbb{N}(\mathbb{R})$, defined by

$$A(x,y) = \sum_{i\in I}\sum_{j\in J} x_i a_{ij} y_j, \quad B(x,y) = \sum_{i\in I}\sum_{j\in J} x_i b_{ij} y_j \qquad (21)$$

(X and Y like in (4) and (5), respectively). Now, if

$$A(x^*,y^*) = \operatorname*{Max}_{x\in X} \operatorname*{Min}_{y\in Y} A(x,y) = \operatorname*{Min}_{y\in Y} \operatorname*{Max}_{x\in X} A(x,y) \qquad (22)$$

then

$$B(x^*,y^*) = \operatorname*{Max}_{x\in X} \operatorname*{Min}_{y\in Y} B(x,y) = \operatorname*{Min}_{y\in Y} \operatorname*{Max}_{x\in X} B(x,y) \qquad (23)$$

and vice versa.

Proof. Suppose $(x^*,y^*)\in X\text{x}Y$ verifies (22). Then there exist two numbers $a^1, a^2\in\mathfrak{A}$ such that,

$$A(x^*,y^*) \in [a^1,a^2], \ a^1 \neq a^2$$

(In the case $a^1 = a^2$, this is trivial).

Since (22) holds, one has

$$A(x,y^*) = \operatorname*{Min}_{y\in Y} A(x,y)$$

but then

$$B(x,y^*) = \operatorname*{Min}_{y\in Y} B(x,y)$$

because, from the hypothesis, $a_{i_1j_1} \circledleq a_{i_2j_2} \Leftrightarrow b_{i_1j_1} \circledleq b_{i_2j_2}$.

In the same way

$$A(x^*,y^*) = \operatorname*{Max}_{x\in X} A(x,y^*) \Rightarrow B(x^*,y^*) = \operatorname*{Max}_{x\in X} B(x,y^*)$$

and a similar reasoning can be followed for

$$\operatorname*{Min}_{y\in Y} \operatorname*{Max}_{x\in X} A(x,y) \text{ and } \operatorname*{Min}_{y\in Y} \operatorname*{Max}_{x\in X} B(x,y)$$

Therefore if (22) is verified, (23) holds.

On the other hand, if $A(x^*,y^*)\in[a^1,a^2]$ and $B(x^*,y^*)\in[b^1,b^2]$ for some $a^1,a^2\in\mathfrak{A}$ and $b^1,b^2\in\mathfrak{B}$, if $a^1\in\mathbb{N}(\mathbb{R})$ is associated with label $\lambda^1\in\Lambda$, it is evident that $b^1\in\mathbb{N}(\mathbb{R})$ will also be associated with the same label $\lambda^1 \in \Lambda$, because if is not so, (23) will not be verified. The same occurs with a^2, b^2 and the corresponding label $\lambda^2 \in \Lambda$. ∎

Thus, given an ordered set of labels defining the possible rewards for an LM game, for each of them one can associate a fuzzy number in such a way that these fuzzy numbers neither affect

a) the label defining the value of the game, if it has a saddle point, nor

b) the interval $[\lambda^1, \lambda^2]$ of labels associated with the interval of fuzzy numbers $[a^1, a^2]$, $a^1, a^2 \in \mathfrak{A}$, to which $A(x^*, y^*)$ belongs.

Now, if the value of the game corresponds to a new label in the interval $[\lambda^1, \lambda^2]$, but it does not belong to Λ, the problem to be solved is: In terms of the labels in Λ, what will be the value of the game?.

To answer this, suppose $f\colon \Lambda \longrightarrow \mathbb{N}(\mathbb{R})$ is a mapping verifying (19), that is, $\forall \lambda^1, \lambda^2 \in \Lambda\colon \lambda^1 \le \lambda^2 \Rightarrow f(\lambda^1) \le f(\lambda^2)$.

Let $v^1 = A(x^*, y^*) \in \mathbb{N}(\mathbb{R})$ be the value of the FM game deduced from the former LM game by means of f.

As it is known, $v^1 \in [f(\lambda^1), f(\lambda^2)]$, for some pair of consecutive labels λ^1, λ^2 in Λ . It will be clear that the membership of v^1 to that interval shall be measured by the supposed relation for ranking fuzzy numbers $\textcircled{\le}$ .Hence the situation may be the one represented in the following figure

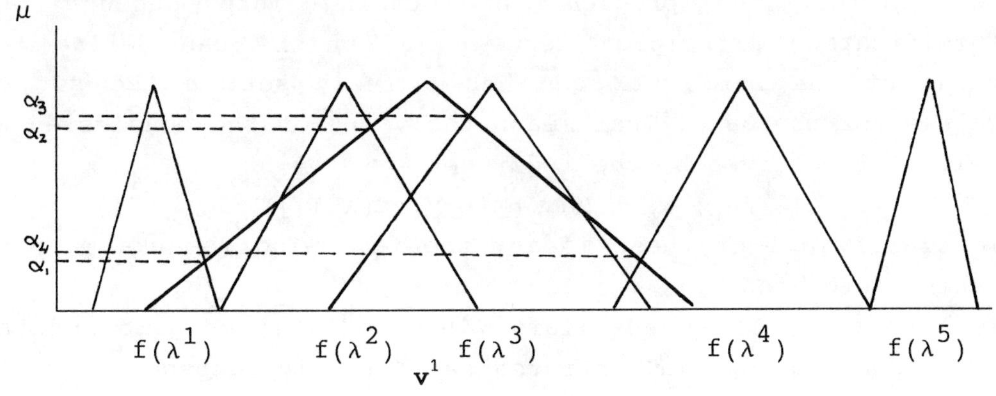

From this example it is seen if $v^1 \in [f(\lambda^2), f(\lambda^3)]$, v^1 also accomplishes the valuation of each label $\lambda^i \in \Lambda$ with an accomplishment degree $\alpha_i \in (0,1]$.

Let now $g: \Lambda \longrightarrow \mathbb{N}(\mathbb{R})$ be another valuation for the labels verifying (19), and let $v^2 \in \mathbb{N}(\mathbb{R})$ be the value of the corresponding FM game deduced from the former one by means of valuation g.

The situation can be represented as in the following figure

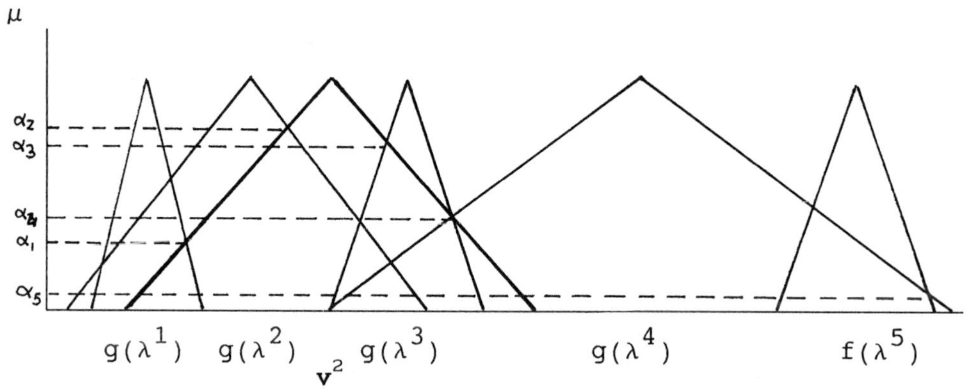

Mapping g also verifies the hypotheses of the above theorem, but now $v^2 \in [g(\lambda^2), g(\lambda^3)]$, that is, as in the case of the last figure, the label that corresponds to v^2 is found again between λ^2 and λ^3, but evidently because of that new valuation g, the degrees $\alpha_i \in (0,1]$ to which v^2 verifies each of the labels $\lambda^i \in \Lambda$ are different from those for v^1.

For solving the problem, and taking into account the fuzzification principle (Goguen, 1967), one can define the value of the former LM game as a fuzzy set on the set of values for the game. That means the value of the considered LM game will be given by the fuzzy set

$$V_L = \{\lambda^k/\alpha_k;\ \alpha_k = \text{Max}\ [\text{Min}(v,\ f(\lambda^k))]\} \qquad (24)$$

To clarify all of these ideas, the following remarks must be taking into account:

1.- The value of an auxiliary (intermediate) FM game and the corresponding optimal strategies directly depend on the relation $\textcircled{\le}$ used for ranking fuzzy numbers. Hence, it will be possible for the same game to get as many different solutions as many relations $\textcircled{\le}$ can be considered.

2.- The valuation associated with each label might be supposed non fuzzy, but the use of fuzzy numbers better reflects an imprecise nature of the rewards.

3.- If the valuations for the labels are changed, the value (24) of the LM game also changes (as it occurs in the classical case if the entries of the payoff matrix are modified by means of some change of scale). But now, due to the nature of payoffs and because of the lack of restrictions on that change of scale, the value V_L is not forced to have a direct functional relation with the new value obtained by some arbitrary change into the valuation.

4.- The problem of obtaining a sole label, finally, as the value of the game, can be trivially solved from (24) by selecting such a label $\lambda^t \in \Lambda$ that,

$$\alpha_t = \text{Max } \{\alpha_k : \lambda^k/\alpha_k\}$$

(But, notice that this would be as if in the classical case for a game with no saddle point, it were required the final value be some entry of the payoff matrix).

This is illustrate with the next example.

5. AN ILLUSTRATIVE EXAMPLE

Let Λ = {VB, B, I, G, VG, E} where the initials correspond to the labels "Very Bad", "Bad", "Indifferent", "Good", "Very Good" and "Excellent", respectively, and verify VB $\le$ B $\le$ I $\le$ $\le$ G $\le$ VG $\le$ E. Let the matrix of payoffs be

$$\begin{pmatrix} \text{Indifferent} & \text{Excellent} & \text{Very Bad} & \text{Good} \\ \text{Very Good} & \text{Bad} & \text{Good} & \text{Bad} \\ \text{Very Bad} & \text{Good} & \text{Indifferent} & \text{Very Good} \end{pmatrix}$$

and suppose with each label in Λ one associates the fuzzy numbers in

$$\mathfrak{A} = \{1,2,3,4,5,6\} \qquad (25)$$

each of them defined by the triangular membership functions (1, 1/2, 3/2), (2, 1, 3), (3, 5/2, 7/2), (4, 2, 6), (5, 3, 7) and (6, 3, 9), respectively. Then one has the following FM game

$$\begin{pmatrix} 3 & 6 & 1 & 4 \\ 5 & 2 & 4 & 2 \\ 1 & 4 & 3 & 5 \end{pmatrix}$$

Suppose now the relation $\textcircled{\le}$ for ranking fuzzy numbers is trivially defined by

$$\forall A,B \in \mathbb{N}(\mathbb{R}): A = (a, \underline{a}, \bar{a}),\ B = (b, \underline{b}, \bar{b}) \Rightarrow A \textcircled{\le} B \Leftrightarrow a \le b$$

Then, in accordance with (8) and (9) we must solve the following parametric LP problems for player 1 and player 2, respectively

$$\begin{array}{ll} \text{Min:} & u_1 + u_2 + u_3 \\ \text{s.t:} & \\ & 3u_1 + 5u_2 + u_3 \ge 1 - p(1-\alpha) \\ & 6u_1 + 2u_2 + 4u_3 \ge 1 - p(1-\alpha) \\ & u_1 + 4u_2 + 3u_3 \ge 1 - p(1-\alpha) \\ & 4u_1 + 2u_2 + 5u_3 \ge 1 - p(1-\alpha) \\ & u_i \ge 0,\ \alpha \in (0,1] \end{array} \tag{26}$$

$$\begin{array}{ll} \text{Max:} & v_1 + v_2 + v_3 + v_4 \\ \text{s.t:} & \\ & 3v_1 + 6v_2 + v_3 + 4v_4 \le 1 + q(1-\alpha) \\ & 5v_1 + 2v_2 + 4v_3 + 2v_4 \le 1 + q(1-\alpha) \\ & v_1 + 4v_2 + 3v_3 + 5v_4 \le 1 + q(1-\alpha) \\ & v_j \ge 0,\ \alpha \in (0,1] \end{array} \tag{27}$$

By taking, for simplicity, $p = q = 0$ and solving them, one obtains $x^* = (.125, .5, .375)$ and $y^* = (.084, .416, .5, 0)$ as optimal strategies for player 1 and player 2, respectively, and (for this easy case) the non fuzzy number,

$$\mu(x) = \begin{cases} 1 & \text{if } x = 3.25 \\ 0 & \text{elsewhere} \end{cases}$$

as a value of the corresponding FM game.

Now, if the degree to which this number verifies each of the labels defined by the fuzzy numbers in $\mathfrak{A}$, is measured one definitely obtains the value of the former LM game which is given by the fuzzy set

$$V_1 = \left\{ \frac{VB}{0}, \frac{B}{0}, \frac{I}{.5}, \frac{G}{.625}, \frac{VG}{.125}, \frac{E}{.083} \right\}$$

Let $\mathfrak{A}'$ be another valuation of the labels in Λ defined by a change of scale, that is

$$\mathfrak{A}' = \{3,4,5,6,7,8\} \qquad (28)$$

with the respective membership functions defined by (3, 2.5, 3.5), (4, 3, 5), (5, 4.5, 5.5), (6, 4, 8), (7, 5, 9) and (8, 5, 11). Solving the corresponding auxiliary FM game one obtains again $x^* = (.125, .5, .375)$ and $y^* =(.084, .416, .5, 0)$ and now the non fuzzy value,

$$\mu(x) = \begin{cases} 1 & \text{if } x = 5.25 \\ 0 & \text{elsewhere} \end{cases}$$

Hence the value of the former LM game is

$$V_2 = \left\{ \frac{VB}{0}, \frac{B}{0}, \frac{I}{.5}, \frac{G}{.625}, \frac{VG}{.125}, \frac{E}{.083} \right\}$$

which evidently coincides with V_1.

Finally, notice that if one supposes $p, q \neq 0$ for solving (26)and (27), as say $p = q = .25$, the value of the auxiliary FM game is the fuzzy number

$$\mu(x) = \begin{cases} 5 - 13/x & \text{if } x \in [2.6, 3.25] \\ 13/x - 3 & \text{if } x \in [3.25, 4.33] \end{cases}$$

and hence the value of the LM game is the fuzzy set

$$V_1^1 = \left\{ \frac{VB}{0}, \frac{B}{.26}, \frac{I}{.80}, \frac{G}{.74}, \frac{VG}{.40}, \frac{E}{.30} \right\}$$

If the new valuation of Λ, (28), is considered and one supposes $p = q = .25$ again in the corresponding parametric LP problems for solving the auxiliary FM game, one obtains as the value for this game the fuzzy number

$$\mu(x) = \begin{cases} 5 - 21/x & \text{if } x \in [4.2, 5.25] \\ 21/x - 3 & \text{if } x \in [5.25, 7] \end{cases}$$

and then the value of the LM game is the fuzzy set

$$V_2^1 = \left\{ \frac{VB}{0}, \frac{B}{.42}, \frac{I}{.86}, \frac{G}{.78}, \frac{VG}{.50}, \frac{E}{.40} \right\}$$

As it is evident, V_2^1 does not exactly coincide with V_1^1 because the margins of the constraints in (8) and (9) have not bee changed in accordance with the corresponding change of scal from $\mathfrak{A}$ to $\mathfrak{A}'$. It is obvious that if one wish to obtain th second value exactly equal to the first one, those margins (and q) must be changed in the same way to the ones in which $\mathfrak{A}$ changes with respect to $\mathfrak{A}$.

A formal relation between the former margins p and q associated with a valuation $\mathfrak{A}$ of Λ, and the corresponding ones associated with another valuation $\mathfrak{A}'$ of Λ will be discussed in forthcoming papers.

6. CONCLUSION

We have presented an approach to solve an LM game with the following features,

a) If the game has a saddle point, its value is one of the labels defining the payoffs of the game.
b) If the matrix defining the game has no saddle point, it can be obtained as solution, by means of a valuation for the labels, a fuzzy set on the set of the former labels.
c) Whatever valuation for the labels, it is always obtained the same pair of consecutive labels among which the value of the game in "terms of labels" is found. This interval never changes, but the membership degrees on the final fuzzy set defining the value of the game do; it depends on both the relation ⓢ used for ranking fuzzy numbers and on the margins that players permit in the accomplishment of the constraints into an auxiliary LP problems used for solving the corresponding FM game.

REFERENCES.

L. Campos (1989): Fuzzy Linear Programming Models to Solve Fuzzy Matrix Games. To be appeared in Fuzzy Sets and Systems.

L. Campos and J.L. Verdegay (1986): On Fuzzy Games. Proc. of III Fall Int. Seminar on Applied Logic, 25-35. Palma (Spain).

L. Campos and J. L. Verdegay (1989): Linear Programming Problems and Ranking of Fuzzy Numbers. Fuzzy Sets and Systems 32, 1-11.

M. Delgado, J.L. Verdegay and M.A. Vila (1989): A General Model for Fuzzy Linear Programming. Fuzzy Sets and Systems, 29, 21-29.

J.A. Goguen (1967): L-Fuzzy Sets. Jour. of Math. Anal. and Appl.,18, 145-174.

J. von Neumann and O. Morgenstern (1947): Theory of Games and Economic Behavior. Princeton University Press (Princeton).

S.A. Orlovski (1977): On Programming with Fuzzy Constraint Sets Kybernetes 6, 197-201.

G. Owen (1982): Game Theory. Academic Press (New York).

FUZZY CONVEXITY AND PERIPHERIAL CORE OF AN EXCHANGE ECONOMY REPRESENTED AS A FUZZY GAME

Antoine BILLOT
Department of Economics, ERMES-CNRS
University of PARIS 2
92, rue d'Assas
75006 Paris, France

Abstract: *One of the theory of games miracles is to have converted the problem of non emptiness of the core into the convexity of the model (sets and functions). But, what happens if the agents' fuzzy preferences are not convex? The peripherial core existence is a possible answer. Defined from a new and weaker notion of convexity, it allows to balance an exchange economy approximately if the usual core in empty and it also permits to distinguish betweeen the usual balancing allocations (when they exist), the one which corresponds to the higher social level of satisfaction.*

Keywords: *Fuzzy preference, core analysis.*

1. INTRODUCTION

The notion of cooperation (**ICHIISHI [1981;1982], SCARF [1967; 1971;1973], SHAPLEY [1973], SCHLEICHER [1979]**), which comes after the development of a coalition, gives birth to what we sometimes call *super-players*, that is groups of players who are going to behave in the same way a single player does. In other words, the super-players rationality is identical to a single agent in a conflict situation. This means that the players are able to **abdicate** their intrinsic power of decision towards a collective institution coming from the coalition to which they belong to. These intermediary agents (middlemen) are theoretical artifacts which symbolize all union forms, trade unions (syndicates), trusts and other diverse associations. Even if the analysis of a syndicate in terms of theory of games makes a large difference between a **simple coalition** and a **syndicate** (see **DREZE & GABSZEWICZ [1971]**) - the last coming from an exogeneous valuation, we can verify that the syndicate really exists while explaining the existence of the coalition with the model -, this analysis is only justified by a theore-

J. Kacprzyk and M. Fedrizzi (eds.), Multiperson Decision Making Using Fuzzy Sets and Possibility Theory, 311–335.

tical willing (and pratically technic) to reduce the number of possible coalitions that can improve their satisfaction, relatively to a given imputation and thus to increase the core's cardinal associated to the cooperative game (see **ROSENMULLER [1981]** and **CORNWALL [1984]**, chap 5).

Literature on cooperative fuzzy games is not very wide. In fact, it gravitates arround two specific poles which are **AUBIN**'s (**[1974;1979;1981;1986]**) studies and **BUTNARIU**'s (**[1979;1980;1985; 1986;1987]**) ones. BUTNARIU continued what he had started with non-cooperative fuzzy games and that with the help of notions like *strategic conception* and *research on the optimal rule of informations exchange.* He generalized the SHAPLEY value concept to a fuzzy game and studied its stability in terms of *individual structure of information.* After specifying where the exact problem in cooperative games was, **AUBIN** writes in **[1986]**: "At first, cooperative games theorists ran into difficulties coming from the fact that coalitions set is finite. The set structure is too poor and its results too trivial or too complex. Many attempts to increase the set of players had been originated; for example, one consisted to take as the players set the segment [0,1] called "players continuum". This proceeding, used for the first time by AUMANN, is what physicists have been using since the "differential calculus" invention. We shall defend a proceeding consisting in keeping a finite number of players and taking a coalitions continuum called "fuzzy coalitions". The cooperative fuzzy games are the one who take count of the fuzzy coalition behaviour".

AUBIN defines a cooperative fuzzy game as an independent game towards the agents preference behaviour, placed on a fuzzy membership to the coalitions. This approach will be similarly inspired by us even if the concepts used are not ours. There transfigures the usual cooperative game model only for the resort to the principles of convexity analysis, which permit to present new results. On the other hand, the direction in which BUTNARIU investigated, proved to be extremely fruitful in terms of *non-constraining cooperation.* It appears to be an interesting generalization of the cooperative game analysis central concepts.

The classic non-vacuity theorems (**SCARF [1967]**, **MUNIER [1973]**, **ISCHIISHI [1981]**, **ROSENMULLER [1981]**, **BORDER [1982]**, **CORNWALL [1984]**) lay on a convexity condition on the fundamental sets (mixed strategies set for pure games, consumption ones in exchange economies) as well as individual functions (convex preorders for payoffs). This condition generates the balanced game structure without which the game's core is empty. A game's core is a set of issues, outlets, imputations or allocations which contains the equilibria and also the issues which can be stabilized by strong enough dissuasive threats.

The property of threat . stabilization was shown by economists even before a much more formal analysis: the theory of games confirmed first intuition. FISCHER had already critisized COURNOT duopoly analysis by saying that no businessman was shortsighted enough not to suppose that his rival would not react to his own hits[1].....

This will lead up economists to think that the *counter* idea, *dissuasion by threat* is formally explained by the *core* notion (**SCHOTTER & SCHOWDIALLER [1980]**). This also means that an eventual betrayal coming from a given usual coalition can be immediately countered by other players reactions. These reactions will then prevent all players belonging to the betrayal coalition to benefit from their membership (to that coalition) and at least dissuade them from participating. The threat notion conducts us to an *arbitration* notion. The core is finally the set of all the acceptable arbitrations by society. Once a way out is proposed, imputation or allocation of the core, no one - that is to say no individuals, nor fuzzy or usual coalitions - is incited to oppose himself, or less he will be abandoned and will not do as good (in payoffs or utility) as he would of been offered. This idea brings us straight to the balanced structure concept.

For a better comprehension of the described phenomenon, we must consider that the agents have the possibility of meeting all engagements. This argument corresponds to the fact that all contracts settled between the agents at the beginning of the exchanges could always be canceled after words if other contracts, proved to be better, would appear. The contracts are described by the ways out, the issues, outlets, imputations or allocations. This *recontract* hypothesis is perfectly analogue to the walrasian vision of the economic market which corresponds to the existence of an auctioneer. One of the theory of games miracles is to have concentred the non-vacuity of the core's problem into the convexity of the model (sets and functions); this convexity interpreted easily, as we shall see, further on even thow quite restrictive. In fact, the

---

1 " *anymore than a chess player assumes that his opponent will not interfere with his effort to capture a knight.*" cf **MOULIN [1981]**, p.91.

question we ask in this paper is: what happens if the agents fuzzy preferences are not convex? At first, we shall redefine fuzzy convexity and try to show that its requirement is weaker. It will be the first part of the paper.

2. ℓ-CONVEXITY AND FUZZY PREORDERS

2.1. The mathematical model

We shall consider a referential set X of any objects. Let's suppose the set convex and thus connected.

Let's consider a *fuzzy relation of preference*, $\mathcal{R}(.,.)$ defined on the Zadeh set [0,1].

The agent will arrange the different elements of X, $\mathcal{R}(.,.)$ being a ℓ-transitive and reflexive (e.f.)[2] fuzzy relation. It means

ℓ-transitive relation.

1) $\forall\ (x,y,z) \in X^3$, if $\mathcal{R}(x,y) \geq \mathcal{R}(y,x)$ and $\mathcal{R}(y,z) \geq \mathcal{R}(z,y)$ then $\mathcal{R}(x,z) \geq \mathcal{R}(z,x)$.

reflexive (e.f.) relation.

2) $\forall\ x \in X$, $\mathcal{R}(x,x) \in [0,1]$.

"$\mathcal{R}(x,y) \geq \mathcal{R}(y,x)$" means that the agent considers that **the relative qualities of the object x are greater than those of the object y**[3].

We assume, to generate preordered fuzzy subset by $\mathcal{R}(.,.)$ noted $\mathcal{X}$, that if $\forall\ (x,y) \in X^2$, $\mathcal{R}(x,y) \geq \mathcal{R}(y,x)$, then $\alpha(x) \geq \alpha(y)$[4]: the agent's local preference ($\mathcal{R}(.,.)$) does not contradict with his global preference ($\alpha(.)$). There is no local irrational (non ℓ-transitive) preference, and thus no agent that would have to later domi-

---

[2] (e.f.) for "*Ecole Française*" (French School); see **PONSARD & FUSTIER [1986]**.

[3] So, $\mathcal{R}(x,x)$ evaluates the **intrinsic qualities of the object x.**

[4] In all the paper the membership functions are symbolized in such a way: the membership function of the fuzzy subset $\mathcal{A}$ is $a(.)$. *All which concerns fuzzy subsets appear in cursives letters.*

nate it with his global rationality[5].

If $\mathcal{R}(.,.)$ is ℓ-transitive and reflexive (e.f.), then $\mathcal{R}(.,.)$ is fuzzy preorder (e.f.). We also know that if the referential set X is convex, therefore connected, under the fuzzy preference continuity hypothesis [6], $\mathcal{X}(.)$ is a continuous function of utility on X (see note 5). We shall suppose that the continuity hypothesis is satisfied.

2.2. Basic propositions

Proposition 1: $[\forall\alpha\in]0,1[: \mathcal{X}(\alpha x+(1-\alpha)y)=0]$
No x and y convex combination belongs to $\mathcal{X}$

Proposition 2: $[\exists\alpha\in]0,1[: \mathcal{X}(\alpha x+(1-\alpha)y)= 0]$
Existence of a x and y convex combination that does not belong to $\mathcal{X}$

Proposition 3: $[\mathcal{X}(x) \neq 0]$
x belongs to $\mathcal{X}$

By convention, we note:
- X(1), the set of objects of X that satisfy proposition 1 and that for all y of X: $X(1) = \{x\in X; \forall y\in X, \forall\alpha\in]0,1[: \mathcal{X}(\alpha x+(1-\alpha)y)=0\}$.

- X(2), the set of objects of X that satisfy proposition 2 and that for all y of X; $X(2)=\{x\in X; \forall y\in X, \exists\alpha\in]0,1[: \mathcal{X}(\alpha x+(1-\alpha)y)=0\}$.

- X(3), the set of objects of X that satisfy proposition 3. It corresponds[7] to the exclusive support of $\mathcal{X}$: $X(3)=\text{supp}\mathcal{X}$. We assume that

---

5 The **Principle of Dominated Irrationality** appears in **BILLOT [1988]**. When the fuzzy relation $\mathcal{R}(.,.)$ is not ℓ-transitive, the agent can dominate this irrationality with the first projection of the relation. He builds his utility function with the Max operator: $\mathcal{X}(x) = \text{Max } \mathcal{R}(x,X)$. Under usual hypothesis (**BILLOT [1986]**), $\mathcal{X}(.)$ is a continuous function of utility.

6 The two following sets are fuzzy **closed** sets: $\forall x'\in X$,
$\mathcal{X}^+= \{x\in X;\mathcal{X}(x)\geq\mathcal{X}(x')\}$ and $\mathcal{X}^-= \{x\in X;\mathcal{X}(x)\leq\mathcal{X}(x')\}$.

7 Do not always assimilate the exclusive support of $\mathcal{X}$ and the set X itself. Nothing prevents an element from X having a zero level of membership to $\mathcal{X}$: $\text{supp}\mathcal{X} = \{x\in X;\mathcal{X}(x)>0\}$. A non exclusive sup-

the **exclusive support of $\mathcal{X}$ is convex** and thus connected.

2.3. Simple propositions

For $x \in X$,

a) It is obvious that: if $[\forall\alpha\in]0,1[$ and $\forall y\in X$, $\mathcal{X}(\alpha x+(1-\alpha)y)=0]$ then $[\exists\alpha\in]0,1[; \forall y\in X, \mathcal{X}(\alpha x+(1-\alpha)y)=0]$.

We can also conclude that: $X(2) \subseteq X(1)$.

b) We know as well that:[8] if $[\neg\exists\alpha\in]0,1[; \forall y\in X:\mathcal{X}(\alpha x+(1-\alpha)y)=0]$[9] then $[\forall\alpha\in]0,1[, \exists y\in X: \mathcal{X}(\alpha x+(1-\alpha)y)\neq 0]$. Because of the convexity of the exclusive support of $\mathcal{X}$, it means: $[\forall\alpha\in]0,1[, \forall y\in X: \mathcal{X}(\alpha x+(1-\alpha)y)\neq 0]$.

Then, the complementary set of X(2) in X and X(1) are disjointed. If we note $X(2)^c$ the complementary set of X(2) in X, we can write: $X(2)^c \cap X(1) = \emptyset$.

2.4. Complex propositions

c) $X(3)^c \subseteq X(1)$.

Proof: If $[\forall\alpha\in]0,1[, \forall y\in X, \mathcal{X}(\alpha x+(1-\alpha)y)=0]$, then, for a given x, $x\in X$, since $\mathcal{X}(.)$ is continuous on X: $\lim_{\alpha \to 1^-} [\mathcal{X}(\alpha x+(1-\alpha)y)]=0$, it means $\mathcal{X}(x) = 0$: $x \notin X(3)$. Thus if $x \in X(1) \Rightarrow x \notin X(3)$; it corresponds to: $X(1) \supseteq X(3)^c$, that is to say, $\text{supp}\mathcal{X}^c \subseteq X(1)$. □

d) $X(2) \subseteq X(3)^c$.

Proof: If $\mathcal{X}(x)=0$, it means, since $\mathcal{X}(.)$ is a continuous function on X: $\lim_{\alpha \to 1^-} [\mathcal{X}(\alpha x+(1-\alpha)y)]=0$. In other words, $\exists\alpha\in]0,1[$ such as, $\forall y\in X$: $\mathcal{X}(\alpha x+(1-\alpha)y)]=0$. i.e.: $x \in X(3)^c \Rightarrow x \in X(2)$; $X(2) \subseteq X(3)^c$. □

e) $X(2) = X(1) = X(3)^c$.

---

port of $\mathcal{X}$ is: $\{x\in X;\mathcal{X}(x)\geq 0\}$. It corresponds trivialy to X.

8 Simple propositions a) et b) allow to write X(2) and X(1) as being the same set; this result appears explicitly in the complex proposition e).

9 "$\neg\exists$" means "there is no..".

Proof: Proposition d) establishes $X(2) \subseteq X(3)^c$, and proposition c) $X(3)^c \subseteq X(1)$. It means: $X(2) \subseteq X(3)^c \subseteq X(1)$.

We know also $X(1) \supseteq X(2)$ and $X(2)^c \cap X(1) = \varnothing$. If the set X(1) is included in X(2), $X(1) \supseteq X(2)$, then $X(2)^c \supseteq X(1)^c \Rightarrow X(2) \cap X(1)^c = \varnothing$. But, we know also that $X(2)^c \cap X(1) = \varnothing$ (with b), that is to say $X(2)^c \cup X(1) = X$ and $X(2) \cup X(1)^c = X$; thus $X(2)^c \cup X(1) = X(2) \cup X(1)^c$. We have just shown, $X(2) \cap X(1)^c = \varnothing$ and, by definition $X(2) \cap X(2)^c = \varnothing$, so, we can conclude: $X(2) = X(1)$. If $X(2) = X(1)$ and $X(2) \subseteq X(3)^c \subseteq X(1)$, so, it is clear that: $X(2) = X(1) = X(3)^c$ or $X(2)^c = X(1)^c = X(3) = \text{supp}\mathcal{X}$.

□

(For all demonstrations, see **QUINE [1972]**, chap 18, 19 and 20.)

What have we shown? If $X(2)^c = X(1)^c = \text{supp}\mathcal{X}$, it means that if an object x belongs to the support of the fss[10] $\mathcal{X}$, coming from the preorder $\mathcal{R}(.,.)$, then $[\forall \alpha \in]0,1[, \exists y \in X, \mathcal{x}(\alpha x + (1-\alpha)y) \neq 0]$. As soon as two objects x and y have a membership degree to $\mathcal{X}$ which is superior to 0, there exists a convex combination which has **also** a membership degree superior to 0. The usual convexity tells us that if x and y belong to the usual set X, and if X is connected, then all the convex combinations of x and y belong to the usual set X. Intuitively, the **weak fuzzy convexity** notion means: **if two elements of the referential set belong to the exclusive support of the fss, then there is a convex combination of these two elements which belongs also to the exclusive support of the fss.** The proposition e) includes that for any exclusive support of a referential set with a continuous membership function, this exclusive support is *convex* in the sense so forth.

Definition 1: A fuzzy subset $\mathcal{X}$ of the convex referential set X is **weak ℓ-convex** iff: $\forall (x,y) \in (\text{supp}\mathcal{X})^2, \exists \alpha \in]0,1[: (\alpha x + (1-\alpha)y) \in \text{supp}\mathcal{X}$.

Definition 2: A fuzzy subset $\mathcal{X}$ of the convex referential set X is **strong ℓ-convex** iff: $\forall (x,y) \in (\text{supp}\mathcal{X})^2, \forall \alpha \in]0,1[: (\alpha x + (1-\alpha)y) \in \text{supp}\mathcal{X}$.

An other way in presenting the strong ℓ-convexity is by considering the two following propositions as equivalent:

$$[\forall \alpha \in]0,1[, \mathcal{x}(\alpha x + (1-\alpha)y) > 0] \Leftrightarrow [\mathcal{x}(x) > 0 \text{ et } \mathcal{x}(y) > 0].$$

In demonstrating the complementary set of X(2) in X is the exclusive support of $\mathcal{X}$, $\text{supp}\mathcal{X}$, we establish the following lemmas:

---

10 "fss" for Fuzzy SubSet.

LEMMA 1: *Let a convex referential set X and $\mathcal{X}$ any fss of X. If $\mathcal{x}(.)$ the membership function is continuous, then $\mathcal{X}$ is weak ℓ-convex.*

LEMMA 1bis: *Let a convex referential set X and $\mathcal{X}$ any fss of X. If $\mathcal{x}(.)$ is continuous and supp$\mathcal{X}$ convex, then $\mathcal{X}$ is strong ℓ-convex.*

This ℓ-convexity is very different from convexity used in fuzzy literature (**LIU [1985]**, **WEISS [1975]**, **KAUFMANN [1973]**, **PREVOT [1977]**, **ZIMMERMANN [1985]**, **PONSARD [1980]** and **CHANG [1968]**), the last one simply coming out of properties of epigraph (see **MOULIN, FOGELMAN-SOULIE [1979]**).

In fact, if we consider a function f(.) defined on a space E, towards $\mathbb{R}$, we call **f(.) epigraph,** noted epi(f), the following set $\{(x,\lambda)\in E\times\mathbb{R};\ f(x)\leq\lambda\}$. A f(.) epigraph is convex if the function is also convex. On the other hand, the complementary set (non exclusive) $\text{epi}(f)^c_\lambda$, defined as following: $\text{epif}(.)^c_\lambda=\{(x,\lambda)\in E\times\mathbb{R};\ f(x)\geq\lambda\}$, is convex if and only if for E convex, $f(\alpha x+(1-\alpha)y)\geq\text{Min}\ \{f(x),f(y)\}$, that is to say, if f(.) is quasi-concave. In other words, a fss $\mathcal{X}$ is convex (in the sense of literature) if the membership function $\mathcal{x}(.)$ is quasi-concave: here, we find again the initial preorders and their usual restrictive convexity.

In our process, the membership function $\mathcal{x}(.)$ comes directly from preorders. The fss comes from the individual preferences and is convex only if the membership function is quasi-concave. This then induces that the fuzzy subset $\mathcal{X}$, fuzzy preordered subset by $\mathcal{R}(.,.)$, is only convex if the membership function $\mathcal{x}(.)$ is quasi-concave, that is to say, if the preorders are convex. That way, there is a link between fundamental set convexity (which concerns the referential set) and the associated preorders: this link is the base of usual core's non vacuity in the cooperative games approach.

Our goal is simple. We shall content ourselfs of the ℓ-convexity, immediately acquired if $\mathcal{x}(.)$ is continuous, i.e. if the referential set X is connected and the choice individual preorder continuous, to show a core and non vacuity conditions free from all preorder usual convexity constraint, since the fss $\mathcal{X}$ is ℓ-convex even if the choice preorder is not usual convex. The **fuzzy core**, we call it the **peripherial core (P.C.)**, is not identical to the usual core or AUBIN fuzzy core. To define PC, we need two references: a planner interested in the PC's non vacuity and the ε-cores of the non fuzzy literature (**WEBER [1979]**, **WOODERS [1983]**).

3. PERIPHERIAL CORE'S NON VACUITY

3.1. ε-cores, game planner and coalitions

Before going any further in the technique as well as its formal developements, we must explain, using a more intuitive analysis, what peripherial core notion means. If the usual core (we call it **intra-muros core** to insist on the fact that it is always included in the peripherial core) is empty - that is to say the game is not balanced - what is pratically happening? Does the economy ignore the decision that will be made by society? What we want to know, is what is going to happen pratically, really. Society must, of course, choose an allocation, a way out, an outlet if it is pure exchange game (the players can stop playing which is not the case of all economic agents). In our approach, in the absence of individual preorder convexity, the intra-muros core is empty. In other words, we have to give satisfying results in the case of individual fuzzy preorders that are no more necessarily convex.

3.1.1. ε-cores

If the game is not in equilibrium, i.e. the preorders are not convex, the intra-muros core is empty. In the exchange economy analysis, the **approached core** concept introduction allowed new condition guaranting its non vacuity (see **WOODERS & ZAME [1984]**). One of the possible formulations of the approached core notion is the ε-core introduced by **SHAPLEY & SHUBIK** in their **[1969]** paper; In the same way, using a balanced structure extension from **SCARF**[11] **[1967]**, **WEBER [1979]** showed that a certain number of games with a players continuum in the sense of AUMANN had non-empty cores at an ε near (ε very little). The approached core analysis for games with a players continuum without a balanced structure was made by **WOODERS [1983]** using a replication hypothesis identical to the one used by **AUBIN [1986]**. The works that have just been quoted as well as AUBIN works (in fuzzy coalitions) show the evident link between the *number of players extension*, by hypothesis or replication, and the *condition of a balanced game.* In other words, we can expect balan-

---

11 or **balancedness**, cf **BONDAVERA [1963]** and **SHAPLEY [1973]**, coming from the convexity of preorders in a game without utility compensations and without side payments, which corresponds to the case of an exchange economy where individual utilities are ordinal.

ced results for a cooperative game - without the preorders convexity - if we extend society's cardinal to the infinite. Nevertheless, in such an economy (or such a game), the core is approached to an ε near which means that the agents in a coalition suffer shortsightness (mesured by ε) that makes them unconcerned by two allocations at an ε near. In terms of utility, they accept loosing ε of satisfaction so that they do not block up an allocation or an outlet. We can then deduce an approached core noted ε-core. One of the shortsightness interpretations proposed by SHAPLEY & SHUBIK explains "ε" as an organisational cost deduced from the members of the coalition.

However, these types of approached-cores are founded on societies *convexification* (replication or direct continuum in the sense of AUMANN) and somewhat replaces individual preorders convexity by the players set convexity. What happens if we do not replicate an economy with a finite number of players? The answer that we shall propose will be founded on the ℓ-convexity of the fuzzy subset associated to the exchanger or player's individual preorder and on a game planner.

3.1.2. The game planner

In Walras's works, there is an agent, the auctionner, whose satisfaction rises with the surplus reduction, the goods can be offered or demanded.

Here, the game planner is a walrasian auctionner's twin brother. His satisfaction increases with the gaps reduction between the peripherial core and the intra-muros core, under the non empty peripherial core's constraint. The last peripherial core (i.e. the one which maximises the planner's utility under the non vacuity constraint) comes fom the same family as **OWEN**'s **Least-Core [1982]**, established by ε-cores regression in a replicated economy.

3.1.3. Coalitions

In an exchange economy: what does "*participate to a coalition*" means? It means *membership* (for free, except for the ε-cores) to an intermediary institution that lets any member realize more profitable exchanges than those proposed in society. AUBIN using the fuzzy coalition notion to *convexify* the coalitions set and so forth to reduce the set of allocations or non blocked imputations (i.e. intra muros core) does not change much to the fact that the argument is, by essence, technic even if an intuitive signification can be attached to it. An economic coalition (i.e. related to an exchange economy with or without production) is founded on the individual interest. Therefore, the interest is defined by an economic rationality that leads to the individual research for a maximum of satisfaction in a fixed environment, even if this satisfaction results from a fuzzy approach of the preferences. Except *endogeneousing* the agents membership to the coalitions, which is not done by either AUBIN or BUTNARIU, nothing let us economically explain the reason why agents belong to different levels for the present coalitions. However, the agent can think that he is represented more or less well by the coalitions and this basic notion is besides the starting point of balanced structures; this means that the agent gives to each coalition to which he participates, a **fraction of representativity**.

In other words, you must not confuse a coalition that can allocate inside (internal market as in the labour theory) and a coalition directly issued from individual choices from which the function is to generate a collective decision and not to proceed to internal allocations that demand an interest solidarity weakly compatible with the existence of an implication of different agents. This conception brings us to reject a fuzzy coalition hypothesis. AUBIN's hypothesis implied a restriction of the intra-muros core but with an evident condition of individual preorder convexity because of a previous non vacuity of the intra-muros core. Our goal is not to restrain the core but, on the contrary, to fill it in situations where it is naturally empty. Our approach pushes us not to add anything that could diminish the core's cardinal.

3.2. The concept of peripherial core

Let us remember the different hypothesis which gouvern our model so there will subsist no ambiguity between AUBIN's approach and our own.

3.2.1. The economic model

1) Let us consider the set of players (or economic agents), noted S. The S set's cardinal, $|S|$, is a finite number, n: $|S|=n$.

2) We call **agents coalition** any usual subset of S. We can then deduce that 2^n coalitions are possible.

Any coalition is possible *a priori*. However, intuitively, we can suppose that some of the 2^n coalitions will not be activated because - as seen before - it is necessary that common interest brings agents together in a solidary group called coalition. It is the reason for which the activated coalitions set, noted AC, will be a usual subset from P(S), the set of parts from S. One calls AC a **coalitions structure** with $AC \subseteq P(S)$. (Thus, $\emptyset \in P(S)$ but trivially $\emptyset \notin AC$.)

3) At any activated coalition C, $C \in AC$, corresponds a family of coefficients (α^i_c), for any agent i belonging to C. α^i_c symbolizes the i's individual *fraction* that he estimates *represented* by the coalition C (while belonging to C at a usual degree 1). That way, a member from family X, assistant at university Y, member of a Z syndicate, will have 3 coefficients α_x, α_y and α_z, with which he will be sure to be **perfectly** described. In other words, for this agent i: $\alpha^i_x + \alpha^i_y + \alpha^i_z = 1$.

4) However, each activated coalition C, $C \in AC$, represents each of its members without any discrimination. He can be assistant or Professor, the agent is represented by the university Y (in which there is two different ranks) in an identic way: $\forall C \in AC$, $\forall (i,j) \in C^2$, $i \neq j$; $\alpha^i_c = \alpha^j_c$.

4bis) The **simultaneous** satisfaction of hypothesis 3) and 4) induces what we call a **balanced structure**.

5) Any subset C from S - especially any element from AC - will be designated by $\mathbb{R}^C$ the set of the vectors of components u_i, where u_i designates the utility level felt by the agent i (for a given allocation), $i \in C$. From this definition, we can deduce[12] that $\mathbb{R}^n = \mathbb{R}^S$.

6) By construction, we will refer to an operator π, such as from any vector of $\mathbb{R}^S$, $u = (u_i)_{i \in S}$, π associates the only components related to the members of a coalition C, which then generates a vector of $\mathbb{R}^C$:

---

12 This writing is a convention which avoids distinguishing the coalition and its cardinal, since we define the referential field of internal utility vectors.

$$\pi: \mathbb{R}^S \longrightarrow \mathbb{R}^C$$
$$(u_i)_{i\in S} \longrightarrow \pi(u) = (u_i)_{i\in c}.$$

By definition, we call vectors of $\mathbb{R}^C$, **C-imputations**. The composition of a cooperative game with n agents (or players) is the data, for any non-empty coalition from S, of a non-empty part of $\mathbb{R}^C$, noted V(C) with:

$$V(C) - \mathbb{R}^C \subset V(C) \qquad (A)$$

What does this all mean?

Let's represent society S ourselves as a collection of n individuals. Let's now consider a vector u from $\mathbb{R}^C$: which gives a certain utility u_i to each member of C. The agent i wants to maximize his utility level u_i (even if fuzzy). The rule of the game permits different alliances. The player's problem as the exchanger's one is to enter a coalition that will guarantee the best level of utility. To do this, the agent i examines every AC coalition and compares them to what's proposed. V(C) is precisely the mathematical translation of the imputation that the coalition C can guarantee to each of its members. This means that if its members decide an imputation $(u_i)_{i\in C} \in \mathbb{R}^C$, then none (i.e. no member of the anti-coalition S-C) will be able to threat this imputation. In other words, there is, for a coalition C, a way to play so that, at the end of the game, each member i of C will be at a certain satisfaction level at least equal to u_i of $(u_i)_{i\in C}$. This explains why coalition C has no interest what so ever in what happens to the other agents than its owns (postulate of selfishness for coalitions and individuals). This is expressed by the equation (A).

3.2.2. The exchange economy

One considers an exchange economy of the same type as AUBIN's. Let's place agents (in a finite number $n = |S|$) in a game situation. They have to share their initial endowments, noted D, composed with ℓ goods divided between n individuals (i.e. $D \in \mathbb{R}^1_+$), with two hypothesis: initial detention of the endowments (H1) and selfishness (H2).

We call **allocation**, any vector x of $\mathbb{R}^{1n}_+$. If the allocation is such as :

$$\sum_{i:1}^{n} x_i = D,$$

the allocation is **feasible**.

3.2.2.1. *The agents specification*

This is where we find again fuzzy preferences. Each agent i, $i \in S$, has fuzzy preference preorder (e.f.), $\mathcal{R}_i(.,.)$, ℓ-transitive and reflexive (e.f.). We suppose that the fuzzy preorder is continuous and each consumption set X_i is convex (therefore connected), $X_i \subset \mathbb{R}^1_+$, which assures the existence of a continuous fuzzy utility function on X_i, taking its values on [0,1]. We write the fuzzy utility function $fu_i(.)$.

We'll say that coalition C (from AC) totally blocks up allocation $x \in \mathbb{R}^{1n}_+$, if there exists an allocation $y \in \mathbb{R}^{1n}_+$, such as:

1) $D = \sum_{i \in S} d_i = \sum_{i \in S} y_i$
2) $\pi_c[fu_i(y)] >> \pi_c[fu_i(x)]$ (">>" means *per component*).

We'll note PR(C) the internal possible redistributions set:
$PR(C) = \{y \in \mathbb{R}^{1n}_+;\ \forall\ i \in C,\ y_i \in \mathbb{R}^1_+ \text{ and } \sum_{i \in C} y_i = \sum_{i \in C} d_i\}.$

3.2.2.2. *The agents behavior*

The set PR(S) is the set of all feasible allocations. By definition, we suppose that $PR(S) = \prod_{i \in S} X_i$. Thus PR(S) is convex and connected. We assume it to be also compact. Each agent, member of the society S is going to generate a fss $\mathcal{PR}_i(S)$ - that **we assume the exclusive support to be convex** - using his fuzzy preorder (e.f.), ℓ-transitive and reflexive (e.f.) by arranging the allocations of PR(S) in this way: $\forall\ (x,y) \in [PR(S)]^2$, if agent i believes that the allocation x's relative qualities (restricted by H1 to x_i) are superior to y's ones, then he'll write $prs_i(x_i) > prs_i(y_i)$, where $prs_i(.)$ is the membership function of the fss $\mathcal{PR}_i(S)$ included in the referential set PR(S). Since $\forall$ i $\in$ S, $\mathcal{R}_i(.,.)$ is a continuous fuzzy preorder (e.f.) and PR(S) is connected, then the membership function $prs_i(.)$ is continuous[13]. We can then establish directly the following result:

LEMMA 2: *$\forall$ i $\in$ S, $\mathcal{PR}_i(S)$ is strong ℓ-convex.*

Proof: It is obvious since $prs_i(.)$ is continuous and supp$\mathcal{PR}_i(S)$ convex. We have just to apply lemma 1bis. □

3.2.2.3. *The peripherial core definition*

By defining a fuzzy utility function $prs_i(.)$ on PR(S), each agent restrains his satisfaction to feasible allocations. The possibilities of blocking - up to now - are only of one type. If allocation x is always less satisfying, in a given coalition, than an internal feasible imputation, this allocation x will be totally blocked. An allocation x will belong to the intra-muros core if no coalition blocks it at all. We can then define the intra-muros core (IMC)[14] as following:

$$IMC = \{x\in PR(S);\ \forall\ C\in AC,\ \exists\ i\in C,\ \forall\ y\in PR(C),\ prs_i(x_i) \geq prs_i(y_i)\}.$$

We know that IMC is a usual set, that is to say:

$$\forall\ x \in IMC;\ imc(x) = 1.$$

But the intra-muros core only describes the allocations which are totally unblocked. This corresponding to an unitary membership degree. Let's now consider allocations that are *a priori* blocked but with a *more or less* level of satisfaction. If a feasible allocation x' is blocked, it does not belong to the intra-muros core:

---

13 It is obvious that if $\prod_{i\in S} X_i$ = PR(S) then $prs_i(.)$ and $fu_i(.)$ are identical. We assume here to be in that case. That is the reason why we can use indifferently $prs_i(.)$ or $fu_i(.)$.

14 By definition; $\forall$ y $\in$ PR(C), $\exists$ y'$\in$ PR(S) such as:
$$\pi_c[prs_i(y'_i)] = [prc_i(y_i)]_{i\in C}.$$
By convention, we note the components $prs_i(y_i)$.

$imc(x') = 0 \Leftrightarrow \exists\ C \in AC;\ \forall\ i \in C,\ \exists\ y \in PR(C) : prs_i(x_i) < prs_i(y_i).$

As soon as a member of each allocation prefers a feasible external allocation, this allocation x' belongs to IMC.

However, by introducing ordinal interpersonal comparisons of utility because of the game planner, there are some rejected allocations that induce a certain level of satisfaction, very closed but inferior to a level of utility corresponding to an element of IMC. To synthetize all of this information, the best tool is the intersection and/or the union of the different fss $\mathcal{PR}_i(S)$. Taking the intersection and union of the different $\mathcal{PR}_i(S)$ is equivalent to confer to a rejected allocation the **minimum level of maximum levels** of satisfaction that it generates at the coalitions members of the AC structure. This principle supposes two hypothesis:

H3: We can make interpersonal ordinal utility comparisons.
H4: $\forall\ i \in C,\ \forall\ C \in AC,\ fu_i(d_i) > 0.$

We can now define a fss, the Peripherial Core, noted $\mathcal{PC}$, (its exclusive support including IMC), as following:

Definition 3: We call **Peripherial Core** the fss of PR(S) such as:

$$\mathcal{PC} = \{x \in PR(S),\ pc(.) \in \underline{M} = [0,1];\ pc(x) = \underset{C \in AC}{\mathrm{Min}}\ [\underset{i \in C}{\mathrm{Max}}\ prs_i(x_i)]\}.$$

Our problem now becomes very easy: every element of $\mathcal{PR}_i(S)$ belongs to $\mathcal{PC}$ ($\forall\ i \in S$).

THEOREM I: *Whatever is the nature (convex or not) of the agents fuzzy preferences, under* H3 *the exchange economy is balanced if* $\forall\ i \in S$, X_i *is convex,* $\mathcal{R}_i(.,.)$ *continuous and* $supp\mathcal{PR}_i(S)$ *convex. (i.e.* $\mathcal{PC} \neq \emptyset$*)*

(For the proof see **BILLOT [1988]**, Part II, chap 2, section 2)

Remark: Traditionally (**SCARF [1967], HILDENBRAND & KIRMAN [1976]**) we proceed, in this kind of proof, by extension of a coalition towards society. Even the less, the reference to a guaranteed level of individual utility is also classic (see **CORNWALL [1984], ROSENMULLER [1981]**).

Anyway, we have established that an exchange economy is always balanced as well as the agents fuzzy preferences are non-convex. To do this, there must exist at least an allocation of $\prod_{i \in S} X_i$ such that the coalitions members have at least a fuzzy utility level superior to 0. This constraint is explained by (H4) which is not very restrictive. It means only that independently of every coalition or coalition structure, the agents attribute to the initial endowments a non-null level of utility. (An other way is defining fss of this section on a membership set $\underline{M}$ =]0,1], which excludes incomparabities, see **BILLOT [1986]**)

This result is very important because it gets rid the core's non-vacuity of the condition of individual preorders convexity. However, we can't be totally satisfied by the fact that the peripherial core is too spreaded out. It contains all of PR(S) allocations that satisfy condition H4 (if we apply the non-null utility constraint) and we can think that they are large. With a strict pragmatic point of view, it is possible to compare the peripherial core's allocations by looking at function $\mathcal{pc}(.)$.

What we have obtained corresponds to the fact that if the IMC is empty, there exists "*approximately balanced*" allocations that can solve the exchange problem in the given economy. To find a ℓ-balanced[15] solution in an exchange economy, it is just necessary that in each coalition, there is an internal allocation which is considered by every member as granting a non-null utility level. It seems natural to assimilate it to the individual endowments but any allocation would do the same thing. To have with certainty a ℓ-balanced solution, there must exits, at least, in each coalition for each member of the coalition, an internal allocation which is not *intrinsically without interest* (cf note 3).

Now that we have a non empty fss that contains all the ℓ-balanced solutions, we must choose the ℓ-balanced allocations that approach the most the IMC because the $\mathcal{PC}$ definition induces for each allocation of $\mathcal{PC}$, a membership degree coming from the utility of the less satisfied agent of the most satisfied agents (cf def3).

The $\mathcal{PC}$'s membership function is defined on $\underline{M}$ = [0,1] (with a possible discontinuity). Because of hypothesis (H3), it is possible

---

15 Henceforth, we note the *approximately balanced* allocations, ℓ-balanced allocations.

for a game planner to compare the different elements composing $\mathcal{PC}$. The essential tool which is going to be used by our planner is the α-cuts[16]. Before, we must prove two results that will confort us in our peripherial core's definition. They tend to show that if IMC exists, it is **included** in supp$\mathcal{PC}$, and that the membership degree of the allocations which belong to IMC is always **superior** to the membership degree of the allocations which only belong to $\mathcal{PC}$ (with a condition on satisfaction restriction inside the coalitions).

LEMMA 3: *If $\mathcal{PC} \neq \emptyset$ and IMC $\neq \emptyset$, then IMC $\subset$ supp$\mathcal{PC}$.*
Proof: If IMC $\neq \emptyset$, it means: $\exists$ x $\in$ PR(S); x $\in$ IMC. That is to say:

$\forall$ C $\in$ AC, $\exists$ i $\in$ C, $\forall$ y $\in$ PR(C): $prs_i(x_i) \geq prs_i(y_i)$.

Thus we can write:

$\forall$ y $\in$ PR(C): $\underset{i \in C}{\text{Max}}\ prs_i(x_i) \geq \underset{i \in C}{\text{Min}}\ prs_i(y_i)$.

As $\mathcal{PC}$ is non-empty, (H4) is satisfied and thus $\forall$ i $\in$ S, $prs_i(d_i) > 0$. So, we can deduce: $\underset{i \in C}{\text{Max}}\ prs_i(x_i) \geq \underset{i \in C}{\text{Min}}\ prs_i(d_i) > 0$ and that $\forall$ C $\in$ AC. It means: $\underset{C \in AC}{\text{Min}}\ \underset{i \in C}{\text{Max}}\ prs_i(x_i) > 0$.

Thus, we can conclude: $pc(x) > 0 \Rightarrow x \in$ supp$\mathcal{PC}$. □

We now bring a new hypothesis that means that no internal allocation exists, and this whatever the coalition, such as the maximum satisfaction that it generates in a given coalition **can be superior** to the maximum of satisfaction generated by every allocation belonging to IMC.

ISP (Internal Satisfaction Postulate): $\forall$ C$\in$AC, $\forall$ y$\in$PR(C), $\forall$ x$\in$IMC,

$$\underset{i \in C}{\text{Max}}\ prs_i(y) \leq \underset{i \in C}{\text{Max}}\ prs_i(x).$$

LEMMA 4: *Let an allocation x $\in$ IMC $\neq \emptyset$. Then, under ISP, $\forall$ x'$\in\mathcal{PC}$' (where $\mathcal{PC}$' is such as supp$\mathcal{PC}$'= supp$\mathcal{PC}$ - IMC): pc(x') < pc(x).*
Proof: if x'$\in\mathcal{PC}$': $\exists$C'$\in$AC; $\forall$i$\in$C', $\exists$y'$\in$PR(C') such as:

$$prs_i(x') < prs_i(y').$$

Then: $\underset{c'}{\text{Max}}\ prs(x'_i) < \underset{c'}{\text{Max}}\ prs(y'_i)$.

We know also, if x'$\in\mathcal{PC}$'$\Rightarrow$ $\forall$ C $\in$ AC - C': $\forall$ y $\in$ PR(C),

---

16 An α-cut of a fss $\mathcal{A}$ is the usual set $A_\alpha = \{x\in\mathcal{A}; a(x)\geq\alpha\}$.

$$\exists\ i \in C;\ prs_i(y) \leq prs_i(x')$$

and $\forall\ j \in C - \{i\};\ prs_j(y) > prs_j(x')$.

Thus we can deduce: $\forall\ C \in AC - C',\ \underset{C}{\text{Max}}\ prs_i(x') \geq \underset{C}{\text{Min}}\ prs_i(y)$.

So, we obtain:

1) $\exists\ C' \in AC,\ \exists\ y' \in PR(C');\ \underset{C'}{\text{Max}}\ prs_i(x') < \underset{C'}{\text{Max}}\ prs_i(y')$

2) $\forall\ C \in AC - C',\ \forall\ y \in PR(C):\ \underset{C}{\text{Max}}\ prs_i(x') \geq \underset{C}{\text{Min}}\ prs_i(y)$.

It means:

$$\underset{AC}{\text{Min}}\ \underset{C}{\text{Max}}\ prs_i(x') < \underset{AC}{\text{Max}} \left[\underset{C'}{\text{Max}}\ prs_i(y'),\ \underset{C}{\text{Min}}\ prs_i(y)\right],$$

that is to say,

$$pc(x') < \underset{AC}{\text{Max}} \left[\underset{C'}{\text{Max}}\ prs_i(y'),\ \underset{C}{\text{Min}}\ prs_i(y)\right].$$

β) Let us show now, that $\forall\ x \in IMC$:

$$pc(x) \geq \underset{AC}{\text{Max}} \left[\underset{C'}{\text{Max}}\ prs_i(y'),\ \underset{C}{\text{Min}}\ prs_i(y)\right].$$

Let us suppose there exists an allocation x belonging to IMC such that:

$$pc(x) < \underset{AC}{\text{Max}} \left[\underset{C'}{\text{Max}}\ prs_i(y'),\ \underset{C}{\text{Min}}\ prs_i(y)\right].$$

It means:

$$\underset{AC}{\text{Min}}\ \underset{C,C'}{\text{Max}}\ prs_i(x) < \underset{AC}{\text{Max}} \left[\underset{C'}{\text{Max}}\ prs_i(y'),\ \underset{C}{\text{Min}}\ prs_i(y)\right].$$

We can deduce two alternative situations:

1) $\exists\ C \in AC - C';\ \underset{C}{\text{Max}}\ prs_i(x) < \underset{C}{\text{Min}}\ prs_i(y) \Rightarrow$

$$\forall\ i \in C:\ prs_i(x) < prs_i(y).$$

So, we can write:

$$\exists\ C \in AC;\ \forall\ i \in C,\ \exists\ y \in PR(C)\ /\ prs_i(x) < prs_i(y).$$

Thus, coalition C blocks up the allocation x. We deduce from it: $x \notin IMC$, which is inconsistent with the beginning hypothesis. (Let us remark that ISP yet forbids this equation.)

2) $\underset{C'}{\text{Max}}\ prs_i(x) < \underset{C'}{\text{Max}}\ prs_i(y')$. This case is trivially prohibited with ISP.

In conclusion, we can now write:

$$\forall\ x \in IMC,\ pc(x) \geq \underset{AC}{\text{Max}} \left[\underset{C'}{\text{Max}}\ prs_i(y'),\ \underset{C}{\text{Min}}\ prs_i(y)\right].$$

Thus: $\forall\ x \in IMC,\ \forall\ x' \in \mathcal{PC}'$:

$$\mathcal{pc}(x) \geq \underset{AC}{\text{Max}} \left[\underset{C'}{\text{Max}}\ \mathcal{prs}_i(y'),\ \underset{C}{\text{Min}}\ \mathcal{prs}_i(y) \right] > \mathcal{pc}(x')$$

$$\Leftrightarrow$$

$$\mathcal{pc}(x) > \mathcal{pc}(x') \qquad \square$$

4. THE α-APPROACHED CORES[17]

Theorem I brings us to consider all solutions which belong to $\mathcal{PC}$ as *approximately balanced*, that is to say the allocations of the peripherial core while IMC is empty. However, even if it is not empty, the membership function of $\mathcal{PC}$, reflecting the less satisfied agent utility of the most satisfied agents, can, there too, distinguish between all of IMC allocations (because of (H3)), the solutions balancing the game which are *most preferable.*

Finally, the agents having demonstrated their preferences and having choosen the coalitions to which they are going to participate in, the game planner disposes of the peripherial core, i.e. a membership function explaining social utility on the approximately balanced allocations. Lemma .4 will encourage the game planner to search for a social allocation more satisfying in the sense that social utility (assimilated to $\mathcal{pc}(.)$) increases when, on one side, we *approach* IMC (if it exists), and on the other side, when we *hierarchize* the $\mathcal{PC}$'s allocations. This social utility, $\mathcal{pc}(.)$, must be, in what is going to follow, a continuous function defined on [0,1], which means supposing $\mathcal{PC}$ is weak ℓ-convex. (cf lemma 1)

4.1. Some results

Let's recall what is an α-cut:

Definition 4: ∀ $\mathcal{A}$ a fss of X, we call A_α "**α-cut of $\mathcal{A}$**", $\alpha \in]0,1]$, the following usual set: $A_\alpha = \{x \in X;\ a(x) \geq \alpha\}$.

Definition 5: We call **α-approached core**, noted PC_α, every α-cut of the peripherial core $\mathcal{PC}$.

---

17 Do not assimilate this notion with α-core's one, issued from the introduction of external effects, (coalition utility depending on the other coalitions utilities). See **ROSENTHAL [1971]**.

We remark that the more α is big, the more the usual obtained set's cardinal is small: let $(\alpha,\alpha') \in]0,1]^2$ with $\alpha > \alpha'$, then $|PC_\alpha| < |PC_{\alpha'}|$.

The game planner's problem becomes simple. He is going to search the level α^* such as $\forall\ \alpha>\alpha^*, |NP_\alpha|=0$ with $|NP_{\alpha^*}|\neq 0$. That is to say, the **last**[18] **peripherial core's α-cut** is non empty.

Definition 6: We call **Last Core**, noted **LC**, the α^*-approached core.

We can then generate certain intuitive results.

LEMMA 5: *If IMC $\neq \emptyset$, under ISP $\Rightarrow$ LC $\subseteq$ IMC.*
Proof: It builds upon lemma 3. If IMC $\neq \emptyset$, we know that for $x' \in$ IMC, $\forall\ x \in \mathcal{PC}'$: $pc(x)>pc(x')$; thus for $\bar{\alpha}$ defined with $\bar{\alpha} = \underset{x\in \mathcal{PC}'}{Max}\ pc(x)$, it is obvious that $\forall\ \alpha > \bar{\alpha}$: $|PC_\alpha| \leq |IMC|$. If it is true for every α, $\alpha>\bar{\alpha}$, it is also true for α^*, which is, by definition, superior to $\bar{\alpha}$ since IMC$\neq\emptyset$. Thus, we can conclude that $|PC_{\alpha^*}| \leq |IMC|$ i.e. LC$\subseteq$IMC.

□

LEMMA 6: *If IMC $= \emptyset \Rightarrow$ LC $= NP_{\alpha^*}$ where $\alpha^* = \underset{x\in \mathcal{PC}'}{Max}\ pc(x)$.*

Proof: If IMC $= \emptyset \Rightarrow \mathcal{PC}' = \mathcal{PC}$. Thus, by definition, the maximum of α such as $|PC_\alpha| \neq \emptyset$ with $\forall\ \beta>\alpha$, $|PC_\beta| = \emptyset$, is obviously defined with $\alpha^* = \underset{x\in \mathcal{PC}}{Max}\ pc(x)$. Then, it induces: LC $= PC_{\alpha^*}$. □

LEMMA 7: *If supp$\mathcal{PC} \neq \emptyset \Rightarrow$ LC $\neq \emptyset$.*
Proof: If supp$\mathcal{PC} \neq \emptyset \Rightarrow \exists\ x \in$ PR(S); $pc(x)>0$ and thus $\exists\ \alpha$, $\alpha=pc(x) \Rightarrow$ LC $\neq \emptyset$. □

Lemma 5 means that the Last Core is included or equal to IMC while the individual preorders are convex. Lemma 6 explains the fact that the upper degree of the membership function $pc(.)$ defines

---

18 To be certain of the α^*'s existence, it is necessary to define $pc(.)$ as a continuous function on a compact set; that is the case, since $\mathcal{PC}$ is weak ℓ-convex and PR(S) compact, by hypothesis.

the Last Core if IMC is non empty. Lemma 7 assures the Last Core's non vacuity as soon as supp$\mathcal{PC}$ is not empty.

4.2. The last core's theorem

We can then conclude by the following theorem which sums up the peripherial core's analysis:

THEOREM II: *The conditions under which the Last Core of an exchange economy is non empty are:*
1) the set of feasible allocations is compact and convex,
2) the Peripherial Core is non empty and weak ℓ-convex,
3) ISP is satisfied.

If we analyze this last result, we remark the disappearance of the convexity hypothesis of individual preorders and the presence of an interpersonal ordinal utility comparison hypothesis equivalent to the existence of a game planner. The introduction of the α-approached core analysis is not here, as usually in current literature, linked to the concept of a replicated economy and/or the agents continuum. Our set of agents is finite and that is what explains the necessity of the interpersonal ordinal utility comparison hypothesis (H3). We must then discuss on the imposed constraint by each of both conditions (preorder's convexity vs (H3)) to be able to define, if possible, which one is less restrictive.

5. CONCLUSION

Our goal was double. We had, on one side, to remove the convexity hypothesis, which is strong, and on the other side, not to affect the coalition structures either by replication or by introduction of fuzzy coalitions. We have reached our goal because of the fss's ℓ-convexity, meaning the introduction of fuzzy unions and intersections that would generate the ordinal comparison principle. Then again, our approach joints our theoretical ambitions. We want to study the practice solutions of a game having no usual solutions. The implicit idea that is in our scheme is *cooperation* even in constitued coalitions and it is formalized by a membership function of the peripherial core, coming from the different $\mathcal{PR}_i(S)$ intersections of agents i the most satisfied of each coalition.

By definition, the peripherial core contains all the allocations socially feasible for which the less satisfied of the most satisfied agents of society has a positive fuzzy utility. However, in a given coalition of non blocking coalitions, the most satisfied agent must prefer them to any possible internal allocation. Then, the most satisfied of all agents of the blocking coalition must not prefer them to any other .internal allocations. In other words, a socially feasible allocation belongs to IMC, on the condition of the non existence of at least one internal allocation in at least one coalition, that is prefered by every member of this coalition.

For the peripherial core, we find these two steps. Every social feasible allocation passes through each coalition. Then we can deduce at the most n possible levels of memberships coming from the most satisfied agents. At least, we'll attribute to the considered allocation, the level of satisfaction coming from the less satisfied agent of the *preselected* agents. These different notions are brought back to the concept of union and intersection of the sets $\mathcal{PR}_i(S)$, thanks to the fss algebra (**ZIMMERMANN [1985]**), and this in each coalition. One of the miracles to which we assist is the exhibition of a balanced solution (thanks to the game planner) corresponding to a generalized concertation of all agents on top of the coalitions which only define a balanced structure. In other words, if we place ourselves in usual conditions (convex preorders of preference), the game planner will generate a **perfectly balanced solution** which will have a better (or equal) social level than the IMC allocations one that we cannot distinguish in the usual case.

REFERENCES

AUBIN J.P. [1974]. "Coeur et Valeur des Jeux Flous", *Compte-rendus de l'Académie des Sciences de Paris,* **279**, 891-894.
AUBIN J.P. [1979]. *Mathematical Methods in Economics and Game Theory*. Amsterdam, North-Holland.
AUBIN J.P. [1981]. "Locally Lipschitz Cooperative Games", *Journal of Mathematical Economics*, **8**, 241-262.
AUBIN J.P. [1986]. *L'Analyse Non-Linéaire et ses Motivations Economiques*. Masson, Paris.
BILLOT A. [1986]. "A Contribution to a Mathematical Theory of Fuzzy Games" in *Fuzzy Economics and Spatial Analysis*, C.Ponsard & B.Fustier Eds, Dijon, Librairie de l'Université, 47-56.
BILLOT A. [1988]. *Comportement Imprecis et Equilibres Economiques: une Analyse Axiomatique*, These de Doctorat es Sciences Economiques Dijon, Novembre 1988. To appear in Librairie de l'Université and for the English translation in Springer Verlag under the new title *Economic Theory of Fuzzy Equilibria.*

BONDAVERA O.N. [1963]. "Some Applications of Linear Programming Methods to the Theory of Cooperative Games", *Problemy Kiberneteki*, **10**, 119-139.
BORDER K.C. [1982]. "The Core of a Coalitional Production Economy", Social Science Working Paper n°461, California Institute of Technology.
BUTNARIU D. [1979]. "Solution Concepts for *n*-Persons Fuzzy Games" in *Advances in Fuzzy Set Theory and Applications*, M.M. Gupta, R.K. Ragade & R.R. Yager Eds, 339-359.
BUTNARIU D. [1980]. "Stability and Shapley Value for *n*-Persons Fuzzy Games", *Fuzzy Sets and Systems*, **7**, 63-72.
BUTNARIU D. [1985]. "Non-Atomic Fuzzy Measures and Games",*Fuzzy Sets ans Systems*, **17**, 39-52.
BUTNARIU D. [1986]. "Fuzzy Measurability and Integrability", *Journal of Mathematical Analysis and Applications*, **117**, 385-410.
BUTNARIU D. [1987]. "Values and Cores of Fuzzy Games with Infinitely Many Players", *International Journal of Game Theory*, **16**, 43-68.
CHANG C.L. [1968]. "Fuzzy Topological Spaces", *Journal of Mathematical analysis and Applications*, **24**, 182-190.
CORNWALL R.R. [1984]. *Introduction to the Use of General Equilibrium Analysis*. Amsterdam, North-Holland.
DREZE J.H. & GABSZEWICZ J.J [1971]. "Syndicates of Traders in an Exchange Economy", in *Differential Games and Related Topics*, North-Holland, Amsterdam.
HILDENBRAND W. & KIRMAN A. [1976]. *Introduction to Equilibrium Analysis*. Amsterdam, North-Holland.
ICHIISHI T. [1981]. "A Social Coalitional Equilibrium Existence Theorem", *Econometrica*, **49**, 369-377.
ICHIISHI T. [1982]. "Non-Cooperation and Cooperation", in *Games, Economic Dynamics and Time Series Analysis*, M. Deistler, E. Furst & G. Schwodiauer Eds, Vienne, Physica-Verlag.
KAUFMANN A. [1973]-[1980]. *Introduction à la Théorie des Sous-Ensembles Flous*. Tomes I et IV, Masson, Paris.
LIU Y.M. [1985]. "Some Properties of Fuzzy Convex Sets", *Journal of Mathematical Analysis and Applications*, **111**, 119-129.
MOULIN H. [1981]. *The Strategy of Social Choice*. Laboratoire d'Econométrie de l'Ecole Polytechnique.
MOULIN H. , FOGELMAN-SOULIE F. [1979]. *La Convexité dans les Mathématiques de la Décision*. Hermann, Paris.
MUNIER B. [1973]. *Jeux et Marchés*. Presses Universitaires de France, Paris.
OWEN G. [1982]. *Game Theory*. 2nd Edition, Academic Press.
PONSARD C. [1980]. "Fuzzy Economic Spaces", First World Regional Science Congress, Harvard University, Cambridge, Massachusetts,

June 15-25.
PONSARD C. & FUSTIER B. Eds [1986]. *Fuzzy Economics and Spatial Analysis.* Dijon, Librairie de l'Université.
PREVOT M. [1977]. *Sous-Ensembles Flous: Une Approche Théorique.* Sirey, Collection de l'Institut de Mathématiques Economiques, Paris.
QUINE W.V.O. [1972,1984]. *Méthodes de Logique.* 3rd Edition. Armand Collin, Paris.
ROSENMULLER J. [1981]. *The Theory of Games and Markets.* North Holland, Amsterdam.
ROSENTHAL R.W. [1971]. "External Economies and Cores", *Journal of Economic Theory*, **3**, 182-188.
SCARF H. [1967]. "The Core of an *n*-Person Game", *Econometrica*, **35**, 50-69.
SCARF H. [1971]. "On the Existence of a Cooperative Solution for a General Class of *n*-person games", *Journal of Economic Theory*, **3**, 169-181.
SCARF H. [1973]. *The Computation of Economic Equilibria.* New Haven, Yale University Press.
SCHLEICHER H. [1979]. Ed, *Jeux, Information et Groupes.* Economica, Paris.
SCHOTTER A. & SCHWÖDIAUER G. [1980]. "Economics and the Theory of Games: A Survey", *Journal of Economic Literature*, **18**, 479-527.
SHAPLEY L.S. [1973]. "On Balanced Games without Side Payments", in *Mathematical Programming*, T.C. Hu & S.M. Robinson Eds, Academic Press, New York, 261-290.
SHAPLEY L.S. & SHUBIK M. [1969]. "On Market Games", *Journal of Economic Theory*, **1**, 9-25.
WEBER S. [1979]. "On ε-Cores of Balanced Games", *International Journal of Game Theory*, **8**, 241-250.
WEISS M.D. [1975]. "Fixed Points, Separation and Induced Topologies for Fuzzy Sets", *Journal of Mathematical Analysis and Applications*, **50**, 142-150.
WOODERS M.H. [1983]. "The Epsilon Core of a Large Replica Game", *Journal of Mathematical Economics*, **11**, 277-300.
WOODERS M.H. & ZAME W.R. [1984]. "Approximate Cores of Larges Games", *Econometrica*, **52**, 1327-1350.
ZIMMERMANN H.J. [1985]. *Fuzzy Sets Theory - and its Applications.* Kluwer, Nijhoff Publishing, Boston.

FUZZY SEQUENCING GAMES

G. Pederzoli* and B. Viscolani**

* Istituto di Informatica, Università di Trento, Via Rosmini 42, I-38100 Trento, Italy.
** Dipartimento di Matematica Applicata ed Informatica, Università di Venezia, Dorsoduro 3825/E, I-30123 Venezia, Italy.

Abstract: Sequencing situations, where uncertain service times are modelled as fuzzy numbers, are considered. This assumption leads to the necessity of dealing with fuzzy costs and fuzzy arrangement of customers waiting in a line. The cooperative behaviour of customers is taken into account in order to establish division rules of possible gains. Fuzzy sequencing games can then be constructed where the primary source of fuzziness is in the cost functions rather than in the coalition formation.

Keywords: Sequencing situations, fuzzy numbers, games

1. INTRODUCTION

In this paper we consider sequencing situations, where a certain number of customers is standing in a queue before a counter waiting for service. Each customer has a cost function, which depends on the waiting time as well as on the service time. When service times are known with certainty, results of scheduling theory, see e.g. Baker (1974), state that the customers should optimally be ordered in a well defined sequence. In practice we are often faced with uncertain service times and therefore this uncertainty is reflected also in the cost structure of the model. Here, in view of a remark made by Prade (1980), we model the uncertainty of service times by means of fuzzy numbers. It follows that instead of an optimal *arrangement* of customers, an optimal fuzzy subset of *arrangements* has to be searched for. The cooperative behaviour of the customers can be represented by a game

J. Kacprzyk and M. Fedrizzi (eds.), Multiperson Decision Making Using Fuzzy Sets and Possibility Theory, 336–341.

where the coalitions are necessarily fuzzy. Such a treatment was introduced by Shapley and Shubik (1969) and pursued by Aubin (1981), where the fuzziness of a coalition is the primary concept. On the contrary, in the present approach the primary source of fuzziness is represented by service times.

2. MOTIVATION AND DEFINITIONS

Curiel, Pederzoli and Tijs (1987) study the possibility of cooperation among customers of a queueing system in a deterministic environment. There each customer (player) is characterized by his cost function and service time, whereas the cooperation entails that some customers original service order be modified. A natural way to render that model more realistic is to allow for the uncertainty of some variables. As far as queueing systems are concerned, Prade (1980) points out that the uncertainty regarding the service time has more reasonably a fuzzy, rather than a random nature. Here we intend to analyze some consequences, on the model of Curiel, Pederzoli and Tijs (1987), of assuming that the service time of a customer be a fuzzy number, rather than a real number.
Following mainly the presentation given by Kaufmann and Gupta (1985) and, more shortly, by Fedrizzi (1987), we say that a fuzzy subset of $\mathbb{R}$ is a *fuzzy number* (in $\mathbb{R}$), if it is *convex* and *normal*. More precisely, if $\mu_A : \mathbb{R} \to [0,1]$ is the membership function of a fuzzy subset A, then A is a fuzzy number iff:

1) $\mu_A(\lambda x + (1-\lambda) y) \geq \mu_A(x) \wedge \mu_A(y)$

for all $x, y \in \mathbb{R}$ and $\lambda \in [0,1]$, and

2) $\mu_A^{-1}(1) \neq \emptyset$

The ordinary real numbers can be considered as special fuzzy numbers. Moreover the usual arithmetic operations on $\mathbb{R}$ are extended to the set of fuzzy numbers (in short f.n.), though not retaining all their original properties. In particular, addition and subtraction of f.n.'s and multiplication of a f.n. by a positive real number are defined as follows in terms of their membership functions.

Addition, A, B f. n.'s:

$$\mu_{A(+)B}(x) = \max\{\mu_A(y) \wedge \mu_B(z),\ y + z = x\},\ x \in \mathbb{R}\ .$$

Subtraction, A, B f. n.'s:

$$\mu_{A(-)B}(x) = \max\{\mu_A(y) \wedge \mu_B(z), y - z = x\}, x \in \mathbb{R} .$$

Multiplication, A f.n., $\alpha \in \mathbb{R}_+$:

$$\mu_{\alpha A}(x) = \mu_A(x/\alpha), x \in \mathbb{R}$$

The addition is commutative, associative and has a neutral element, which is the real 0, but no fuzzy number has an opposite, except for (ordinary) real numbers.
A structure of distributive lattice is then imposed by the definitions of maximum and minimum of fuzzy numbers.

Minimum:

$$\mu_{A(\wedge)B}(x) = \max\{\mu_A(y) \wedge \mu_B(z), y \wedge z = x\}, x \in \mathbb{R} .$$

Maximum:

$$\mu_{A(\vee)B}(x) = \max\{\mu_A(y) \wedge \mu_B(z), y \vee z = x\}, x \in \mathbb{R} .$$

Consequently, a partial order, $\leq$, of fuzzy numbers, which is consistent with the order of real numbers, is defined as follows:

$$A \leq B \text{ iff } A\,(\wedge)\,B = A \qquad \text{(i.e. iff } A\,(\vee)\,B = B).$$

It is immediate to notice that $\leq$ is not complete. We say that a f.n. A is non-negative if $A \geq 0$, where 0 is the real 0, i.e. if

$$\mu_A(x) = 0, \qquad \text{for all } x < 0 .$$

In the following we will use both upper and lower case letters to indicate fuzzy numbers and write +, -, $\wedge$, $\vee$ instead of (+), (-), ($\wedge$), ($\vee$).

3. SEQUENCING SITUATIONS

Paraphrasing Curiel, Pederzoli and Tijs (1987), we consider a queue consisting of n

customers who are waiting to be served. The set of customers is $N = \{1, 2, \dots n\}$ and any ordering of them in the queue is described by a permutation σ of N, $\sigma \in \Pi_N$. For any customer i, $i \in N$, let $s_i \geq 0$ be his (fuzzy) service time and $c_i : \mathbb{R}_+ \to \mathbb{R}_+$, $c_i(t) = \alpha_i t + \beta_i$, his linear cost function, $\alpha_i, \beta_i \in \mathbb{R}_+$. $c_i(t)$ denotes the cost for customer i if his waiting time plus service time is equal to t: if t is a fuzzy time, then $c_i(t)$ is a fuzzy cost. Last, non-negativity of the service times is a natural assumption.
A *fuzzy sequencing situation* is an ordered triple (σ, α, s), where $\sigma \in \Pi_N$; $\alpha \in \mathbf{R}^n_+$, with α_i being the leading coefficient in c_i; s is an ordered n-tuple of non -negative fuzzy numbers (service times). We call a sequencing situation (σ, α, s) *real* if $s \in \mathbf{R}^n_+$, so that the service times are certain. The (fuzzy) *total cost* for the customers in N if served according to σ is given by

$$C_\sigma = \sum_{i=1}^{n} \alpha_i \Big(s_i + \sum_{j \in P(\sigma, i)} s_j\Big) + \sum_{i=1}^{n} \beta_i .$$

We are interested in the possibility for the customers to decrease the total cost by rearranging their positions in the queue.
We define the *patience-index* p_i of customer i as

$$p_i = s_i / \alpha_i$$

Equivalently we might define an "urgency-index" to be a_i / s_i, as Curiel, Pederzoli and Tijs (1987) do for the real case. Nevertheless we prefer the former choice in this fuzzy context for simplicity of computation.
In case s were a real n-tuple, the following proposition, adapted from Curiel, Pederzoli and Tijs (1987), provides the way for minimizing the total cost.

Proposition 1

Let (σ, α, s) be a real sequencing situation, then

$$C_\sigma = \min_{\pi \in \Pi_N} C_\Pi \quad \text{iff} \quad p_{\sigma^{-1}(i)} \leq p_{\sigma^{-1}(i+1)}, \qquad 0 \leq i < n .$$

Unfortunately in the opposite case, when s is not a real n-tuple, the inequality $p_i \leq p_j$ loses its meaning in general.
Proposition 1 still holds only if the set of patience indexes $\{p_1, p_2 \dots p_n\}$ is totally ordered by $\leq$. Then, with Proposition 1, every other result of Curiel, Pederzoli and Tijs (1987) on the "sequencing games" should easily be proved to hold.

On the contrary, if $\{p_1, p_2 \dots p_n\}$ is only partially ordered by ≤, then the fuzzy nature of the problem becomes more significant. Instead of looking for a permutation minimizing the total cost C_σ, we are restricted to looking for a minimizing fuzzy subset of Π_N. This is the typical situation of an optimization problem with fuzzy state, as described and exemplified in Dubois and Prade (1979).
A statement similar to Proposition 1 should characterize the support of the minimizing fuzzy subset of permutations, i.e. the permutations at which the membership function of the minimizing subset is positive.

Claim 1
Let (σ, α, s) be a fuzzy sequencing situation and let σ belong to the support of the minimizing subset of Π_N, with respect to the total cost C_σ. Then

$$p_1 \le p_j \Rightarrow \sigma(i) \le \sigma(j), \quad \text{for all} \quad i, j, \in N .$$

Sketch proof

a) Claim 1 is implied by Proposition 1 in the case that $\{p_1, p_2 \dots p_n\}$ is totally ordered by ≤; that is true in particular if (σ, α, s) is a real sequencing situation.

b) If $N^1 \subset N$ is any subset of customers such that $\{p_1 : i \in N^1\}$ is totally ordered by ≤ and we want to minimize the total cost with the positions of the remaining customers in $N \setminus N^1$ fixed, then Claim 1 must be satisfied for $i, j \in N^1$.

Therefore we see that the fuzziness introduced in the service times induces a fuzziness in the optimal service ordering of customers.

4. GAIN DIVISION RULES

Knowing an optimal arrangement σ, Curiel, Pederzoli and Tijs (1987) calculate the cost savings which is obtained by changing the original service order into σ and discuss how to divide such savings among the customers. This problem brings about the definition of a *sequencing game* and has an important solution in the *equal gain splitting rule*, (EGS in short).
The crucial point of the EGS-rule lies in the fact that the cost savings C_σ-C_τ, corresponding to a change from an arrangement σ to an arrangement τ, is expressed, in the real case, as a sum of a few binomials of the type $\alpha_j s_i - \alpha_i s_j$.
In view of the fact that ≤ is not a total order for the fuzzy numbers and that the subtraction

of two f.n. 's. cannot be 0, the cost savings $C_\sigma - C_\tau$ cannot be represented in a simple way in the fuzzy case, like it was in the real one.
Therefore we expect that:

i) there should be several division rules equivalent to the EGS-rule and that could be represented as a *fuzzy division rule*;
ii) the definition of the value of coalitions in a *fuzzy sequencing game,* being essentially a cost savings due to a suitable rearrangement, encounters the same difficulties of representation cited above.

The intriguing aspects of fuzzy sequencing games require to be tackled by studying special classes of them. A possible interesting choice is to have all the service times to be identical fuzzy numbers.

REFERENCES

Aubin, J. - P. (1981) *Cooperative fuzzy games*, Math. of Op. Res., 6, 1-13.

Baker, K.R. (1974) *Introduction to Sequencing and Scheduling*, Wiley, New York.

Curiel I., G. Pederzoli and S. Tijs (1987) *Sequencing games.* EJOR 40, 344-355.

Dubois D. and H. Prade (1979) *Decision making under fuzziness*, in M.M. Gupta, R.K. Ragade and R.R. Yager (Eds.): *Advances in Fuzzy Set Theory and Applications.* North-Holland, Amsterdam, pp. 279-302.

Fedrizzi M. (1987) *Introduction to fuzzy sets and possibility theory,* in J. Kacprzyk and S.A. Orlovski (Eds.): *Optimization Models Using Fuzzy Sets and Possibility Theory.* D. Reidel Publishing Company, Dordrecht, pp. 13-26.

Kaufmann A. and M.M. Gupta (1985) *Introduction to Fuzzy Arithmetic.* Van Nostrand, New York.

Prade H.M. (1980) *An outline of fuzzy or possibilistic models for queueing systems,.* in P.P. Wang and S.R. Chang (Eds.) *Fuzzy Sets, Theory and Applications to Policy Analysis and Information System.* Plenum Press, New York, pp. 147-153.

Shapley, L.S. and M. Shubik (1969) *On market games.* J. of Ec. Theory, 1, 9-25.

INDEX